Symbian OS C++
for Mobile Phones
Volume 2

TITLES PUBLISHED BY SYMBIAN PRESS

- Wireless Java for Symbian Devices
Jonathan Allin
0471 486841   512pp   2001   Paperback

- Symbian OS Communications Programming
Michael J Jipping
0470 844302   418pp   2002   Paperback

- Programming for the Series 60 Platform and Symbian OS
Digia
0470 849487   550pp   2002   Paperback

- Symbian OS C++ for Mobile Phones, Volume 1
Richard Harrison
0470 856114   826pp   2003   Paperback

- Programming Java 2 Micro Edition on Symbian OS
Martin de Jode
0470 092238   498pp   2004   Paperback

- Symbian OS C++ for Mobile Phones, Volume 2
Richard Harrison
0470 871083   448pp   2004   Paperback

- Symbian OS Explained
Jo Stichbury
0470 021306   448pp   2004   Paperback

Symbian OS C++ for Mobile Phones Volume 2

Programming with extended functionality and advanced features

Richard Harrison

With
Alan Robinson, Arwel Hughes, Dominic Pinkman, Elisabeth Måwe, Gregory Zaoui, Nick Johnson, Richard Potter

Reviewed by
Alex Peckover, Alex Wilbur, Chris Trick, Dan Handley, John Roe, Leon Bovett, Murray Read, Nick Tait, Paul Hateley

Managing editor
Phil Northam

Assistant editor
Freddie Gjertsen

John Wiley & Sons, Ltd

Published by John Wiley & Sons Ltd, The Atrium, Southern Gate, Chichester,
West Sussex PO19 8SQ, England
Telephone (+44) 1243 779777

Email (for orders and customer service enquiries): cs-books@wiley.co.uk
Visit our Home Page on www.wileyeurope.com or www.wiley.com

Other Wiley Editorial Offices

John Wiley & Sons Inc., 111 River Street, Hoboken, NJ 07030, USA

Jossey-Bass, 989 Market Street, San Francisco, CA 94103-1741, USA

Wiley-VCH Verlag GmbH, Boschstr. 12, D-69469 Weinheim, Germany

John Wiley & Sons Australia Ltd, 33 Park Road, Milton, Queensland 4064, Australia

John Wiley & Sons (Asia) Pte Ltd, 2 Clementi Loop #02-01, Jin Xing Distripark, Singapore 129809

John Wiley & Sons Canada Ltd, 22 Worcester Road, Etobicoke, Ontario,
Canada M9W 1L1

Wiley also publishes its books in a variety of electronic formats. Some content that
appears in print may not be available in electronic books.

Library of Congress Cataloging-in-Publication Data

Harrison, Richard.
Symbian OS C++ for mobile phones / By Richard Harrison.
p. cm.
Includes bibliographical references and index.
ISBN 0-470-85611-4 (Paper : alk. paper)
1. Cellular telephone systems – Computer programs. 2. Operating
systems (Computers) I. Title.
TK6570.M6H295 2003
621.3845′6 – dc21

03006223

British Library Cataloguing in Publication Data

A catalogue record for this book is available from the British Library

ISBN 0-470-87108-3

Typeset in 10/12pt Optima by Laserwords Private Limited, Chennai, India
Printed and bound in Great Britain by Biddles Ltd, King's Lynn
This book is printed on acid-free paper responsibly manufactured from sustainable
forestry in which at least two trees are planted for each one used for paper production.

Contents

Foreword

David Wood, Executive Vice President, Research, Symbian

Less than eighteen months have passed since the appearance of the first volume of *Symbian OS C++ for Mobile Phones*. These eighteen months have seen giant strides of progress for the Symbian ecosystem. In 2003 alone, the number of commercially available add-on applications for Symbian OS phones tripled. In the fourth quarter of that year, an unprecedented number of distinct new Symbian OS phone models – eight – reached the market. And in December of that year, for the first time, over a million phones running Symbian OS shipped in a single month. Looking ahead, five different 3G Symbian OS phones have recently reached shops around the world, underscoring Symbian's leadership position for the emerging generation of mobile phones. New licensing deals have been announced with premier companies in Japan, China, Korea and Taiwan, highlighting the global interest in the capabilities of the Symbian ecosystem. Last but not least, the Symbian Enterprise Advisory Council has been formed, in which leading providers of mobile business solutions are actively collaborating to promote the rapid take-up of Symbian OS phones for business use.

The good news for Symbian OS developers is that, despite these dramatic changes, the basics of the Symbian development world remain the same. Applications written to run on Symbian OS phones in 2003 will also run on Symbian OS phones reaching the market in 2004 and 2005, in most cases with very few changes and optimizations (and in many cases with no changes required at all). Symbian OS is written in a style of C++ that holds consistently throughout all levels of the software, and throughout all versions of the operating system. Once you learn the rules, you find they apply far and wide. Symbian OS was deliberately designed to be future-proof – to 'expect the unexpected'. As the first waves of the 3G future reach us, it is reassuring to see how well the programming framework thrives despite all the changes.

Over the last eighteen months, Symbian's 500-strong team of in-house software engineers has considerably extended the scope and functionality of Symbian OS. Volume 2 of *Symbian OS C++ for Mobile Phones* is your chance to boost your own understanding of the resulting prodigious software suite. This book builds on the foundations of its predecessor, covering some of the pivotal features of Symbian OS in more detail, and goes on to describe the key new software features which are now appearing in the latest breakthrough phones.

Symbian provides the platform that enables innovation through openness; developers such as the readers of this book provide the ingenuity and the diverse domain knowledge to create myriad solutions. It is my fervent wish that software which you write, with this book as your guide, becomes dear to millions of users of Symbian OS phones.

About This Book

Symbian OS C++ for Mobile Phones Volume 2 provides information in three main areas:

1. It provides a comprehensive review of the basic techniques needed to program a Symbian OS application. The descriptions are supplemented by many straightforward and easy-to-follow examples, which range from code snippets to full applications.

2. It promotes further understanding of Symbian OS, by describing the interaction between an application and the underlying system code. This theme pervades the whole book, but a particular example is Chapter 3, which provides an illuminating walk-through of the lifecycle of a typical application, from startup to closedown.

3. It describes some of the significant new features that are introduced in Symbian OS v7.0s. This aspect is particularly significant in the discussion of multimedia services in Chapter 7, and in Chapter 8, which provides an up-to-date description of the use of Symbian OS communications and messaging services.

Symbian OS is used in a variety of phones with widely differing screen sizes. Some have full alphanumeric keyboards, some have touch-sensitive screens and some have neither. As far as possible, the material in this book is independent of any particular user interface. However, real applications run on real phones so, where necessary, we have chosen to use the Series 60 user interface and the Nokia 6600 phone as concrete examples. Wherever relevant, the text explains the principal differences between the Series 60 and UIQ user interfaces. This kind of information is invaluable for anyone who wishes to create versions of an application to run on a variety of Symbian OS phones.

Symbian OS C++ for Mobile Phones Volume 2 complements the Symbian OS software development kits. When you've put this book down, the exclusive Symbian OS v7.0s TechView SDK supplied will be your first resource for reference information on the central Symbian OS APIs. For more specialized and up-to-date information relating to a specific mobile phone, you will need to refer to a product-specific SDK, available from the relevant manufacturer.

These SDKs contain valuable guide material, examples and source code, which together add up to an essential developer resource. As a general rule, if you have a query, look first at the SDK: you'll usually find the additional information you need that takes things further than we could in just one book.

Conventions

To help you get the most from the text and keep track of what's happening, we've used a number of conventions throughout the book.

> **These boxes hold important, not-to-be-forgotten information that is directly relevant to the surrounding text.**

While this style is used for asides to the current discussion.

We use several different fonts in the text of this book:

- When we refer to words you use in your code, such as variables, classes and functions, or refer to the name of a file, we use this style: `iEikonEnv`, `ConstructL()`, or `e32base.h`.

- URLs are written like this: ***www.symbiandevnet.com***.

- And when we list code, or the contents of files, we'll use the following convention:

```
Lines that show concepts directly related to the surrounding text are
shown on a gray background
```

```
But lines which do not introduce anything new, or which we have seen
before, are shown on a white background.
```

- We show commands typed at the command line like this:

```
abld build winscw udeb
```

Innovation Through Openness

The success of an open operating system for smartphones is closely linked to the degree to which the functionality of lower levels of software and hardware can be accessed, modified, and augmented by add-on software and hardware. As Symbian OS smartphones ship in volume, we are witnessing the arrival of a third wave of mobile phones.

The first wave was voice-centric mobile phones. Mobile phone manufacturers have performed wonders of optimization on the core feature of these phones – their ability to provide great voice communications. Successive generations of products improved their portability, battery life, reliability, signal handling, voice quality, ergonomics, price, and usability. In the process, mobile phones became the most successful consumer electronics product in history.

The second wave was rich-experience mobile phones. Instead of just conveying voice conversations between mouth and ear, these phones provided a much richer sensory experience than their predecessors. High-resolution color screens conveyed data vividly and graphically. High-fidelity audio systems played quality music through such things as ringtones and audio files. These phones combined multimedia with information and communications, to dramatic effect.

But the best was still to come. The primary characteristic of the third wave of mobile phones is their openness. Openness is an abstract concept, but one with huge and tangible consequences for developers. The key driver is that the growing on-board intelligence in modern phones – the smartness of the hardware and software – can now be readily accessed by add-on hardware and software. The range of applications and services that can be used on a phone is not fixed at the time of manufacture, meaning new applications and services can be added afterwards. The phone can be tailored by an operator to suit its customers and these customers can then add further customizations, reflecting specific needs or interests.

The Symbian Ecosystem

Open phones allow a much wider range of companies and individuals to contribute to the value and attractiveness of smartphones. The attractiveness of a phone to an end-user is no longer determined only by the various parties involved in the creation of that phone. Over-the-air downloads and other late-binding mechanisms allow software engineers to try out new ideas, delivering their applications and services directly to end-users. Many of these ideas may seem unviable at time of manufacture. However, the advantage of open phones is that there is more time, and more opportunity, for all these new and innovative ideas to mature into advantageous, usable applications that can make a user's life easier – whether it be over-the-air synchronization with a PC, checking traffic or having fun with 3D games or photo editing.

The real power of open phones arises when add-on services developed for a phone are reused for add-on services on other phones. This allows an enormous third-party development ecosystem to flourish. These third parties are no longer tied to the fortunes of any one phone, or any one phone manufacturer. Moreover, applications that start their lives as add-ons for one phone can find themselves incorporated at time of manufacture in subsequent phones, and be included in phones from other manufacturers. Such opportunities depend on the commonality of the underlying operating system. Open standards drive a virtuous cycle of research and development: numerous companies that can leverage the prowess, skills, experience and success of the Symbian ecosystem.

Symbian OS Phones

This book focuses on Symbian OS v7.0s, and the additional technologies it brings to mobile phones, as well as expanding on the core programming techniques explored in *Symbian OS C++ for Mobile Phones Volume 1*. We use the first Symbian OS v7.0s phone, the Nokia 6600, to illustrate most of the examples in the text, but these should also be demonstrable in Symbian OS v7.0 as well as other Version 7.0s phones. Phones following on from the Nokia 6600 include the Nokia 7700, 7610 and 9500, Panasonic X700 and Samsung SGH-D710.

Symbian OS phones are currently based on the following user interfaces open to C++ and Java programmers: the Series 80 Platform (Nokia 9200/9500 Communicator series), the Series 90 Platform (Nokia 7700), the Series 60 Platform (Nokia 7610, 6600, 6620, 7650, 3650, 3660, 3620, N-Gage, Siemens SX1, Sendo X, Panasonic X700 and Samsung SGH-D710), and UIQ (Sony Ericsson P800, P900, BenQ P30, Motorola A920, A925, A1000). The Nokia 6600 was the first smartphone to include

Java MIDP 2.0. Read on for a brief summary of the user interface families now available.

Mobile Phones with a Numeric Keypad

These phones are designed for one-handed use and require a flexible UI that is simple to navigate with a joystick, softkeys, jogdial, or any combination of these. Examples of this come from the Series 60 Platform. Fujitsu produces a user interface for a range of phones including the F2102v, F2051 and F900i for NTT DoCoMo's FOMA network. Pictured is the Siemens SX1.

Mobile Phones with Portrait Touch Screens

These mobile phones tend to have larger screens than those in the previous category and can dispense with a numeric keypad altogether. A larger screen is ideal for viewing content or working on the move, and pen-based interaction gives new opportunities to users and developers. The best current example of this form factor is UIQ, which is the platform for the Sony Ericsson P800 and P900, as well as BenQ P30 and Motorola's A920, A925 and A1000. The P800, P900 and P30 actually combine elements of full screen access and more traditional mobile phone use by including a numeric keypad, while the Motorola smartphones dispense with a keypad altogether. Pictured is the Sony Ericsson P900.

Mobile Phones with Landscape Screens

These mobile phones have the largest screens of all Symbian OS phones and can have a full keyboard and could also include a touch screen. With this type of mobile phone developers may find enterprise applications particularly attractive. A current example of the keyboard form factor is the

Series 80 Platform. This UI is the basis of the Nokia 9200 Communicator series, which has been used in the Nokia 9210i and Nokia 9290 and will be used in the Nokia 9500. Based on Series 90, the Nokia 7700 is an example of a touch screen mobile phone without keyboard aimed more at high multimedia usage.

When you're ready to use the Symbian OS C++ programming skills you've learned in this book, you'll want an up-to-the-minute overview of available phones, user interfaces and tools. For the latest information, start at ***www.symbian.com/developer*** for pointers to partner websites, other books, white papers and sample code. If you're developing technology that could be used on any Symbian OS phone, you can find more information about partnering with Symbian at ***www.symbian.com/partners***.

We wish you an enjoyable experience programming with Symbian OS and lots of success.

About the Authors

Richard Harrison, Lead Author

Richard joined Symbian (then known as Psion) in 1983 after several years teaching maths, physics and computer science. During that time he wrote a Forth language implementation for Acorn Computers, and wrote accompanying user manuals for the Acorn Atom and BBC Micro.

He has spent the majority of his time in system integration (SI), building and leading the SI team. He has produced user documentation for software for the Sinclair QL, the PC application software for the Psion Organiser I and the source code translator for the original version of OPL. Joint author of the Organiser II spreadsheet and principal designer and author of the Psion Series 3 and 3a word processors, he was also lead author of the Psion SIBO SDK team.

Educated at Balliol College, Oxford, with an MA in Natural Science (Physics), Richard also graduated from Sussex University with an MSc in Astronomy, and spent a further two years of postgraduate research in the Astronomy Group at Imperial College.

Alan Robinson

Alan Robinson joined Symbian shortly after its formation in 1998 and has mostly worked on documentation and examples in messaging and communications. Alan previously contributed to *Wireless Java for Symbian Devices* (Wiley, 2001) and *Symbian OS C++ for Mobile Phones Vol 1* (Wiley, 2003).

A graduate of Cambridge University with a BA in literature and philosophy, he became interested in applying logical theory and took a Computing MSc at Middlesex University. He has worked on developer kits for a startup company's messaging middleware platform, and for IBM's MQ Series.

Arwel Hughes

Arwel joined Symbian (then Psion) in 1993, working on documentation for the Series 3a and also some software development. Since the formation of Symbian, he has contributed documentation and examples for Symbian OS. This is rather like painting the famous Forth Bridge: just when you think you can see the end...

Arwel previously worked on IBM mainframes in roles including programmer and systems programmer for a number of companies including GKN, Prudential Assurance, Shell and Chase Manhattan Bank. He has a BSc in Applied Mathematics from Sheffield University.

Dominic Pinkman

Dominic joined Psion in October 1995 as a technical author. He has written and maintained documentation for APIs throughout Symbian OS, and was a co-author of the book *Symbian OS C++ for Mobile Phones Vol 1* (Wiley, 2003).

He has an MSc in Computer Science from the University of Kent and a BA in Modern Language studies from Leicester University.

Elisabeth Måwe

Elisabeth joined the system documentation team in 2000 and has since been designing and writing the Symbian Developer Library, specializing in operating system customization, kits, emulators, test, build and release tools. She has also been involved in training and usability management.

Elisabeth has a BA in Technical Communication/Information Design from Mälardalens Högskola and Coventry University, as well as an MA in Contemporary English Language and Linguistics from Reading University. After graduating in 1996 she worked as a technical author, information designer and web editor for various IT companies in the UK, producing documentation for both network management and market research software. She would like to thank Alex Peckover and Murray Read for providing both example code and technical expertise.

Greg Zaoui

Gregory Zaoui first joined Symbian in 1998, as a graduate software engineer with a 'Licence de Mathématiques' from the University of Strasbourg. He has been working on various projects for System Integration on build tools and release management. He then joined the newly created Test Solutions group in 2002, as a technical architect for TechView and other test tools.

His interests range from skiing and windsurfing to talmudic studies. Gregory would like to thank Richard Harrison and Paul Treacy for their excellent mentoring, as well as Clare Oakley (Test Solutions manager) without whom it would be impossible to talk consistently about test tools for Symbian OS. He also would like to acknowledge Elisabeth Måwe for her very active participation to the chapter, Konstantin Michaelov for his very useful example cases, Andrew Thoelke for the profiler bits, and all Test Solutions developers for their contribution. Gregory would also like to add special thanks to his dear wife Tamar for her constant encouragement and most precious help.

Nick Johnson

Since joining Symbian, Nick worked for a year in the Multimedia team helping implement next-generation Multimedia APIs and frameworks on Symbian OS and then subsequently transferred to Symbian's Marketing department, where he is now working as a developer consultant assisting Symbian partners with their Multimedia troubles.

Previous to this, Nick first spent three years studying Computer Science and Cybernetics at Reading University before subsequently spending two years working in 3D sound research at Sensaura Ltd. Here he spent time both developing new 3D sound algorithms and implementing the Xbox and GameCube ports of their cross-platform 3D audio middleware library. After leaving Sensaura, Nick spent a few weeks in the games industry working on 'Microsoft Train Simulator 2' before deciding that it just wasn't for him and instead joined Symbian.

Outside work, Nick enjoys learning Japanese and about Japanese culture, is a home cinema/film enthusiast, enjoys collecting/drinking rare liquors and vodkas and spends large amounts of time trying to convince friends that LaserDiscs are still the way forward . . .

Richard Potter

Richard joined Symbian in the summer of 2002 as a technical author. He works on documentation for the Security and Networking subsystems and has also written some Perl and Python scripts to aid the team.

Richard's unusual route to becoming a technical author includes advertising photography, being a singer/guitarist in a rock band, an MSc in Astrophysics, and an MPhil in Experimental Particle Physics working at the Stanford Linear Accelerator Center, Palo Alto, California. Many thanks to Jelte Liebrand for his advice.

Acknowledgements

Many thanks, in no particular order, to Marit Doving, Ian Weston, Omid Rezvani, Jason Dodd, Ade Steward, Ski Club, Iain Dowie, Sander Siezen, Nick 'I, Robot' Tait, Colin Turfus, Martin de Jode, Dave Jobling, Bart Govaert, Phil 'Ooc Clavdivs' Spencer, Karen Mosman, Colin Anthony, and System Management Group for the Symbian OS system model. Their contributions and support have all been very much appreciated. Much respect to the Laughing Gravy and Dingo Dave at the Stage Door for providing vital fuel. Original cover concept by Jono Tastard.

About the Cover

The mobile phone has traditionally connected the mouth to the ear – at Symbian's Exposium 2003, Symbian introduced the concept of Symbian OS enabling a new generation of connected communications devices by connecting the mouth to the ear to the eye. To realize this vision, the mobile phone industry is working together through Symbian to develop the latest technologies, support open industry standards, and ensure interoperability between advanced mobile phones as networks evolve from 2.5G to 3G and beyond . . .

Symbian licenses, develops and supports Symbian OS, the platform for next-generation data-enabled mobile phones. Symbian is headquartered in London, with offices worldwide. For more information see the Symbian website, ***http://www.symbian.com/***. 'Symbian', 'Symbian OS' and other associated Symbian marks are all trademarks of Symbian Software Ltd. Symbian acknowledges the trademark rights of all third parties referred to in this material. © Copyright Symbian Software Ltd 2004. All rights reserved. No part of this material may be reproduced without the express written permission of Symbian Software Ltd.

1

Symbian OS Fundamentals

Before we head into the deeper aspects of Symbian OS, we need to spend some time looking at some of the basic operations, programming patterns and classes that are common to all applications, and indeed, to almost all code that runs in the system.

What we are going to see here are the basic patterns that are used over and over again: building blocks that allow you to build safe and efficient code.

Symbian OS uses object-orientation, and is written in C++, with a tiny bit of assembler thrown in at the lowest levels. This means that the vast majority of applications are written in C++.

The use of C++ in Symbian OS is not exactly the same as C++ in other environments:

- C++ does more than Symbian OS requires – for example, full-blown multiple inheritance.

- C++ does less than Symbian OS requires – for example, it doesn't insist on the number of bits used to represent the basic types, and it knows nothing about DLLs.

- Different C++ communities do things differently because their requirements are different. In Symbian OS, large-scale system design is combined with focus on error handling and cleanup, and efficiency in terms of ROM and RAM budgets.

1.1 Object Creation and Destruction

One of the fundamental characteristics about object-oriented systems is the creation and destruction of objects. Objects are created, have a finite lifetime, and are then destroyed.

Symbian OS C++ for Mobile Phones, Volume 2. Edited by Richard Harrison
© 2004 Symbian Software Ltd   ISBN: 0-470-87108-3

Object creation and destruction is intimately tied up with the issue of cleanup, making sure that your applications are coded in such a way that they do not leak memory – a real issue for systems that may not be rebooted for long periods, if at all.

We'll first just look at the very basics of object creation and destruction in Symbian OS. In some ways this may give a misleading picture – the full picture will only emerge once we've looked at error handling and cleanup. This is because object creation, object destruction, error handling and cleanup are all intimately tied together with the aim of ensuring that objects, once created, are always destroyed when no longer needed.

There are two places in Symbian OS where you can create objects: the heap and the program stack.

1.1.1   The Heap (Dynamic Objects)

All threads have an associated heap, termed the default heap, from which memory can be allocated at runtime. This is where you put large objects, and objects that can only be built at runtime, including dynamic variable length strings. This is also where you put objects whose lifetimes don't coincide with the function that creates them – typically such objects become data members of the parent or owning object, with the relationship expressed as a pointer from owning object to owned object.

Memory is allocated from the thread's default heap, as and when required, using the C++ operator `new` and, very rarely, using user library functions such as `User::Alloc()`. If there is insufficient free memory, then an allocation attempt fails with an out-of-memory error.

In Symbian OS, classes that are intended to be instantiated on the heap are nearly always derived from the `CBase` class. This class gives you two things:

- zero initialization, so that all data members, member pointers and handles are initially zero

- a virtual destructor, so that the object can be properly destroyed. This is an important point when we come to look at cleanup issues later.

Strictly speaking, `CBase` is a base class for all classes that own resources, for example other bits of heap, server sessions, etc. What this means is that all `CBase` derived classes must be placed on the heap, but that not all heap objects are necessarily `CBase` derived.

The following code fragment shows a simple way of creating a heap-based object:

```
class CMyClass : public CBase
    {
public:
    CMyClass();
    ~CMyClass();
    void Foo();
private:
    TInt      iNumber;
    TBuf<32> iBuffer;
    }
```

```
CMyClass* myPtr = new CMyClass;
if (myPtr)
    {
    myPtr->Foo();  // can safely access member data & functions
    }
delete myPtr;
```

If there is insufficient memory to allocate the CMyClass object, then myptr is NULL. If allocation succeeds, myPtr points to the new CMy-Class object, and further, the data members iNumber and iBuffer are *guaranteed* to be binary zeroes. Conversely, the delete operator causes the object's destructor to be called before the memory for the object itself is released back to the heap.

There's one very important variation on this. Take a look at the following code:

```
CMyClass* myPtr = new (ELeave) CMyClass;
myPtr->Foo(); // can safely access member data & functions
...
delete myPtr;
```

The main difference here is that we have specified ELeave as part of the new operation. What this means is that instead of returning a NULL value when there isn't enough memory in which to create the CMyClass object, the operation 'leaves'. We'll explore what leaving means in more detail later when we investigate error handling and cleanup, but for the moment, think of it as an operation where the function returns immediately.

If the new operation doesn't leave, then it means that memory allocation for the new object has succeeded, the object has been created, and program control flows to the next C++ instruction, that is, the instruction myPtr->Foo(). It also means that there's no need to check the value of myPtr – the fact that the new operation returns means that myPtr will have a sensible value.

1.1.1.1   Ownership of Objects

In a typical object-oriented system such as Symbian OS, where objects are created dynamically, the concept of ownership is important. All objects need to be unambiguously owned so that it is clear who has responsibility for destroying them.

Use a destructor to destroy objects that you own.

1.1.1.2   Don't Forget About Objects – Even by Accident

Don't allocate objects twice. It sounds obvious, but allocating an object a second time, and putting the address into the same pointer variable into which you put the address of the first allocated object, means that you lose all knowledge of that first object. There is no way that a class destructor – or any other part of the C++ system – can find this object, and it represents a memory leak.

1.1.1.3   Deleting Objects

As we've seen, deleting an object is simply a matter of using the `delete` operator on a pointer to the object to be deleted. If a pointer is already zero, then calling `delete` on it is harmless. However, you must be aware that `delete` does not set the pointer itself to zero. While this does not matter if you are deleting an object from within a destructor, it is very important if the deletion occurs anywhere else. Double deletion doesn't always cause an immediate crash, and sometimes it leaves side-effects that only surface a long time after the real problem – the double delete – occurred. As a result, double deletes are very hard to debug.

On the other hand, double deletes are easy to avoid – just follow this little discipline:

> **C++ `delete` does not set the pointer to zero. If you delete any member object from outside its class's destructor, you must set the member pointer to NULL.**

1.1.2   The Program Stack (Automatic Objects)

The stack is used to hold the C++ automatic variables for each function. It is suitable for fixed size objects whose lifetimes coincide with the function that creates them. In Symbian OS, the stack is a limited resource. A thread's stack cannot grow after a thread has been launched; the thread is panicked – terminated abruptly – if it overflows its stack. This means that stack objects in Symbian OS shouldn't be too big, and they should only be used for *small* data items – for example, strings of a few tens of characters, say. Taking string data as an example, a good rule of thumb is to put anything larger than a file name on to the heap. However, it's quite acceptable to put pointers (and references) onto the stack – even pointers to very large objects.

You can control the stacksize in a `.exe`, through the use of the `epocstacksize` keyword of the `.mmp` file used to create the `.exe`. However, this only applies to console programs, servers or programs with no GUI – and not to GUI programs as they are launched with `apprun.exe`. GUI programs have a small program stack, of the order of 20k, and must be considered a valuable resource.

You can control the stacksize when you launch a thread explicitly from your program. However, avoid the temptation to create a large stack as this will eat into valuable resources.

We put built-in types, or classes that don't need a destructor, on to the program stack. They don't need a destructor because they own no data. This means that they can be safely discarded, without the need for any kind of cleanup. You simply exit from the function in which the automatic variable was declared. The type of objects that can go on to the stack are:

- any built-in type: these are given `typedefs`, such as `TInt` for a signed integer.

- any enumerated type, such as `TAmPm`, which indicates whether a formatted time-of-day is am or pm. Note that all enumeration types have names starting with a `T`, though enumerated constants such as `EAm` or `EPm` begin with `E`.

- class types that do not need a destructor, such as `TBuf<40>` (a buffer for a maximum of 40 characters) or `TPtrC` (a pointer to data, or to a string of any number of characters). `TPtrC` contains a pointer, but it only *uses* (rather than *has*) the characters it points to, and so it does not need a destructor.

For example, given a function `Foo()` in class `CMyClass`, we can create a `TInt` and a `TBufC<16>` type as automatic variables, use them in the body of the function, and then simply discard them, without doing any kind of cleanup when the function exits.

```
void CMyClass::Foo()
    {
    TInt myInteger;
    TBufC<16> buffer;
    ...
    // main body of the function
    } // variables are 'lost' on exit from the function.
```

1.2   Error Handling and Cleanup

In machines with limited memory and resources, such as those that Symbian OS is designed for, error handling is of fundamental importance. Errors are going to happen, and you can't afford not to handle them correctly.

Symbian OS provides a framework for error handling and cleanup and is a vital part of the system with which you need to become familiar. Every line of code that you write – or read – will be influenced by thinking about cleanup. No other Symbian OS framework has so much impact; cleanup is a fundamental aspect of Symbian OS programming. Because of this, we've made sure that error handling and cleanup are effective and very easy to do.

1.2.1   What Kinds of Error?

The easiest way to approach this is by focusing on out-of-memory errors.

These days, desktop PCs come with at least 256 MB of RAM, virtual memory swapping out on to 20 GB or more of hard disk, and users who expect to perform frequent reboots. In this environment, running out of memory is rare, so you can be quite cavalier about memory and resource management. You try *fairly* hard to release all the resources you can, but if you forget then it doesn't matter too much: things will get cleaned up when you close the application, or when you reboot. That's life in the desktop world.

By contrast, Symbian OS phones have as little as 4 MB of RAM, and often no more than 16 MB, although there are now devices with 32 MB. Nevertheless, by comparison with a PC, this is small; there is no disk-backed virtual memory. Remember that your users are *not* used to having to reboot frequently.

You have to face some key issues here – issues that don't trouble modern desktop software developers:

- You have to program efficiently, so that your programs don't use RAM unnecessarily.

- You have to release resources as soon as possible, because you can't afford to have a running program gobble up more and more RAM without ever releasing it.

- You have to cope with out-of-memory errors. In fact, you have to cope with potential out-of-memory for *every single operation* that can allocate memory, because an out-of-memory condition can arise in any such operation.

- When an out-of-memory situation arises that stops some operation from happening, you must not lose any data, but must roll back to an acceptable and consistent state.

- When an out-of-memory situation occurs part way through an operation that involves the allocation of several resources, you must clean up all those resources as part of the process of rolling back.

For example, consider the case where you are keying in data into an application for taking notes. Each key you press potentially expands the buffers used to store the information. If you press a key that requires the buffers to expand, but there is insufficient memory available, the operation will fail. In this case, it would clearly be quite wrong for the application to terminate – all your typing would be lost. Instead, the document must roll back to the state it was in before the key was processed, and any memory that was allocated successfully during the partially performed operation must be freed.

In fact, the Symbian OS error handling and cleanup framework is good for more than out-of-memory errors. Many operations can fail because of other environmental conditions – reading and writing to files, opening files, sending and receiving over communications sessions. The error handling and cleanup framework can make it easier to deal with those kinds of error too.

Even user input errors can be handled using the cleanup framework; as an example, code that processes the OK button on a dialog can allocate many resources before finding that an error has occurred. Dialog code can use the cleanup framework to flag an error and free the resources with a single function call.

There's just one kind of error that the cleanup framework can't deal with: programming errors. If you write a program with an error, you have to fix it. The best service Symbian OS can do for you (and your users) is to kill your program as soon as possible when the error is detected, with enough diagnostics to give you a chance to identify the error and fix it – hopefully, before you release the program. In Symbian OS, this is called a **panic**.

1.2.2 Panics

The basic function to use here is `User::Panic()`. This takes a string that we call the panic category, and a 32-bit integer number. The category and the number in combination serve as a way of identifying the programming error.

On real hardware, a panic simply displays a dialog titled *program closed*, citing the process name, the panic category and the number you identified.

The panic category should be no longer than 16 characters, but if it is longer, then only the first 16 characters are used.

The typical pattern is to code a global function (or a class static function, which is better) that behaves like this:

```
static void Panic(TInt aPanic)
    {
    _LIT(KPanicCategory, "MY-APP");
    User::Panic(KPanicCategory, aPanic);
    }
```

You can call this with different integer values.

The reason for raising a panic, as we have said, is to deal with programming errors, and not environmental errors such as out-of-memory. A common usage is to deal with the parameters passed to a function call. For some parameters, there may be a defined range of valid values, and accepting a value outside that range could cause your code to misbehave. You therefore check that the parameter is within range, and raise a panic if it's not. Another common case is where you want to ensure that a function is called only if your program is currently in a particular state, and raise a panic if not.

A commonly used way of raising panics is to use **assert macros**, of which there are two: __ASSERT_DEBUG and __ASSERT_ALWAYS. The first is compiled only into debug versions of your program, while the latter is compiled into both debug versions and production versions. As a general rule, put as many as you can into your debug code, and as few as you can into your release code. Do your own debugging; don't let your users do it for you.

Take the member function Foo() of a class CMyClass as an example:

```
enum TMyAppPAnic
    {
    ...
    EMyAppIndexNegative = 12,
    ...
    }
void CMyClass::Foo(TInt aIndex)
    {
    __ASSERT_ALWAYS(aIndex > 0, Panic(EMyAppIndexNegative));
    ...
    }
```

The pattern here is __ASSERT_ALWAYS(*condition*, *expression*), where the *expression* is evaluated if the *condition* is not true. Before attempting to execute the code, we make sure that the parameter aIndex is positive, and panic with code EMyAppIndexNegative if not. This gets handled by the Panic() function above, so that if this code were taken on a production machine, it would show **Program closed** with a category of

MY-APP and a code of 12. The number 12 comes from the enumeration TMyAppPanic containing a list of panic codes.

1.2.3   Leave and the Trap Harness

The majority of errors suffered by a typical program are environment errors, typically caused by a lack of suitable resources at some critical time, the most common of which is lack of memory. In such circumstances, it is usually impossible to proceed, and the program must roll back to some earlier known state, cleaning up resources that are now no longer needed.

Leave or *leaving* is part of that mechanism.

A leave is invoked by calling the static function User::Leave(), and this is called from within a *trap harness*. The trap harness catches errors – more precisely, it catches any functions that leave. If you're familiar with the exception handling in standard C++ or Java, TRAP() is like try and catch all in one, User::Leave() is like throw, and a function with L at the end of its name is like a function with throws in its prototype. It's very important to know what these functions are: the Symbian OS convention is to give them a name ending in L().

In a typical program, code can often enter a long sequence of function calls, that is, where one function calls another, which in turn calls another. If the function at the 'bottom' of this nested set of calls can't allocate memory (for example) for some object that it needs to create, then you might expect that function to return some error value back to its caller. The calling function then has to check the returned value, and it in turn needs to return this error value back to its caller. If each calling function needs to check for various return values, the code quickly becomes intractably complex.

Use of the trap harness and leave can cut through all this. User::Leave() causes execution of the active function to terminate, and to return through all calling functions, until the first one is found that contains a TRAP() or TRAPD() macro. Just as important, it rolls back the stack frames as it goes. You can find the TRAP() and TRAPD() macros in e32std.h. While their implementations are not meant to be understood, the main points are that TRAP() calls a function (its second parameter) and returns its leave code in a 32-bit integer (its first parameter). If the function returns normally, without leaving, then the leave code will be KErrNone (which is defined as zero). TRAPD() defines the leave code variable first, saving you a line of source code, and then essentially calls TRAP().

For example, in an application having word processing type behavior, a key press might cause many things to happen, such as the allocation of undo buffers and the expansion of a document to take the new character or digit. If anything goes wrong, you might want to undo the operation completely, using code such as:

```
TRAPD(error, HandleKeyL());
if(error)
    {
    RevertUndoBuffer();

    // Any other special cleanup
    User::Leave(error);
    }
```

If the `HandleKeyL()` function leaves, `error` contains the leave value; this is always negative. If the function does not leave, `error` contains 0.

While this can save a lot of hard coding effort, it does present a problem. A typical program allocates objects on the heap, quite often lots of them. It is also common behavior for a function to store a pointer to a heap object in an automatic variable, even if only temporarily. If, at the time a leave is taken, the automatic variable is the *only* pointer to that heap object, then knowledge of this heap object will be lost when a leave is taken – an effect known as a memory leak. As Symbian OS doesn't use C++ exceptions, then it needs to use its own mechanisms to ensure that this does not happen.

1.2.4   The Cleanup Stack

The cleanup stack addresses the problem of cleaning up objects that have been allocated on the heap, but to which the owning pointer is an automatic variable. If the function that has allocated the object leaves, then the object needs to be cleaned up.

Take the following code from the `cleanup` example; the **Use 3** menu item is handled by the following code:

```
case ECleanupCmdUse3:
    {
    CX x*=new (ELeave) CX;
    x->UseL();
    delete x;
    }
```

The `UseL()` function allocates memory and might leave. It's coded:

```
void CX::UseL()
    {
    TInt* pi new (Eleave) TInt;
    delete pi;
    }
```

This code can go wrong in two places. First, the allocation of `CX` might fail – if so, the code leaves immediately with no harm done.

Second, if the allocation of the TInt fails in UseL(), then UseL() leaves, and the CX object, which is pointed to only by the automatic variable x in HandleCommandL(), can never be deleted. The memory is orphaned – a memory leak has occurred.

1.2.4.1 Use the Cleanup Stack if Needed

What's actually happening here is that after the line

```
CX x*=new (ELeave) CX;
```

has been executed, the automatic x points to a cell on the heap, but after the leave, the stack frame containing x is abandoned without deleting x. That means that the CX object is on the heap, but no pointer can reach it, and it will never be destroyed.

The solution is to make use of the cleanup stack. In a console application, the cleanup stack must be explicitly provided; in a standard GUI application, it's provided by the UI framework. As its name implies, the cleanup stack is a stack of entries that represent objects that are to be cleaned up if a leave occurs. It is accessed through the CleanupStack class. The class has push and pop type functions that allow you to put items on to the stack.

One thing we should note here is that stack frame object destructors are not called if a leave happens. Only objects on the cleanup stack are cleaned up. Objects on the stack frame are not cleaned up.

Here's how we should have coded it:

```
case ECleanupCmdUse3:
    {
    CX x*=new (ELeave) CX;
    CleanupStack::PushL(x);
    x->UseL();
    CleanupStack::PopandDestroy(x);
    }
```

The cleanup stack class, CleanupStack, is defined in e32base.h. With these changes in place, here's what's happening:

- Immediately after we have allocated the CX and stored its pointer in x, we also push a copy of this pointer onto the cleanup stack.

- We then call UseL().

- If this doesn't fail (and leave), our code pops the pointer from the cleanup stack and deletes the object. We could have used two lines of code for this (CleanupStack::Pop(), followed by delete x), but this is such a common pattern that the cleanup stack provides a single function to do both.

- If `UseL()` *does* fail, then *as part of leave processing*, all objects on the cleanup stack in the current trap harness are popped and destroyed anyway.

We could have done this without the aid of the cleanup stack by using code like this:

```
case ECleanupCmdUse3:
    {
    CX* x = new(ELeave) CX;
    TRAPD(error, x->UseL());
    if(error)
        {
        delete x;
        User::Leave(error);
        }
    delete x;
    }
```

However, this is much less elegant. The cleanup stack works particularly well for a long sequence of operations, such as:

```
case ECleanupCmdUse3:
    {
    CX* x = new(ELeave) CX;
    CleanupStack::PushL(x);
    x->UseL();
    x->UseL();
    x->UseL();
    CleanupStack::PopAndDestroy();
    }
```

Any one of the calls to `UseL()` may fail, and it would begin to look very messy if we had to surround *every* `L` function with a trap harness just to address cleanup. Indeed it would reduce efficiency and increase the size of the code to unacceptable levels.

The native exception handling of C++ addresses the problem of automatics on the stack by calling their destructors explicitly, so that a separate cleanup stack isn't needed. C++ exception handling was not available at all on GCC, or reliably on Microsoft Visual C++, when Symbian OS was designed, so it wasn't an option to use it. Stack-based objects are not cleaned up in Symbian OS; only cleanup stack objects are.

The cleanup stack is, as the name implies, a stack, and you can add more than one item to it. So, for example, we can have:

```
case ECleanupCmdUse3:
    {
    CX* x1 = new(ELeave) CX;
```

```
CleanupStack::PushL(x1);
CX* x2 = new(ELeave) CX;
CleanupStack::PushL(x2);
x1->UseL();
CleanupStack::PopAndDestroy(2);
}
```

The 2 in the call to `PopAndDestroy(2)` causes the last two items to be removed from the stack and destroyed. When you do this, you must be careful not to remove more items from the cleanup stack than you put on to it, otherwise your program will panic.

Don't use the cleanup stack when you don't need it.

A common mistake when using the cleanup stack is to be overenthusiastic, putting all possible pointers to heap objects on to the cleanup stack. This is *wrong*. You only need to put a pointer to an object on to the cleanup stack to prevent an object's destructor from being bypassed. If the object's destructor is going to be called anyway, then you must not use the cleanup stack.

If an object is a member variable of another class (rather than an automatic like x), then it will be destroyed by the class's destructor, so you should never push a member variable to the cleanup stack.

Member variables are indicated by an `i` prefix, so code like this is *always* wrong:

```
CleanupStack::PushL(iMember);
```

This is likely to result in a double deletion, once from the cleanup stack and once from the class's destructor.

Another way of thinking about this is that once a pointer to a heap object has been copied into a member of some existing class instance, then you no longer need that pointer on the cleanup stack. This is because the class destructor takes responsibility for the object's destruction (assuming, of course, that the destructor is correctly coded to do this).

1.2.4.2 *What if* `CleanupStack::PushL()` *Itself Fails?*

Pushing to the cleanup stack may potentially allocate memory, and therefore may itself fail! You don't have to worry about this, because such a failure will be handled properly. But for reassurance, here's what happens under the covers.

Symbian OS addresses this possibility by always keeping at least one spare slot on the cleanup stack. When you do a `PushL()`, the object you are pushing is first placed on to the cleanup stack (which is guaranteed to work, because there was a spare slot). Then, a new slot is allocated. If *that* fails, then the object you just pushed is popped and destroyed.

The cleanup stack actually allocates more than one slot at once, and doesn't throw away slots that have been allocated when they are popped. So pushing and popping from the cleanup stack are very efficient operations.

Since the cleanup stack is used to hold fairly temporary objects, or objects whose pointers have not yet been stored as member pointers in their parent object during the parent object's construction, the number of cleanup stack slots ever needed by a practical program is not too high. More than 10 would be very rare. So the cleanup stack itself is very unlikely to be a contributor to out-of-memory errors.

1.2.4.3   *CBase and the Cleanup Stack*

In the earlier examples, when we push x to the cleanup stack, we actually invoke the function `CleanupStack::PushL(CBase* aPtr)`, because `CX` is derived from `CBase`.

When a subsequent `PopAndDestroy()` happens, this function can only call the destructor of a `CBase`-derived object. When we looked at the basics of object creation and destruction, we saw that `CBase`'s destructor was a virtual function. What this means is that any object derived from `CBase` can be pushed onto the cleanup stack and, when it is popped and destroyed, its destructor is called (as you would expect).

> **The cleanup stack and C++ destructors make it very easy for a programmer to handle cleanup. Use the cleanup stack for objects pointed to only by C++ automatics. Use the destructor for objects pointed to by member variables. It just works. You very rarely need to use TRAP(). The resulting code is easy to write, compact, and efficient.**

One thing we should say here about `CBase`-derived classes is that they should not be allocated on the stack as automatics. The zero initialization will not work, so the class may not behave as expected. These classes are designed exclusively to be used on the heap; their behavior is undefined if used in stack context. During a leave, the destructor of an object may not be called. The methods used to resolve this all assume that the object is heap based. Most `CBase` classes have private constructors, which are called from static factory functions usually called `NewL()` and/or `NewLC()`.

1.2.4.4 Two-phase Construction

The cleanup stack is used to hold pointers to heap-based objects so that they can be cleaned up if a leave occurs. This means you must have the opportunity to push objects on to the cleanup stack. One key situation in which this would not be possible when using normal C++ conventions is in between the allocation performed by new and the invocation of a C++ constructor that follows the allocation.

This problem requires us to invent a new rule: that C++ constructors cannot leave. We also need a work-around: two-phase construction.

To see why C++ constructors must not leave, take a look at the code taken from cleanup. The example has a class called CY that contains a member variable which points to a CX. Using conventional C++ techniques, we allocate the CY from the constructor:

```
class CY : public CBase
    {
public:
    CY();
    ~CY();
public:
    CX* iX;
    };
```

```
CY::CY()
    {
    iX = new(ELeave) CX;
    }
CY::~CY()
    {
    delete iX;
    }
```

The cleanup example's **Use 4** menu item calls cleanup-friendly code, as follows:

```
case ECleanupCmdUse4:
    {
    CY* y = new(ELeave) CY;
    CleanupStack::PushL(y);
    y->iX->UseL();
    CleanupStack::PopAndDestroy();
    }
```

Looks good, doesn't it? We have used C++ constructors in the usual way, and we've used the cleanup stack properly too. Even so, this code *isn't* cleanup safe. It makes *three* allocations as it runs through:

- The command handler allocates the CY: if this fails, everything leaves and there's no problem.

- The CY constructor allocates the CX: if this fails, *the code leaves, but there is no* CY *on the cleanup stack*, and the CY object is orphaned.

- CX::UseL() allocates a TInt: by this time, the CY *is* on the cleanup stack, and the CX will be looked after by CY's destructor, so if this allocation fails, everything gets cleaned up nicely.

The trouble here is that the C++ constructor is called at a time when no pointer to the object itself is accessible to the program. The code

```
CY* y = new(ELeave) CY;
```

is effectively expanded by C++ to:

```
CY* y;
CY* temp = User::AllocL(sizeof(CY));   // Allocate memory
temp->CY::CY();                        // C++ constructor
y = temp;
```

The problem is that we get no opportunity to push the CY on to the cleanup stack between allocating the memory for the CY and the C++ constructor, which might leave. There's nothing we can do about this.

> **It's a fundamental rule of Symbian OS programming that no C++ constructor should contain any functions that can leave.**

To get around this, we need to provide a separate function to do any initialization that might leave. We call this function the **second-phase constructor**, and our convention is to call it ConstructL().

In the cleanup example, the class CZ is like CY but uses a second-phase constructor:

```
class CZ : public CBase
    {
public:
    static CZ* NewL();
    static CZ* NewLC();
    void ConstructL();
    ~CZ();
public:
    CX* iX;
    };
```

```
void CZ::ConstructL()
    {
    iX=new(ELeave) CX;
    }
```

`CZ::ConstructL()` performs the same task as `CY::CY()`, but the leaving function `new(ELeave)` is now in the second-phase constructor `ConstructL()`. This is now cleanup safe.

1.2.4.5 *Wrapping up `ConstructL()` in `NewL()` and `NewLC()`*

Working with the two-phase constructor pattern can be inconvenient because the user of the class has to remember to call the second-phase constructor explicitly.

To make this easier, and transparent, we use the `NewL()` and `NewLC()` patterns. The `CZ` class has a static `NewLC()` function that's coded:

```
CZ* CZ::NewLC()
    {
    CZ* self = new(ELeave) CZ;
    CleanupStack::PushL(self);
    self->ConstructL();
    return self;
    }
```

Because `CZ::NewLC()` is static, you can call it without any existing instance of `CZ`. The function allocates a `CZ` with `new(ELeave)` and then pushes it on to the cleanup stack so that the second-phase constructor can then safely be called. If the second-phase constructor fails then the object is popped and destroyed by the rules of the cleanup stack. If all goes well, it leaves it on the cleanup stack – that's what the C in `NewLC()` stands for. `NewLC()` is useful if, on returning, we want to refer to the new `CZ` from an automatic variable.

Often, however, we don't want to keep the new `CZ` on the cleanup stack, and we use the `NewL()` static function that can be coded:

```
CZ* CZ::NewL()
    {
    CZ* self = new(ELeave) CZ;
    CleanupStack::PushL(self);
    self->ConstructL();
    CleanupStack::Pop();
    return self;
    }
```

This is exactly the same as `NewLC()` except that we pop the `CZ` object off the cleanup stack *after* the second-phase constructor has completed successfully, that is, after the second-phase constructor has returned.

As you can see, the two implementations are almost the same, and you will commonly see `NewL()` implemented in terms of `NewLC()` like this:

```
CZ* CZ::NewL()
    {
    CZ* self = CZ::NewLC();
    CleanupStack::Pop();
    return self;
    }
```

One interesting thing to note in the `NewL()` implementation is that we have popped an item from the cleanup stack without destroying it. All we're doing here is putting a pointer on to the cleanup stack only for as long as there is a risk of leaving. The implication is that on return from the `NewL()`, ownership of the object is passed back to the caller of `NewL()` who then has to take responsibility for it.

`CZ::NewLC()` and `CZ::NewL()` operate as **factory functions** – static functions that act as a kind of constructor.

1.2.4.6   Other Forms of Cleanup

The cleanup mechanism is more general purpose than we've shown so far. In fact, not only can you delete an object, but you can perform many other cleanup operations, two of which we label as close and release. Closing is simply the result of calling `Close()` on the object to be cleaned up, while releasing is the result of calling `Release()`.

The reason for this is that sometimes you need to create and use an R type object as an automatic variable rather than as a member of a C class.

For example, you would call `Close()` if your object were a handle; you would call `Release()` to release some resource that you might be holding. It's important, of course, that the class instance that needs cleanup defines and implements a `Close()` function (or a `Release()`). Let's look at `Close()`, for example: instead of using `CleanupStack::PushL()` as we've done before, we use the templated function `CleanupClosePushL()`, a global non-member function, passing a reference to the class instance which is to be the subject of the cleanup operation.

```
...
RTestTwo two;
CleanupClosePushL(two);
...
CleanupStack::PopAndDestroy();
...
```

What's actually happening here is that `CleanupClosePushL()` pushes a `TCleanupItem` on to the cleanup stack. A `TCleanupItem` is a general purpose object that encapsulates an object to be cleaned up, and a static function that will close it. When `CleanupStack::PopAnd-Destroy()` is called, the cleanup operation is performed on that object. In this particular example, the object is `RTestTwo`, and the cleanup operation is the function `Close()` that is called on it.

The same logic applies to `Release()` except that we use `Cleanup-ReleasePushL()` instead, and `Release()` is called on the object to be cleaned up.

You can look up the `TCleanupItem` class in `e32base.h`. Anything pushed to the cleanup stack is actually a cleanup item. `Cleanup-Stack::PushL(CBase*)`, `CleanupStack::PushL(TAny*)`, `Cle-anupClosePushL()`, and `CleanupReleasePushL()` simply create appropriate cleanup items and push them on to the cleanup stack.

There is also a `CleanupDeletePushL()` function which is the same as `CleanupStack::PushL(TAny*)` except that in this case the class destructor is called. This is often used when we have a pointer to an `M` class object that needs to be placed on to the cleanup stack.

You can also create your own `TCleanupItem` and push it if the cleanup functions offered by these facilities are insufficient.

1.3 Naming Conventions

Like any system, Symbian OS uses naming conventions to indicate what is important, and to make understanding Symbian OS source code easier. Symbian OS source code and all SDKs adhere to these conventions.

One point to note is that APIs take on *American English spelling* so expect to see `Color`, `Center` and `Gray` rather than `Colour`, `Centre` and `Grey`.

1.3.1 Class Names

Classes use an initial letter to indicate the basic properties of the class; the main ones are as follows:

Category	Examples	Description
`T` classes, types	`TDesC`, `TPoint`, `TFileName`	`T` classes don't have a destructor. They act like built-in types. That's why the `typedefs` for all built-in types begin with `T`. `T` classes can be allocated as automatics (if they're not too big), as members of other classes, or on the heap.

(continued overleaf)

Category	Examples	Description
C classes	`CActive`, `CBase`	Any class derived from `CBase`. C classes are *always* allocated on the default heap. `CBase`'s `operator new()` initializes all member data to zero when an object is allocated. `CBase` also includes a virtual destructor, so that by calling `delete` on a `CBase*` pointer any C object it points to is properly destroyed.
R classes	`RFile`, `RTimer`, `RWriteStream`, `RWindow`	Any class that owns resources other than on the default heap. Usually allocated as member variables or automatics: in a few cases, can be allocated on the default heap. Most R classes use `Close()` to free their associated resources.
M classes, interfaces	`MGraphicsDeviceMap`, `MGameViewCmdHandler`, `MEikMenuObserver`	An interface consisting of virtual functions. A class implementing this interface should derive from it. M classes are the only approved use of multiple inheritance in Symbian OS: they act similarly to `interfaces` in Java. The old technical term was 'mixin', hence the use of M.
Static classes	`User`, `Math`, `Mem`, `ConeUtils`	A class consisting purely of `static` functions that can't be instantiated into an object. Such classes are useful containers of library functions.
Structs	`SEikControlInfo`	A C-style `struct`, without any member functions. There are only a few of these in Symbian OS: most later code uses T classes even for `structs`.

Some other prefixes are occasionally used for classes, in rare circumstances. The only one we'll encounter in this book is HBufC, for heap-based descriptors. Kernel-side programming uses D for kernel-side CBase-derived classes.

The distinction between T, C and R is very important in relation to cleanup. In general, T classes don't require cleanup because they allocate no resources, R classes acquire resources that need to be closed or released, and C classes need to be deleted.

One thing we need to be honest about here is that some M type classes are not pure interfaces, because some Symbian classes implement code. Nevertheless the majority of these types are pure interfaces.

1.3.2 Data Names

These also use an initial letter, excepting automatics.

Category	Examples	Description
Enumerated constant	EMonday, ESolidBrush	Constants or values in an enumeration. If it has a name at all, the enumeration itself should have a T prefix, so that EMonday is a member of TDayOfWeek. You will find that some #defined constants use an E prefix, in circumstances where the constants belong to a logically distinct set.
Constant	KMaxFileName, KRgbWhite	Constants of the #define type or const TInt type. KMax-type constants tend to be associated with length or size limits: KMaxFileName, for instance, is 256 (characters).
Member variable	iDevice, iX, iOppFleetView	Any non-static member variable should have an i prefix. The i refers to an 'instance' of a class.
Arguments	aDevice, aX, aOppFleetView	Any variable declared as an argument. The a stands for 'argument', not the English indefinite article. Don't use an for words that begin with a vowel!
Automatics	device, x, oppFleetView	Any variable declared as an automatic.

Static members aren't used in native Symbian OS code. Global variables, such as console, are sometimes used in .exes (though not in

DLLs). Globals have no prefix. Some authors use initial capitals for globals, to distinguish them from automatics; other authors use an initial lower case g. We have no strong views on the right way to do things here, preferring to avoid the issue by not using globals.

The `i` convention is important for cleanup. The C++ destructor takes care of member variables, so you can spot over-zealous cleanup code, such as `CleanupStack::PushL(iMember)` by using this naming convention.

As with class names, you should use nouns for value names, since they are objects, not functions.

1.3.3 Function Names

It's not the initial letter that matters so much here, as the final letter.

Category	Examples	Description
Non-leaving function	`Draw()`, `Intersects()`	Use initial capital. Since functions do things, use a verb, rather than a noun.
Leaving function	`CreateL()`, `AllocL()`, `NewL()`, `RunL()`	Use final `L`. A leaving function may need to allocate memory, open a file, etc. – generally, to do some operation which might fail because there are insufficient resources or for other environment-related conditions (not programmer errors). When you call a leaving function, you must always consider what happens both when it succeeds and when it leaves. You must ensure that both cases are handled. The cleanup framework of Symbian OS is designed to allow you to do this. This is the most important naming convention in Symbian OS.
LC functions	`AllocLC()`, `CreateLC()`, `OpenLC()`, `NewLC()`	An LC function leaves something on the cleanup stack when it returns. Commonly, this is a pointer to an allocated object, but more generally it's a cleanup item. If the function fails, then it leaves.

Category	Examples	Description
Simple getter	`Size()`, `Device()`, `ComponentControl()`	Get some property or member data of an object. Often getters are used when the member is private. Use a noun, corresponding with the member name.
Complex getter	`GetTextL()`	Get some property that requires more work, and perhaps even resource allocation. Resource-allocating getters should certainly use `Get` as a prefix; other than that, the boundary between simple and complex getters is not hard-and-fast.
Setter	`SetSize()`, `SetDevice()`, `SetCommandHandler()`, `SetCharFormatL()`	Set some property. Some setters simply set a member. Some involve resource allocation, which may fail, and are therefore also an `L` function.

1.3.4 Macro Names

Symbian OS uses the usual conventions for C preprocessor macro names:

- Use only upper case, and split words with underscores, creating names like `IMPORT_C` and `EXPORT_C`.

- For build-dependent symbols, use two leading and trailing underscores (`__SYMBIAN32__`, `__WINS__`). The symbols `_DEBUG` and `_UNICODE` are notable exceptions to this rule. Double underscores are also used to indicate guards.

1.4 Descriptors

One of the most common and recurring programming requirements in computing is the handling of strings.

In Symbian OS, strings are handled using descriptors. Descriptors are one of the most fundamental and heavily used entities in Symbian OS. They offer a safe and consistent mechanism for dealing with both strings and general binary data regardless of the type of memory in which they reside.

Although string handling was one of the motivators for descriptors, they are also useful for handling more general data in a safe and convenient manner.

Descriptors have had a reputation for being awkward and difficult to use, but in reality they are straightforward once you understand the concept behind them.

1.4.1 What are Descriptors?

The fundamental idea behind a descriptor is that it is an object that represents data in a defined location in memory. It represents data by encapsulating a pointer to the data together with the length of that data. All access to data is done through the member functions of the descriptor, and this means that any operation on the data always guarantees that no memory location outside the bounds represented by the descriptor can be overwritten. It is possible for your code to attempt an illegal access, but this is treated as bad programming rather than an environment or resource problem and raises an exception known as a panic. This is what makes descriptors a safe vehicle for manipulating strings and general data.

The area occupied by data represented by a descriptor is considered to be non-expandable, even though the length of data actually represented can shrink, or expand to fill that area. The area is created, or defined, when the descriptor is created.

One point we should note here – by length, we mean a logical length, for example the number of characters. This is different from the size of the descriptor, which is the number of bytes occupied by the data; length and size are not necessarily the same. For example, the length of a descriptor containing a string of five Unicode characters is 5, but the size is 10 bytes.

The other important idea is that a descriptor gives you a neat way of passing data between functions. We'll look at this later. First let's look at some of the basic ideas behind descriptors and get a feel for them.

There are three types of descriptor that you can create in your code:

- Buffer descriptors
- Pointer descriptors
- Heap descriptors.

1.4.1.1 Buffer Descriptors

In a buffer descriptor, the data is part of the descriptor object (Figure 1.1), and the whole descriptor object can either go on to the program stack, or it can be a data member of some class. As well as containing the data, the object also contains the length of the data.

A buffer descriptor is a `TBuf` type or a `TBufC` type, and they can be used where a `char []` would be used in C. These two types differ slightly in their behavior, but this is something we'll look at later.

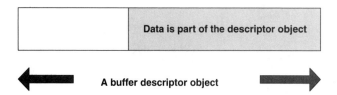

Figure 1.1   Buffer descriptor

There are a number of ways of creating a `TBuf`, but here's one.

```
{
...
_LIT(KTxtHelloWorld,"Hello World!");

TBuf<16> buffer(KTextHelloWorld);
...
}
```

As you can see, the class is templated and takes an integer value. This value defines the maximum length of data that the descriptor can hold; in effect it defines the size of the data area. In this case, the size of the data area is such that it can hold 16 characters. Remember that this does not necessarily mean that the data area occupies 16 bytes.

After construction, the buffer object holds the string 'Hello World!', and has a length of 12. There is no need for a trailing NULL character to flag the end of your string. The data and the length of that data are neatly encapsulated by the descriptor object.

As we said earlier, the program stack is a very limited resource in Symbian OS, so you do need to be aware of the size of buffer descriptors that you create.

1.4.1.2 *Pointer Descriptors*

In a pointer descriptor (Figure 1.2), the descriptor object is separate from the data it represents. The descriptor object itself can either go on to the program stack, or be a data member of some class, but the data itself can be anywhere that is addressable by your code.

A pointer descriptor is a `TPtr` type or a `TPtrC` type; a `TPtrC` can be used where a `const char*` would be used in C. Again, there are slight differences between a `TPtrC` and a `TPtr`, which we'll come to later.

You can create a `TPtrC` in a number of ways, for example by taking a pointer to a memory location and a length value that represents the length of that data.

```
Void foo(TUint16* aBuf, TInt aLength)
   {
   TPtrC myPtr(aBuf,aLength);
   ...
   }
```

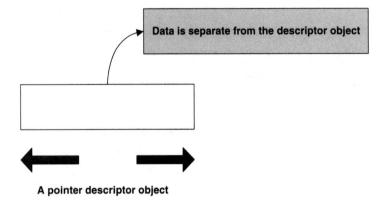

A pointer descriptor object

Figure 1.2   Pointer descriptor

Another common usage is to create a `TPtrC` to represent data already held by an existing descriptor. In this case the length of the data is the length of the data in the source descriptor.

```
{
_LIT(KTxtHelloWorld,"Hello World!");

TBufC<16> buffer(KTextHelloWorld);
TPtrC myPtr(buffer);
...
}
```

A pointer descriptor has the useful property that it can be passed into a function without the need to pass the data itself. Pointer descriptors are also commonly used to represent fragments of other strings.

1.4.1.3   Heap Descriptors

The last descriptor type is the heap descriptor (Figure 1.3). As its name suggests, a heap descriptor is one that provides a fixed length buffer, just like the buffer descriptor we saw earlier where the data is part of the descriptor object, but here the whole descriptor object lives on the heap.

A heap descriptor is an `HBufC` type, and you use this where a `malloc()`'d cell would be used in C. This is the only class in Symbian OS whose name begins with `H`.

These types are useful when the amount of data may be too big for a buffer descriptor, or where the data to be held by the descriptor cannot be determined until runtime.

As you can see from the code fragment, you need an `HBufC*` type in your code, and this can either go on to the program stack or be a data member of some class. There are a number of functions available for creating a heap descriptor, but the static `HBufC::NewL()` is a common

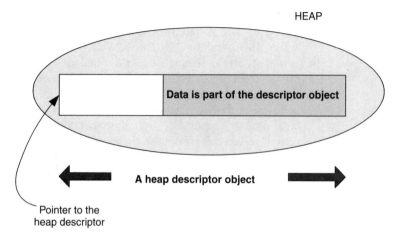

Figure 1.3 Heap descriptor

one. If you use the program stack, always be aware of the cleanup issues we raised earlier to ensure that there is no risk of memory leaks.

```
{
_LIT(KTxtHelloWorld,"Hello World!");

HBufC* buffer = HBufC::NewL(256);
*buffer = KTxtHelloWorld;
...
}
```

Although these three types of descriptor seem to be different, the underlying class structure makes such differences transparent, allowing them to be treated in the same way.

1.4.2 Modifiable and Non-modifiable Descriptors

You may have noticed that buffer descriptors and pointer descriptors both come in two types: `TBufC` and `TBuf`, `TPtrC` and `TPtr`.

We use the `C` in the type names to mean constant or non-modifiable. The idea is that with `TBufC` and `TPtrC` descriptors, you can initialize them with data, but you can't modify that data *through the descriptor*. For example, if you have a `TBufC` containing a string, you can get the length of the string, find substrings, compare it with other strings and so on. What you can't do is change part of the content, for example by appending a second string or replacing the first two characters with something completely different. In fact the only way to change the content of this type of descriptor is to reassign new data. These descriptor types are not constant in the usual C++ sense; they just have a limited set of functions, none of which allow you to make modifications to the data.

These descriptors give you a useful vehicle for passing data to functions and accessing data within those functions.

On the other hand, the `TBuf` and `TPtr` types do allow you to change their content. The classes give you a much richer set of functions that let you append characters, delete characters, format numbers and so on. The list is quite long, and we'll look at some examples later to give you a flavor. We refer to these as modifiable descriptors.

The `TBuf/TPtr` types differ from the `TBufC/TPtrC` types by having an extra machine word that contains the maximum length that the descriptor data can expand to. The size of the data area is fixed when the descriptor is created, but the length of the content can vary, provided it remains within the confines of this data area.

In a sense, `TBuf` offers a superset of behavior over and above that provided by `TBufC`.

The heap descriptor `HBufC` is a non-modifiable type like `TBufC`, except that, as we have seen, it lives on the heap. However, there are other ways of modifying the `HBufC` contents, as we shall see.

1.4.3   More on Buffer and Heap Descriptors

The buffer descriptors, `TBuf` and `TBufC`, are templated, and the template value determines the length of the data area. Try to assign data that's longer than the template value and the descriptor will protect itself by panicking.

The heap descriptor is slightly different. It is allocated as a single cell from the heap. To save memory, the length of the data area is implied by the length of the underlying heap cell. When memory is allocated from the heap, the length of a heap cell is usually rounded up to a value that depends on the underlying hardware. This means that you can end up with a descriptor that's slightly bigger than one you asked for.

1.4.4   The Underlying Class Structure

Having briefly seen the concrete descriptor objects, we now need to look at the class structure underlying them so that you can get a fuller understanding of what's going on.

The concrete descriptor classes we saw earlier are all derived from a set of base classes. The base classes are abstract, which means that you can't instantiate them. They implement nearly all the behavior of descriptors. The most important base classes are `TDesC` and `TDes` (Figure 1.4). We won't bother looking at `TBufCBase`, as this just provides some implementation detail.

TDesC
This is the most fundamental class; it defines and implements the const functions such as `Length()`, `Size()`, `Find()`, etc., that all descriptors

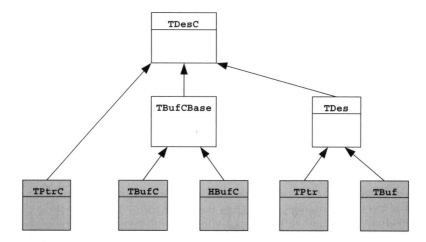

Figure 1.4 Descriptor classes

can use. It also contains a data member that holds the current length of the data that a descriptor represents.

The three concrete descriptor classes TPtrC, TBufC and HBufC are all derived from TDesC, and they only inherit TDesC's const functions and, therefore, cannot make changes to their own content. The concrete descriptor classes themselves just add the functionality that allows you to initialize them with data, either at construction time or later through the Set() functions.

TDes

This class is derived from TDesC, but adds to TDesC by defining and implementing all the functionality that lets you modify the data, for example Copy(), Append(), Format(), Replace(), Trim() and many, many more. In addition it has a data member that is used to hold the maximum length of data that the descriptor can hold.

The two concrete descriptor classes TPtr and TBuf are derived from TDes, which means that they inherit both TDes *and* TDesC behavior. The concrete descriptor classes themselves just add the functionality that allows you to initialize them with data.

The following code shows TBufC<5> and TBuf<12> descriptors both initialized with the string 'Hello'.

```
{
_LIT(KTxtHello,"Hello");
TBufC<5> buffer1(KTextHello);
TBuf<12> buffer2(KTextHello);
...
}
```

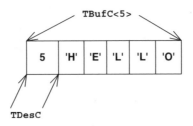

Figure 1.5   TBufC<5> object layout

The TBufC<5> object layout (Figure 1.5) looks like this, showing the TDesC contribution to the object. The template parameter value defines the length of the data area available. The length of the data is also 5; it could have been less than 5, but definitely not more than 5.

Compare this with the layout for the TBuf<12> object (Figure 1.6), showing the TDesC and TDes contributions. The length is still 5, but the maximum length is 12, meaning that there is room to insert, or append, another seven items.

The differences between a TPtrC and a TPtr are fundamentally the same, except that both objects refer to the data through a pointer. Figures 1.7 and 1.8 show the equivalent layouts.

An HBufC type (Figure 1.9) is similar to the TPtrC type except that the HBufC object is on the heap.

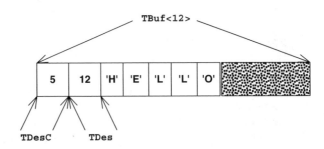

Figure 1.6   TBuf<12> object layout

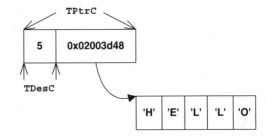

Figure 1.7   Pointer descriptor – TPtrC

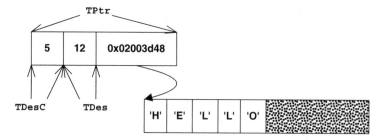

Figure 1.8 Pointer descriptor – TPtr

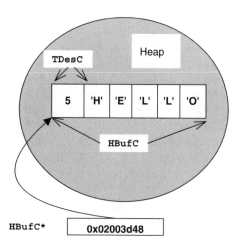

Figure 1.9 HBufC type

1.4.5 Using the Abstract Classes in Interfaces

You cannot instantiate TDesC and TDes because they have protected constructors. In this sense they are abstract classes. But in functions designed to manipulate strings, and indeed any other kind of data that you may be using your descriptors for, you should always use the base classes as arguments. Use:

- const TDesC& to pass a descriptor for data that you will read from, but not attempt to change

- TDes& to pass a descriptor for data that you want to change.

Take this global function called greetEntity() as an example:

```
void greetEntity(TDes& aGreeting, const TDesC& aEntity)
    {
    aGreeting.Append(aEntity);
```

You pass one modifiable descriptor that contains the greeting text ('Hello', perhaps), and you pass a non-modifiable descriptor that contains the entity to be greeted ('World', maybe). The function simply appends the entity on to the greeting. You can call this function using code like this:

```
_LIT(KTxtHello,"Hello");
_LIT(KTxtWorld,"World");
...
TBuf<12> helloworld(KTxtHello);
greetEntity(helloworld,KTxtWorld);
```

When the call to greetEntity() returns, helloworld contains the text 'Hello World'. This is a very simple example that nevertheless shows a heavily used pattern in Symbian OS programming. The important point to note is that you can pass any of the concrete descriptor types into this function. If we had prototyped the function to take, say, a TBuf and a TBufC type instead, then we would have needed to add another five variants of the function to cope with all possible permutations, which is clearly wasteful.

You may be wondering how a TDes type can know that the object passed is a TBuf, say, rather than a TPtr. The answer is that the descriptor classes are not totally honest object oriented citizens. The top four bits of the iLength member of the base class TDesC, which holds the length of descriptor data, is reserved for use as a class identifier. In effect it gives a TDesC type knowledge of the concrete descriptor type of which it is part.

1.4.6   The Literal _LIT

In some of the code fragments we saw earlier, we used a macro called _LIT.

```
_LIT(KTxtHelloWorld,"Hello World!");
```

This is what we call a literal. _LIT is a macro that produces a TLitC object that associates a symbol with a literal value. Although this is not a TDesC type, it behaves as though it were.

The macro builds into the program binary an object of the type TLitC, which stores the required string. The binary layout of TLitC is designed to be identical to a TBufC; this allows a TLitC to be treated (using the proper casts) as a TDesC.

So, if you code

```
_LIT(KHelloRom,"Hello");
const TDesC& helloDesc = KHelloRom;
```

then the TDesC& reference looks exactly like a reference to a TBufC. More importantly, compared to an older deprecated literal form, _L, no runtime temporary and no inline constructor code are generated, so ROM budget is significantly reduced for components that contain many string literals.

Literals have two operators – & and () – that return a const TDesC* and a const TDesC& type respectively. Here's how you might use them:

```
_LIT(KTxtMatchString,"Hello");
_LIT(KFormat2,"Text is %S");
...
TInt length;
length = KTxtMatchString().Length();
...
TBuf<256> x;
x.Format(KFormat2,&KTxtMatchString);
```

In older versions of Symbian OS the only literal available was _L, which produces a TPtrC from a literal value and can be used without a name, so that you can write in your program:

```
const TDesC& helloRef = _L("hello");
```

While this works, it is *inefficient* for a number of reasons, including the following:

- When a program that includes this code is built, the string, including the trailing \0, is built into the program code. We can't avoid the trailing \0, because we're using the C compiler's usual facilities to build the string.

- When the code is executed, a temporary TPtrC is constructed as an invisible automatic variable with its pointer set to the address of the first byte of the string and its length set to 5.

- The address of this temporary TPtrC is then assigned to the const TDesC& reference, helloRef, a fully-fledged automatic variable.

- The TPtrC constructor has to count all of the characters in the string to see how long it is – this is a particular cause of inefficiency.

The code works provided that the reference is used only during the lifetime of the temporary TPtrC or the TPtrC pointer and length are copied if the descriptor is required outside the lifetime.

Another issue is that in the second step above, constructing a TPtrC, including the pointer, the length and the four-bit descriptor class identifier are always required. Code has to be built and run, wasting code bytes and execution time.

For this reason, _L is *deprecated* for shipped applications; it is much better to use _LIT. It is nevertheless still acceptable to use it in test programs, or other programs where you know that memory use is not critical.

Programs should use resource files to contain their language-dependent literal strings; we'll cover resource files in a later chapter. However, many system components require literals for system purposes that don't get translated for different locales.

1.4.7   Some Standard Descriptor Functions

We mentioned that descriptors contain many convenience functions – const functions in TDesC and both const and non-const functions in TDes. This is a quick tour to give you a taster. For more information, see any standard product SDK. You'll also see many descriptor functions throughout this book.

If you wish, you can write your own descriptor functions; just put them into another class (or, exceptionally, no class at all) and pass descriptors to your functions as const TDesC& or TDes& parameters. Symbian OS itself provides utility classes that work in this way; perhaps the most fundamental is TLex, which provides string scanning functions and includes code to convert a string into a number.

1.4.7.1   Basics

You can get basic information about descriptor data: use Length() to find the logical length of the data, for example how many characters it is, and Size() to find out how many bytes the data occupies. Be careful not to use Size() where you really mean Length().

TDes provides Maxlength(), which tells you the maximum length of the data area; if the descriptor is intended to hold a string, then this is the maximum number of characters that the descriptor can hold. Any manipulation function that causes this to be exceeded causes a panic.

If you write your own string handling functions, you would normally construct them using the descriptor library functions. In some circumstances, you may wish to access descriptor contents directly and manipulate them using C pointers – Ptr() allows you to do this – *but be careful*. In particular, make sure that you honor MaxLength() if you're modifying descriptor data, and make sure that any function you write panics if asked to do something that would overflow the allocated MaxLength() – the standard way of ensuring this is to call SetLength() before modifying the content, which will panic if the length exceeds MaxLength().

Manipulating Data

TDes's Copy() function copies data to the descriptor starting at the beginning, while Append() can copy additional data to the descriptor, starting where the existing data stops. There are variants of the Copy() function that perform case or accent folding when copying character data:

- Insert() inserts data into any position, pushing subsequent data towards the end of the descriptor data area.

- Delete() deletes any sequence of data, moving down subsequent data to close the gap.

- Replace() overwrites data in the middle of the descriptor data area.

1.4.7.2 Substring Functions

The TDesC functions Left(), Right(), and Mid() are especially applicable to strings. They let you construct TPtrC descriptors that represent portions of existing descriptor data. For example, you might have a buffer descriptor containing text that needs parsing. Following the parse operation you might have a number of TPtrC descriptors, each of which represents syntactically significant parts of the buffer content. The original descriptor data will not have been changed, deleted, or copied.

1.4.7.3 Formatting

TDes::Format() is a bit like sprintf(): it takes a format string and a variable argument list and saves the result in the descriptor. Functions such as AppendFormat() are similar but append the result to existing descriptor data. Functions such as Format() are implemented in terms of AppendFormat().

Many lower-level functions exist to support AppendFormat(): various Num() functions convert numbers into text, and corresponding AppendNum() functions append converted numbers on to an existing descriptor. For simple conversions, the AppendNum() functions are much more efficient than using AppendFormat() with a suitable format string.

In C, scanning functions are provided by sscanf() and packaged variants such as scanf(), fscanf() and so on. Similar functions are available in Symbian OS through TLex and associated classes which scan data held in descriptors. These functions are relatively specialized and it was not thought appropriate to implement them directly in TDesC.

1.4.8 Representing Binary Data

If you look through the header files defining these descriptor types in any of the product SDKs, you will find that they are conditionally defined in terms of more fundamental types; for example, TDesC is defined as either a TDesC16 or a TDesC8.

Descriptors have two implementations: one where the data is represented as a 16-bit quantity and the other where the data is represented by an eight-bit quantity. By default, descriptors are defined in terms of 16-bit quantities, reflecting the fact that today strings are defined in terms of Unicode characters, which are two-byte entities. This means that if you use, say, TDesC in your code, it really means TDesC16.

This is all well and good if you intend to use your descriptors for handling strings. However, if you intend your descriptor to handle binary data, for example to act as a buffer for data passed across a communications link, then it is usually easier to consider this as a sequence of bytes, plain and simple, and you would use the eight-bit version.

So, for example: instead of declaring a buffer TBuf<128>, you would explicitly use TBuf8<256>. All the functionality of descriptors is still available to you, but it operates on eight-bit data rather than 16-bit data.

1.4.9 More on HBufC

There are some features about heap descriptors that are useful, but also some things that are not obvious at first sight.

1.4.9.1 The Des() Function

You can't directly modify an HBufC type; however, you can use the Des() function to create a TPtr through which you can change the HBufC, and this can prove extremely convenient. For example:

```
{
_LIT(KTxtHelloWorld,"Hello World!");
_LIT(KTxtFriends,"Friends!");

HBufC* buffer = HBufC::NewL(256);
*buffer = KTxtHelloWorld;
TPtr temp = buffer->Des();
temp.Replace(6,6,KTxtFriends);
...
}
```

Prior to calling Replace(), the length of both the TPtr and the HBufC is 12, as you might expect; after the call the length is 14. The value returned by calling Length() on the TPtr is the same as the value returned by calling Length() on the HBufC. The maximum length of the TPtr is the length of the HBufC data area (not its size!).

There is a common mistake that many people tend to make once they become familiar with the Des() function. When they have an HBufC that is to be passed to a function that takes a const TDesC& type, they assume that they first have to create a `TPtr` using `Des()` and pass this into the function. While this works, it is far simpler and much more efficient to simply dereference the heap descriptor; for example, if you have a function `foo()` prototyped as

```
void foo(const TDesC& aDescriptor);
```

then the following code will work fine:

```
HBufC* buffer = HBufC::NewL(256);
...
foo(*buffer);
```

1.4.9.2 Creating an *HBufC* from Another Descriptor

Programs often need to create an `HBufC` from an existing descriptor. For example, you might want to pass data into a function and then do some processing on that data. Typically you do not know the length of the data being passed to your function, so the easiest way to handle this is to create an `HBufC` descriptor within your function. We use the descriptor function `Alloc()`, or more usually one of its variants: `AllocL()` and `AllocLC()`. Here's a simple example.

```
_LIT(KTxtHelloWorld,"Hello World!");

myFunctionL(KtxtHelloWorld);
```

```
void myFunctionL(const TDesC& aBuffer)
    {
    HBufC* mydata = aBuffer.AllocLC();
    // do something with this
    }
```

`AllocL()` (and the `Alloc()` and `AllocLC()` variants) creates an `HBufC` that is big enough to contain `aBuffer`'s data, and then copies the data from `aBuffer` into `mydata`. `AllocL()` does three jobs in one: it works out the length of the source descriptor, it allocates the heap descriptor, and then it copies the data into the heap descriptor.

A *common mistake* is to do this explicitly (calling `Length()` on `aBuffer`, then calling `NewL()` to create the `HBufC`, and then calling `Copy()` on the `HBufC`), which is wasteful and inefficient.

1.4.9.3 How Big is an *HBufC*?

As we saw earlier, the length of a heap descriptor's data area is implied by the length of the underlying heap cell that hosts the descriptor. Heap cells have a granularity that depends on the specific hardware to which Symbian OS has been ported. When allocating memory, Symbian OS always rounds up to the next nearest granularity boundary, so if you create an HBufC of a specific length, you may end up with a bigger descriptor than you asked for.

This becomes important in cases where, for example, you pass the descriptor to some function whose purpose is to return data in that descriptor. Don't assume that the length of data returned in the descriptor is the same as the length you used to allocate the descriptor.

1.4.9.4 Changing the Size of the Descriptor

Although descriptors are not dynamically extensible, you *can* reallocate a heap descriptor so that it is bigger or smaller than the original. You do this using ReAllocL() (or the ReAlloc() and ReAllocLC() variants). Take the following code fragment as an example:

```
_LIT(KTxtHelloWorld,"Hello World!");

HBufC* myData = HBufC::NewL(16);
*myData = KTxtHelloWorld;
myData = myData->ReAllocL(256);
```

We first create a heap descriptor with a requested maximum length of 16, and then initialize it with the string 'Hello World!'. The call to ReAllocL() creates a new heap descriptor and, if successful, copies the original data into it, and returns the *new address*. The original copy is deleted. Of course, if there is insufficient memory, then the function leaves. After the reallocation has taken place, myData is likely to point to a different heap cell. There may be times when it points to the original heap cell, but it is safer to assume that it doesn't. This is particularly important if you use the Des() function, in which case you need to re-Set() any TPtr returned by Des().

1.5   Active Objects

In the old days, programs were written such that, every so often, the program would decide to check for user input and would then process it. With GUI systems, however, the user is in control – their input is the focus of the application's existence.

In GUI systems, application programs *wait for* events, for example keyboard input, pointer input, completion of an I/O request, timer events, etc. These events and the services associated with them are provided by *asynchronous service providers*. Application programs then simply *respond* to these events.

Symbian OS provides some fundamental building blocks for event-handling systems known as active objects. Here, we'll try to give a flavor of how they work and why they are important.

1.5.1 The Asynchronous Service

At the heart of what we call event-driven programming is the concept of the asynchronous service.

When a program requests a service of some other component or part of the system, that service can be performed either *synchronously* or *asynchronously*. A synchronous service is the 'usual' pattern for function calls: when the function returns, either the service requested has been performed, or some kind of error has been returned. An asynchronous service is *requested* by a function call, but completion occurs later and is indicated by some kind of signal. Such a service request may be handled by another thread. Between the issue of the request and the receipt of the signal, the request is said to be *pending*. The requesting program may do other processing while the request is pending, but ultimately it will wait. The operating system wakes up the program when completion of any of its pending requests is signaled.

This is the crux of a typical event-driven program: it makes requests for keyboard input, pen events, timer events, etc., and once it has nothing else to do, it sits and waits. When an event occurs, the program comes out of its wait state, responds to the event in whatever manner is appropriate, typically issues a request for another similar event, and then waits again.

Figure 1.10 shows the typical pattern in a simplistic way. Ignoring startup, shutdown, and many other issues, there's a central core that waits

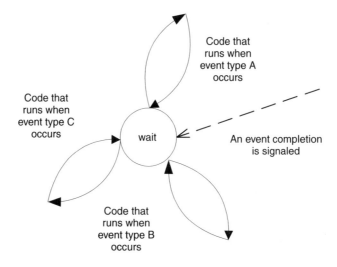

Figure 1.10 Understanding an event-driven program

and, when an event occurs, calls the code that can deal with that event. Control then flows back to the central core.

Consider that Figure 1.10 represents code running in a single thread. If only one event-handling section of code can run at any one time, and once that code runs nothing else in that thread can run until control flows back to the central core, then you begin to see the idea behind active objects, i.e. **non-preemptive multitasking** within the context of a single thread.

The code that runs when an event occurs is represented by an active object. The decision as to which active object should run next is taken by what we call the active scheduler.

Asynchronous requests are more often than not handled by servers. While it is useful to know this, it is not a prerequisite to understanding the way in which active objects work.

1.5.2 Multitasking and Preemption

Symbian OS implements **preemptive multitasking** so that it can run multiple applications and servers concurrently. Active objects, however, are used to implement **non-preemptive multitasking** within the context of a single thread.

In our simple visualization earlier, we assumed that events, or request completions, occur in some kind of predictable order, and that our thread gets a chance to deal with an event before the next event occurs. Of course, it doesn't happen like that. The timing of events is unpredictable, and when our thread gets a chance to run, there may be more than one event ready to be handled. This is why active objects, like threads, have priorities that affect their scheduling. On completion of an active object's execution, control returns to the active scheduler, which then schedules active objects according to the following rules:

- If there is just one object now eligible to run, run it now.

- If there is more than one eligible object, choose the one with the highest priority.

- If there are no eligible objects, wait for the next event and then decide what to do based on these rules.

The function that gets called to handle an event is the active object's RunL(). We'll refer to this frequently.

Some events are more important than others. It's much better to handle events in priority order than in a simple first-in, first-out order (FIFO). Events that control the thread (key events to an application, for example) can be handled with higher priority than others (for example, some types of animation). But once a RunL() has started – even for a low priority event – it runs to completion. No other `RunL()` can be called until the current one has finished. That's okay provided all your event handlers are short and efficient.

Non-preemptive multitasking is surprisingly powerful. Actually, there should be no surprise about this; it's the natural paradigm to use for event handling. For example, the window server handles key and pointer events, screen drawing, requests from every GUI-based application in the system, and animations including a flashing text cursor, sprites and self-updating clocks. It delivers all this sophistication using a single thread with active-object-based multitasking.

> **In many systems, the preferred way to multitask is to multithread. In Symbian OS, the preferred way to multitask is to use active objects.**

In a truly event-handling context, using active objects is pure win–win over using threads: you lose no functionality over threads, as events occur sequentially and can be handled in priority order.

You gain convenience over threads because you know you can't be preempted; you don't need to use mutexes, semaphores, critical sections, or any kind of synchronization to protect against the activities of other active objects in your thread. Your `RunL()` is guaranteed to be an atomic operation. It's also more efficient: you don't incur the overhead of a **context switch** when switching between active objects.

Non-preemptive multitasking in Symbian OS is not the same as **cooperative multitasking**. The 'cooperation' in cooperative multitasking is where one task says: 'I am now prepared for another task to run', for instance by using `Yield()` or some similar function. What this really means is: 'I am a long-running task, but I now wish to yield control to the system so that it can get any outstanding events handled if it needs to'. Active objects don't work like that – during `RunL()` you have the system to yourself until your `RunL()` has finished.

All multitasking systems require a degree of cooperation so that tasks can communicate with each other where necessary. Active objects require less cooperation than threads because they are not preemptively scheduled. They can be just as independent as threads. A thread's active scheduler manages active objects independently of one another, just as the kernel scheduler manages threads independently of one another.

1.5.3 A More In-depth Look at Active Objects

An event-handling thread has a single **active scheduler** which is responsible for deciding the order in which events are handled. It also has one or more **active objects**, derived from the `CActive` class, which are responsible both for issuing requests (which will later result in an event happening) and for handling the event when the request completes. The active scheduler calls `RunL()` on the active object associated with the completed event.

The active scheduler is, in effect, a 'wait loop' which waits for events and, on receipt of a signal marking the occurrence of an event, decides which event has occurred of the many it may be expecting, and then dispatches the `RunL()` function of the appropriate active object.

In fact, the UI framework maintains an outstanding request to the window server for user input and other system events. When this request completes, it calls the appropriate application UI and control functions to ensure that your application handles the event properly. So, ultimately, all application code is handled under the control of an active-object `RunL()`.

It's because this framework is provided for you that you can get started with Symbian OS application programming without knowing exactly how active objects work. But for more advanced GUI programming, and for anything to do with servers, or indeed any service that is requested as an asynchronous request, we do need to know how they work in detail.

1.5.4 Some Simple Active Objects

The best way to understand how to use active objects is to look at a simple example. The `activehello` example demonstrates two active objects in use. Figure 1.11 shows what it looks like when you launch it and press the 'Options' key to display its menu.

The **Set Hello** menu item triggers a Symbian OS timer. Three seconds later, that timer completes, causing an event. This event is handled by an active object, which puts an info-message saying **Hello world!** on the

Figure 1.11 `activehello`, demonstrating two active objects

screen. During that three-second period, you can select the **Cancel** menu item to cancel the timer, so that the info-message never appears.

The **Start flashing** menu item starts the **Hello world!** text in the center of the display flashing, and **Stop flashing** stops it. The flashing is implemented by an active object that creates regular timer events, and then handles them. Its event handling changes the visibility of the **Hello world!** text, and redraws the view.

The two active objects represent two different styles of usage: the first, where a one-off event is generated and its completion handled; and the second, where an event is generated, and as part of the handling of the completion of that event, a further event is generated so that we have a continuous cycle, broken only by selecting the stop flashing menu item.

1.5.4.1 The Set Hello Example

The behavior that we see when the **Set Hello** menu item is selected is handled by the CDelayedHello class. This is how the class is defined:

```
class CDelayedHello : public CActive
    {
public:
    // Construct/destruct
    static CDelayedHello* NewL();
    ~CDelayedHello();

    // Request
    void SetHello(TTimeIntervalMicroSeconds32 aDelay);

private:
    // Construct/destruct
    CDelayedHello();
    void ConstructL();

    // from CActive
    void RunL();
    void DoCancel();
    TInt RunError(TInt aError);

private:
    RTimer iTimer;          // Has
    CEikonEnv* iEnv;        // Uses
    };
```

From the declaration, you can see that CDelayedHello:

- *is* an active object, derived from CActive

- *has* an event generator, an RTimer object

- *defines* a request function, SetHello(), that requests an event from the RTimer

- *implements* the `RunL()` function to handle the event generated when the timer request completes

- *implements* the `DoCancel()` function to cancel an outstanding timer request

- *overrides* the default implementation of `RunError()` to handle the case if `RunL()` leaves.

All active object classes share this basic pattern, shown in Figure 1.12. They are derived from `CActive` and implement its `RunL()`, `DoCancel()`, and `RunError()` functions. They include an event generator, and at least one request function.

Construction

Construction of a `CDelayedHello` active object is implemented by the `NewL()` static function, which follows the standard pattern that we saw earlier when we looked at error handling and cleanup. Here it is:

```
CDelayedHello* CDelayedHello::NewL()
    {
    CDelayedHello* self = new (ELeave) CDelayedHello();
    CleanupStack::PushL(self);
    self->ConstructL();
    CleanupStack::Pop(self);
    return self;
    }

CDelayedHello::CDelayedHello()
: CActive(CActive::EPriorityStandard)
    {
    CActiveScheduler::Add(this);
    }
```

The C++ constructor is required for any derived active object class. Inside it, you call `CActive`'s constructor to specify the active object's **priority.** The active scheduler uses this value to decide which active object is to be dispatched first if it finds that more than one event is ready to be handled.

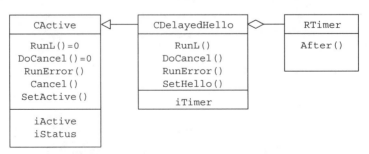

Figure 1.12 The basic pattern of an active object

You should specify `CActive::EPriorityStandard` here unless there are good reasons to specify something lower, or something higher. We'll look at those reasons later. The new object adds itself to the active scheduler, so that the active scheduler can include it in event handling. Adding the object to the active scheduler does not allocate memory, which is why we can do this in the C++ constructor – it cannot leave.

```
void CDelayedHello::ConstructL()
    {
    iEnv = CEikonEnv::Static();
    User::LeaveIfError(iTimer.CreateLocal());
    }
```

The second-phase constructor, the `ConstructL()` function, gets a pointer to the UI environment, and then uses `iTimer.CreateLocal()` to request that the kernel create a kernel-side timer object, which we access through the `RTimer` handle. If there is any problem here, we leave.

Destruction

```
CDelayedHello::~CDelayedHello()
    {
    Cancel();
    iTimer.Close();
    }
```

Before we destroy the active object, the destructor *must* cancel any outstanding requests for timer events. We do this by calling the standard `CActive` function `Cancel()`. It works by checking to see whether a request for a timer event is outstanding and, if so, calls `DoCancel()` to handle it.

> An active object must implement a `DoCancel()` function, and must also call `Cancel()` in its destructor.

The destructor closes the `RTimer` object, which destroys the corresponding kernel-side object. After this, the base `CActive` destructor is invoked and this removes the active object from the active scheduler.

Requesting and Handling Events
`SetHello()` requests a timer event after a given delay:

```
void CDelayedHello::SetHello(TTimeIntervalMicroSeconds32 aDelay)
    {
    _LIT(KDelayedHelloPanic, "CDelayedHello");
```

```
    __ASSERT_ALWAYS(!IsActive(), User::Panic(KDelayedHelloPanic, 1));

    iTimer.After(iStatus, aDelay);
    SetActive();
    }
```

Every line in this function is important:

- First, we assert that no request is already outstanding (that is, that `IsActive()` is false). The client program must ensure that this is the case, either by refusing to issue another request when one is already outstanding, or by canceling the previous request.

- Then, we request the timer to generate an event after aDelay microseconds. The first parameter to `iTimer.After()` is a `TRequestStatus&` which refers to the `iStatus` member which we inherit from `CActive`. As we'll explain below, `TRequestStatus` plays a key role in event handling.

- Finally, we indicate that a request is outstanding by calling `SetActive()`.

This is the invariable pattern for active object request functions. Assert that no request is already active (or, in rare cases, cancel it). Then issue a request, passing your `iStatus` to some function that will later generate an event. Then call `SetActive()` to indicate that the request has been issued.

You can deduce from this that an active object can be responsible for only one outstanding request at a time. You can also deduce that all request functions take a `TRequestStatus&` parameter. A `TRequestStatus` object is an important partner, and we'll look at this in more detail later.

Our UI program calls this function from `HandleCommandL()` using:

```
case EActiveHelloCmdSetHello:
    iDelayedHello->Cancel();              // just in case
    iDelayedHello->SetHello(3000000);  // 3-second delay
    break;
```

In other words, it cancels any request so that the assertion in `SetHello()` is guaranteed to succeed, and then requests a delayed info-message to appear after three seconds.

When the timer event occurs, it is handled by the active object framework and results in the active object's `RunL()` being called:

```
void CDelayedHello::RunL()
    {
    iEnv->InfoMsg(R_ACTIVEHELLO_TEXT_HELLO);
    }
```

Clearly, this code is very simple; it's a one-line function that produces an info-message with the usual greeting text.

Note that if you run this in a Series 60 UI, the code only works in a debug build on the emulator.

The degree of sophistication in an active object's `RunL()` function can vary enormously from one active object to another. The application framework's `CCoeEnv::RunL()` function initiates an extremely sophisticated chain of processing. In contrast, the function above is a simple one-liner.

Error Handling for RunL ()

`CActive()` provides the `RunError()` virtual function, which the active scheduler calls if `RunL()` leaves. You implement this to handle the leave. The protocol here is that your function must return `KErrNone`, which tells the active scheduler that the leave has been handled. The active scheduler interprets any other return value as meaning that the leave has not been handled, and calls its own `Error()` function (`CActiveScheduler::Error()`) which by default panics.

```
TInt CDelayedHello::RunError(TInt aError)
    {
    return KErrNone;
    }
```

Clearly, in this example `RunL()` cannot leave, and all we do is return `KErrNone` to illustrate the point. However, in more sophisticated applications, it is entirely possible that the function could leave, and your `RunError()` would then need to do some real work.

If an active object's `RunL()` calls a leaving function, then that active object should provide an override of the `RunError()` function.

Canceling a Request

If your active object can issue requests, it *must* also be able to cancel them. `CActive` provides a `Cancel()` function that checks whether a request is active and, if so, calls `DoCancel()` to cancel it. As the implementor of the active object, you have to implement `DoCancel()`:

```
void CDelayedHello::DoCancel()
    {
    iTimer.Cancel();
    }
```

There is no need to check whether a timer request is outstanding, because `CActive` has already checked this by the time `DoCancel()` is called.

> **There is an obligation on any class with request functions to provide corresponding cancel functions also.**

1.5.4.2   The Start Flashing Example

The **Start flashing** menu item starts the **Hello world!** text in the center of the display flashing, and **Stop flashing** stops it. The flashing is implemented by an active object that creates regular timer events and then handles them. Its event handling changes the visibility of the **Hello world!** text, and redraws the view.

The behavior that we see when the **Start flashing** menu item is selected is handled by the `CFlashingHello` class. This is slightly more complex than `CDelayedHello` because the `RunL()` function not only handles the completion of a timer event but must also reissue the timer request to ensure that a subsequent timer event occurs and `RunL()` is called again. There is also added complexity because it has drawing code.

We'll look at the essential differences.

Construction

This is the same as in the **Set Hello** example, except that now we need to supply a pointer to the app view object, because we will be doing some drawing.

The second-phase constructor, the `ConstructL()` function, saves the pointer to the app view object and creates the kernel-side timer object in exactly the same way as we did in the **Set Hello** example.

Requesting and Handling Events

The `Start()`, `RunL()` and `DoCancel()` functions of `CFlashing-Hello` show how you can maintain an outstanding request. Here's `Start()`:

```
void CFlashingHello::Start(TTimeIntervalMicroSeconds32 aHalfPeriod)
    {
```

```
_LIT(KFlashingHelloPeriodPanic, "CFlashingHello");
__ASSERT_ALWAYS(!IsActive(),
User::Panic(KFlashingHelloPeriodPanic, 1));
// Remember half-period
iHalfPeriod=aHalfPeriod;

// Hide the text, to begin with
ShowText(EFalse);

// Issue request
iTimer.After(iStatus, iHalfPeriod);
SetActive();
}
```

This is called from `HandleCommandL()` when you select the **Start flashing** menu item. It begins by asserting that a request is not already active, and ends by issuing a request – just as before. Because a whole series of requests will be issued, `Start()` doesn't merely pass the half-period parameter to the `iTimer.After()` but stores it as a member variable for later use.

`Start()` also starts off the visible aspect of the flashing process, by immediately hiding the text (which is visible until `Start()` is called).

When the timer completes, `RunL()` is called:

```
void CFlashingHello::RunL()
{
// Change visibility of app view text
ShowText(!iAppView->iShowText);

// Re-issue request
iTimer.After(iStatus, iHalfPeriod);
SetActive();
}
```

`RunL()` changes the visibility of the text, to implement the flashing effect. Then, it simply renews the request to the timer with the same `iHalfPeriod` parameter as before. As always, the renewed request is followed by `SetActive()`.

In this way, events are generated and handled, and then generated again in a never-ending cycle. The only way to stop the flashing is to issue `Cancel()` which, as usual, checks whether a request is outstanding and, if so, calls our `DoCancel()` implementation:

```
void CFlashingHello::DoCancel()
{
// Ensure text is showing
ShowText(ETrue);

// Cancel timer
iTimer.Cancel();
}
```

We make sure the text is showing, and then cancel the timer. Show-Text() is a simple utility function that sets the visibility of the text; it simply changes the iShowText in the app view, and then redraws the app view.

Because the DoCancel() contains drawing code, it must be executed when there is still an environment in which drawing is possible. This means that we must cancel the flashing hello *before* calling CEikAppUI::Exit(). We *destroy* the active object from its owning class's destructor – this is the right place to destroy it, since the destructor gets called in cleanup situations, while the command handler does not.

The general rule is that you should always be careful about doing anything fancy from an active object's DoCancel(). Nothing in a DoCancel() should leave or allocate resources, and DoCancel() should complete very quickly.

1.5.4.3 *How it Works*

We'll briefly look at the way an event-handling thread, with an active scheduler and active objects, works under the covers.

A thread can have many active objects. Each active object is associated with just one object that has request functions – functions that take a TRequestStatus& parameter. When a program calls an active object's request function (SetHello() in the delayed hello example we saw earlier), the active object passes the request on to the asynchronous service provider. It passes its own iStatus member as the TRequestStatus& parameter to the request function and, having called the request function, it immediately calls SetActive().

The TRequestStatus is a 32-bit object, intended to take a completion code. Before starting to execute the request function, the asynchronous service provider sets the value of the TRequestStatus to KRequestPending, which is defined as 0x80000001.

When the asynchronous service provider finishes processing the request, it generates an event. This means it signals the requesting thread's **request semaphore**, and *also* posts a completion code (such as KErrNone or any other standard error code – anything except KRequestPending is permissible) into the TRequestStatus.

The active scheduler is responsible for detecting the occurrence of an event, so as to associate it with the active object which requested it, and call RunL() on that active object.

The active scheduler calls User::WaitForAnyRequest() to detect an event. This function suspends the thread until one or more requests have completed. The active scheduler then scans through all its active objects, searching for one which has issued a request (iActive is set) and for which the request has completed (iStatus is some value *other* than KRequestPending). It clears that object's iActive and calls its `RunL()`. When the `RunL()` has completed, the scheduler issues `User::WaitForAnyRequest()` again.

So the scheduler handles precisely one event per `User::WaitForAnyRequest()`. If more than one event is outstanding, there's no problem: the next `User::WaitForAnyRequest()` will complete immediately without suspending the thread, and the scheduler will find the active object associated with the completed event.

If the scheduler can't find the active object associated with an event, this indicates a programming error known as a **stray signal**. The active scheduler panics the thread.

Given the delicacy of this description, you might expect writing an active object to be difficult. In fact, as we've already seen with `CDelayedHello`, it's not. You simply have to:

- issue request functions to an asynchronous service provider, remembering to call `SetActive()` after you have done so
- handle completed requests with `RunL()`
- be able to cancel requests with `DoCancel()`
- set an appropriate priority
- handle leaving from the `RunL()` with `RunError()`.

Indeed, if the active scheduler is signaled more than once before it can dispatch an active object's `RunL()`, it has to decide which active object to service first. This is the reason for assigning priorities to active objects. The one with the highest priority value is dispatched first.

The `CActive` class defines a basic set of priority values, encapsulated by the `TPriority` enum, and the application framework defines an additional set, `TActivePriority`. Most applications will use `CActive::EPriorityStandard` as their standard priority value, but if you need to use non-zero values for your own active objects, then you'll need to look at the enum values so that you can ensure you fit in with them.

1.6 Summary

In this chapter, we have reviewed the basics of Symbian OS programming, including:

- object creation and destruction
- naming conventions
- error handling and cleanup
- descriptors
- active objects.

2
Symbian OS User Interfaces

2.1 Introduction

UIQ and Series 60 are the only two user interfaces available for Symbian OS for which third-party developers can write C++ applications. Both provide a framework, built on top of Symbian OS, that can be reused by application writers, and a set of standard applications, for instance PIM applications, multimedia and email. The purpose of this chapter is to compare and contrast the two UIs, in particular their application framework APIs.

Series 60 is one of a range of developer platforms created by Nokia, and it has several versions. Versions 1.0, 1.1 and 1.2 (collectively referred to as the Series 60 Platform 1.x) are based on Symbian OS v6.1. Series 60 Platform 2.0, which this chapter describes, is based on Symbian OS v7.0s. The main features introduced in Platform 2.0 that affect application UIs are skins and bidirectional text support. Skins allow users to customize the UI, by changing the background bitmap, icons and color scheme. Skins are described later in this chapter. Bidirectional text support allows languages that are written from right to left, for instance Hebrew and Arabic, to be edited and displayed. It also affects the ordering and alignment of controls throughout the UI.

All Series 60 phones use a navigation controller that allows navigation in four directions, a confirmation key, and two hardware buttons, called softkeys, beneath the screen. These buttons make Series 60 phones easy to use with one hand. Users can input text using the phone's keypad and can optionally use a predictive text input system.

UIQ is produced by UIQ Technology AB, a subsidiary of Symbian Ltd. UIQ 1.0 was released in September 2000, and there have been several releases since. This chapter describes version 2.1, which runs on Symbian OS v7.0.

Symbian OS C++ for Mobile Phones, Volume 2. Edited by Richard Harrison
© 2004 Symbian Software Ltd   ISBN: 0-470-87108-3

UIQ phones have a large touch-sensitive screen (either 6×8 cm or 4×6 cm), and use a pen as their main input device. However, like Series 60, all UIQ phones provide a hardware confirmation key and other hardware keys for navigation, minimally up and down, and optionally left and right, which make it possible to browse the phone's contents with one hand. The touch-sensitive screen allows text input methods like handwriting recognition and an on-screen virtual keyboard.

To a user these UIs have major differences, but to a programmer they have a lot in common. They have the same underlying framework, which means that application UIs written for both UIs have the same structure, based around application, document, app UI and view classes.

2.2   The Common Framework

Qikon and Avkon are the names of the UI-specific application framework layers implemented on top of the common Symbian OS UI framework, which is called Uikon. Qikon and Avkon implement framework classes defined in Uikon which must be overridden further by application writers. They also define many UI-specific controls, like dialogs, list boxes and editors that can be reused in applications.

Uikon provides the base classes for the three fundamental UI classes: the application (`CEikApplication`), the document (`CEikDocument`) and the application UI (`CEikAppUi`). All Series 60 and UIQ applications minimally need to define their own classes that derive from these. However, they must not derive directly from the Uikon base classes, but from the UI-specific implementations of them.

These implementations have the same name as the Uikon classes, but in UIQ the `CEik` prefix is replaced with `CQik`, and Series 60 uses `CAkn`:

	Application	**Document**	**App UI**
Symbian OS (Uikon)	CEikApplication	CEikDocument	CEikAppUi
Series 60 (Avkon)	CAknApplication	CAknDocument	CAknAppUi/ CAknViewAppUi
UIQ (Qikon)	CQikApplication	CQikDocument	CQikAppUi

The convention of using a `Qik` or `Akn` prefix is used throughout UIQ and Series 60 to identify UI-specific classes, headers and libraries. In general, a type, header or library without such a prefix is part of Symbian OS, and is present in both UIs. There are some exceptions to this, notably avkon, which appears in a few filenames like `avkon.lib` which is the main Series 60 UI library, and `avkon.hrh` which holds the Series 60 resource constants.

With the exception of the app UI, there are very few differences between the Uikon, Qikon and Avkon implementations of these three classes. The next three subsections describe these differences.

2.2.1 The Application

This represents the application's properties, including its UID, capabilities and caption in an appropriate language. Minimally, it must implement two required functions: `CreateDocumentL()` to create the document, and `AppDllUid()` to return the application's UID.

2.2.1.1 *ini Files*

An `ini` file can be used by an application to store its initialization information. For instance, in UIQ it might store the current folder and the zoom level. If one exists, the `ini` file is opened and read each time the application is launched. To avoid the impact this has on application startup time, Series 60 by default does not support them.

The UI framework opens an application's `ini` file by calling `CEikApplication::OpenIniFileLC()`. `CAknApplication` overrides this function to leave with `KErrNotSupported`. If a Series 60 application needs to use an `ini` file, then it must implement `OpenIniFileLC()` itself.

2.2.1.2 *Application Switching*

In both UIQ and Series 60, applications can be left running in the background when a new application is launched. In UIQ this is enforced by the lack of a **Close** or **Exit** menu option in all applications. In Series 60 users have a choice: all applications should provide an **Exit** option, but users can switch between applications instead, for example by using the fast swap window, which is launched by holding down the menu key.

When an application is launched, `CEikApplication::PreDocConstructL()` is called by the UI framework to initialize the application. Both `CAknApplication` and `CQikApplication` override this to first check whether a new instance of the application really should be launched. If there is already a running instance, then instead of launching a new one, the existing instance is brought to the foreground.

2.2.2 The Document

This is responsible for storing and restoring the modifiable data that applies to a particular instance of the application. Minimally, it must instantiate the application UI by implementing a function called `CreateAppUiL()`. Because of this responsibility, all applications, even those that do not need to store any data, must create an instance of a document class.

2.2.2.1 Responding to Low Memory

Because UIQ applications cannot be closed down by users, it is up to the OS to free up RAM when a shortage occurs.

UIQ does this by calling the document's `SaveL()` function with a notification code of `MSaveObserver::EReleaseRAM`, for all running applications except the one in the foreground. `CQikDocument::SaveL()` saves the application's data, then sends an `EEikCmdExit` command to the app UI, which should respond by exiting the application.

An application's derived document class should override `SaveL()` if this default behavior is inappropriate. For instance, UIQ also calls `SaveL()` when data storage has run out, with an `MSaveObserver::EReleaseDisk` notification code. `CQikDocument::SaveL()` ignores these messages. You may choose to respond to this, for instance by deleting any data your application no longer needs or by canceling any operation that is filling up the disk.

2.2.2.2 Document File Creation

To further reduce application startup time, Series 60 by default disables document file creation and opening. In other words, `CAknDocument::OpenFileL()` returns NULL. `OpenFileL()` is called by the framework during application initialization, and returns the file store containing the application's document. Series 60 applications must override `CAknDocument::OpenFileL()` if they need to store and restore data.

2.2.3 The Application UI (app UI)

The main job of the app UI is to handle commands. Commands are generated from a variety of sources, for instance the softkeys and the **Options** menu in Series 60, and the menu bar and the toolbar in UIQ. The app UI also handles key, pointer and other events, and owns and manages the controls in the UI, including the views. Input handling is described in Section 2.7.

2.2.3.1 Views

In both UIQ and Series 60, many applications use more than one view to display their data in different ways. In multiple-view applications, the app UI is responsible for:

- creating and destroying views
- registering and deregistering views with the view server
- activating views

- adding views to the control stack and removing them (for views that need to handle input)

- setting a default view

- passing view-specific events to the correct view for handling.

There are several differences between the UIs in the way these tasks are carried out.

Views in UIQ are derived from `CCoeControl` and `MCoeView`. In multiple-view applications in Series 60, views are derived from `CAknView` and the app UI should be derived from `CAknViewAppUi` rather than from `CAknAppUi`. A `CAknView` is constructed from an `AVKON_VIEW` resource, and defines its own softkeys and **Options** menu pane.

In both UIs, view creation and registration are carried out in the derived app UI's `ConstructL()` function. In UIQ this function must also include a call to `CQikAppUi::BaseConstructL()` and in Series 60 to `CAknAppUi::BaseConstructL()`. `BaseConstructL()` reads the resource file and initializes various UI-specific elements of the app UI, for instance in Series 60 a key sounds object, which plays a sound when the app UI receives key events, and in UIQ the toolbar.

In UIQ, the derived app UI's `ConstructL()` should also include a call to `CCoeAppUi::RegisterViewL()` to register views (and there should be matching calls to `DeregisterView()` in its destructor). In Series 60, `CAknViewAppUi::AddViewL()` is called instead of `RegisterViewL()`. `AddViewL()` also causes the app UI to take ownership of the view; `CAknViewAppUi`'s destructor takes care of view deregistration and destruction.

In both UIs, if an application has multiple views, one of them should be set as the default view using `CCoeAppUi::SetDefaultViewL()`. This also should be called in the derived app UI's `ConstructL()`, because after calling `ConstructL()` the framework activates the default view, and if you haven't set one, the first registered view will be activated.

The default view is the view that is activated when the application is launched. In UIQ, it is also the view that is activated each time the application is brought to the foreground by being switched to. When an application is switched away from in UIQ, its active view is deactivated, and therefore it should save all outstanding changes (or discard any unimportant ones). The next time the application is brought to the foreground, the default view is restored.

In contrast, in Series 60, when an application goes into the background and is then brought to the foreground again, no view switch occurs. The application's previously active view remains active. Only if the application has no active view (this is the case when the application is launched) is the default view activated. Series 60 views therefore do not need to save their changes when the application is switched away from.

There is more information about views in Chapter 6.

2.2.3.2 Data Management

In both UIs it is possible in principle for applications to create multiple document files, although none of the built-in applications do this. In UIQ, categories (called **folders** in the UI) are the recommended way for users to manage data. Series 60 provides standard dialogs for file/folder and memory selection (CAknFileSelectionDialog and CAknMemorySelectionDialog) and for file creation (CAknFile-NamePromptDialog) which can be used in applications. They can be defined using resource structs declared in commondialogs.rh.

A principle of UIQ is that applications, not users, are responsible for saving and managing data. Data management is done using categories. Each application can have its own set of categories. UIQ's standard applications use **Business**, **Personal** and **Unfiled** categories, but users can edit these names, add new ones or remove them. Categories are managed using the app UI's category model, CQikCategoryModel.

2.2.3.3 Access to UI Components

As well as handling commands, the app UI gives access to the UI components that generate commands.

CQikAppUi::SetToolbarL() and RemoveToolbarL() are used to add and remove the toolbar. CAknAppUi::StatusPane() gives access to the Series 60 status pane, and CAknAppUi's Cba() function (CBA is short for control button array) gives access to the button group container that holds the left and right softkeys.

Toolbars, status panes and softkeys are described in more detail later in this chapter.

2.2.3.4 Zooming

In UIQ, **Zoom** is recommended as a standard **Edit** menu option for views that allow text input. When **Zoom** is selected, the standard UIQ zoom settings dialog should be launched by calling CQikZoomDialog::RunDlgLD(). This supports small, medium and large zoom levels.

Each zoom level corresponds to a zoom factor, which is the factor by which font sizes are reduced or increased. A 100% magnification has a zoom factor value of 1000, 200% has the value 2000 and so on. An application's three zoom factor values may be persisted using an ini file called zoom.ini, located in the same folder as the application. If no such file exists, the zoom factor is read from a resource, which contains default zoom factors.

CQikAppUi::ZoomFactorL() gets the application's zoom factors.

In Series 60 it is up to the application whether or not to support zooming. A zoom option is not a requirement and the app UI does not provide any special support for it.

2.2.3.5 *SetMopParent()*

App UIs and views in Series 60 should call a function called `SetMop-Parent()` on any controls they own. `SetMopParent()` is defined in `CCoeControl` so is part of the core OS, but because it was added in v7.0s it is unavailable in UIQ. Its purpose is to inform controls of their owning component. Not calling it may cause scroll indicators, skins and other features to work incorrectly.

2.3 The Screen Layout

Series 60 and UIQ each support a wide variety of devices. Phone manufacturers can customize the UI's look and feel, for instance the icons, color schemes, fonts and text, but all phones running on the same platform have the same basic screen layout and UI components.

2.3.1 Customizing the Look and Feel

Symbian OS includes an abstract look and feel layer that is implemented by the UI in a component called Uiklaf. This component is used by Uikon to get information about the UI. For instance, Uiklaf specifies the UI's standard fonts and the appearance of borders around controls. Also, in some phones, applications briefly fade just before moving into the background. This behavior is specified in Uiklaf.

All Uiklaf classes have a `Laf` prefix. They are only intended to be used internally by Uikon.

As an aside, fading is used in other situations. For instance in both UIs, the foreground application is faded when a dialog is displayed, and in Series 60 only, it is faded when a menu pane is displayed.

2.3.2 Series 60

In Series 60, the screen is divided into three areas or panes (Figure 2.1). These are referred to as the status pane, main pane, and control (or softkey) pane.

2.3.2.1 *Status Pane*

The status pane, usually displayed at the top of the screen, displays information about the foreground application including its title and

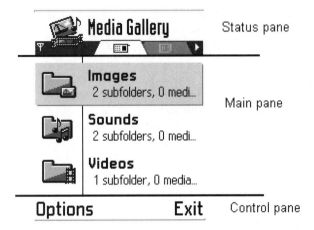

Figure 2.1   Series 60 screen layout

icon, as well as general information about the phone, for instance the signal strength.

It consists of six subpanes. These are laid out from left to right (or from right to left in some cases, for instance in Arabic and Hebrew layouts, in which the ordering of many controls is reversed):

- signal pane (signal strength indicator)

- context pane (displays the application's icon)

- title pane (displays the application's title)

- navigation pane (may contain tabs, images, or text, or can be empty)

- battery pane (battery strength indicator)

- small indicator pane (contains connectivity and some other indicators).

Of these, only the title pane, context pane and navigation pane can be customized by an application writer.

By default, the navigation pane is empty, but it can contain a tab group, as shown in Figure 2.1. This indicates the currently active view in a multiple view application, or the current page in a multi-page dialog and whether more views or pages exist.

In single view applications, the navigation pane can contain a label or image instead of a tab group. This may also be useful when the item in the view is part of a sequence. In this case, left and right arrow bitmaps can show that there is a next or previous item. For instance, the Calendar application displays the date in a navigation label and uses this for navigation.

When the application is in a text editing state, the navigation pane automatically displays an edit indicator which shows whether the input mode is numeric or alphabetic, and upper or lower case (phones that use non-Western languages have different input modes).

The application's initial status pane is specified in the application's `EIK_APP_INFO` resource that is required in all applications' resource files.

A status pane is defined by a `STATUS_PANE_APP_MODEL` resource and its subpanes by `SPANE_PANE`s. The following types of subpane resource exist:

- `TITLE_PANE`. This can optionally contain an icon instead of text, but not both. The default is the application's caption read from its `aif` file.

- `CONTEXT_PANE`. This contains an icon, representing the application. This icon is displayed alongside the title. The default is the application's icon read from its `aif` file.

- `NAVI_DECORATOR`. This can contain different types of controls, including a tab group (`TAB_GROUP`), a text label (`NAVI_LABEL`), or an image (`NAVI_IMAGE`).

In some situations, you may want to hide the status pane, for instance in games that require a full screen display. You can access it using `CAknAppUi::StatusPane()`, and it can be hidden or exposed by calling `CEikStatusPane::MakeVisible()`.

2.3.2.2 Control Pane

The control pane contains the softkey labels and the scroll indicator, if required. A softkey label is a string, usually a single short word identifying the action associated with the softkey. The labels change depending on the state of the application. Often, the left-hand softkey is labelled **Options** and activates a menu pane. It can also be used for issuing affirmative commands, for instance **Ok, Select** and **Yes**, while the right-hand softkey is used for **No, Back, Cancel** and **Exit** commands.

The two softkeys are defined using a `CBA` resource struct containing an array of two `CBA_BUTTON`s which define the left- and right-hand softkeys. These structs are declared in `uikon.rh`.

Series 60 declares many standard `CBA` resources in `avkon.rsg`, with names beginning with `R_AVKON_SOFTKEYS`. You can reuse these or define your own.

For example, here is Series 60's resource definition for **Yes** and **No** softkeys:

```
RESOURCE CBA r_avkon_softkeys_yes_no
    {
    buttons =
```

```
{
CBA_BUTTON {id=EAknSoftkeyYes; txt=text_softkey_yes;},
CBA_BUTTON {id=EAknSoftkeyNo; txt=text_softkey_no;}
} ;
}
```

The `txt` and `id` fields are used to define the softkey label and the command ID it invokes. Note that the **Yes/No** text is defined in a separate file containing the localized text strings.

The command IDs issued by the softkeys are defined in `avkon.hrh`. Some common ones are:

- `EAknSoftkeyOk`

- `EAknSoftkeyCancel`

- `EAknSoftkeySelect`

- `EAknSoftkeyOptions`

- `EAknSoftkeyBack`

- `EAknSoftkeyYes`

- `EAknSoftkeyNo`

- `EAknSoftkeyDone`

- `EAknSoftkeyClose`

- `EAknSoftkeyExit`.

Softkeys can be defined in several places:

- the `cba` field in an `EIK_APP_INFO` resource

- the `buttons` field in a `DIALOG` resource

- the `cba` field in an `AVKON_VIEW` resource

- the `softkeys` field in an `AVKON_LIST_QUERY` resource

- the `softkeys` field in an `AVKON_MULTISELECTION_LIST_QUERY` resource.

The application's initial softkeys are specified in the `EIK_APP_INFO` resource; often, these are **Options** and **Exit** (`R_AVKON_SOFTKEYS_OPTIONS_EXIT`). The **Options** menu pane is defined by `EIK_APP_INFO`'s `menubar` field.

2.3.2.3   *Main Pane*

This is the area between the status pane and the control pane and is generally the area available for the application to draw to – it is the area returned by the app UI's `ClientRect()` function.

2.3.3 UIQ

The screen in UIQ is divided into five areas (Figure 2.2), one of which, the toolbar, is optional.

2.3.3.1 *Application Picker*

This is used to switch between applications. The rightmost icon switches to the application launcher, which gives access to all installed applications. Users can change the applications that are displayed in the Application picker.

2.3.3.2 *Menu Bar*

UIQ has no control pane or softkeys. Instead it uses a menu bar that is always visible beneath the application picker. Each view typically has its own menu bar. Most UIQ menu bars have two left-aligned menu titles and a right-aligned **Folders** menu. The leftmost title should be the name of the application. This provides the principal menu pane with standard functions like **New**, **Find** and **Send as**. The next menu title to the right, which can be omitted if not required, normally provides standard **Edit**

Figure 2.2 UIQ screen layout

commands like **Cut**, **Copy**, **Paste** and **Zoom**. UIQ menu panes should be kept short and cascading menu panes are deprecated. If the menu pane contains more items than will fit on the screen, a scroll bar is added to it, although this should be avoided.

Note that UIQ menus do not include an **Exit** or **Close** option, except in debug builds, where it can be useful to check that the application frees all resources when closed. They also should not provide a **Save** option; UIQ applications should save their data without user intervention.

For more guidelines on how to design menus, see the UIQ Style Guide in the UIQ SDK documentation.

2.3.3.3   *Toolbar*

The toolbar is an optional view-specific bar at the bottom of the screen that holds frequently used controls. These are often command buttons, so an alternative name for it is the button bar.

The application's initial toolbar is specified in the `toolbar` field in the application's `EIK_APP_INFO` resource. It can be changed using `CQikAppUi::SetToolBarL()`, or removed using `CQikAppUi::RemoveToolbarL()`.

The toolbar is defined in a resource file using a `QIK_TOOLBAR` resource, and toolbar buttons by `QIK_TBAR_BUTTON`s. By default these are command buttons, but this can be changed using `QIK_TBAR_BUTTON`'s `type` field. You can customize their behavior, alignment, and spacing (although these have default values) and they can contain text or a bitmap or both. The following code defines a toolbar button containing a standard 'Go back' bitmap (Figure 2.3):

```
QIK_TBAR_BUTTON
    {
    id=EMyAppDone;
    flags=EEikToolBarCtrlHasSetMinLength;
    alignment=EQikToolbarRight;
    bmpfile="z:\\System\\Data\\quartz.mbm";
    bmpid=EMbmQuartzBackarrow;
    bmpmask=EMbmQuartzBackarrowmask;
    }
```

Controls other than buttons can be used in the toolbar by using `QIK_TBAR_CTRL` resources instead.

Note that an optional UI component called the tab screen can be displayed in the same area as the toolbar, but is defined separately. This is described in Section 2.4.7.

Figure 2.3   UIQ toolbar with standard Go back button

2.3.3.4 Status Bar

In general the purpose of the status bar is to display status information such as the battery level and signal strength, and to hold the button that activates the virtual keyboard. Its contents are defined by the phone manufacturer; it cannot be modified by third-party developers.

2.4 Common UI Components

This section describes how UI components that exist in both UIs are defined, created and used, including the differences between the two implementations.

2.4.1 Menu Bars

Menus are defined in resource files, and consist of the following four components:

- The menu bar, which in UIQ is a horizontal bar containing the menu titles. In Series 60 it is never displayed. It is defined using a `MENU_BAR` resource (`CEikMenuBar` in C++).

- Menu titles. These specify a menu pane and in UIQ the text to display in the menu bar. They are defined by `MENU_TITLE` resources. Note that in Series 60, the labels that are displayed in the control pane are not defined in the `MENU_TITLE` resource but in the `CBA` resource that defines the softkeys.

- Menu panes, which are vertical lists of menu items displayed when the user selects a menu title in UIQ or the **Options** softkey in Series 60. They are defined by `MENU_PANE` resources (`CEikMenuPane` in C++). Both Series 60 and UIQ support cascading menu panes, although they are deprecated in UIQ.

- Menu items, which are items in the menu pane that may be selected by the user. They are defined by `MENU_ITEM` resources, and are associated with a label that appears in the menu pane and a command ID that is issued when the item is selected.

The menu bar in UIQ is always visible. It supports multiple menu titles, each of which is associated with a menu pane. In Series 60, the control pane is used instead of a menu bar. Nevertheless, a menu bar still needs to be specified in most Series 60 applications, because it defines the menu pane that is activated by the **Options** softkey.

In both Series 60 and UIQ, the initial menu bar that is used when the application is launched is specified in the `menubar` field in the `EIK_APP_INFO` resource struct.

Different views typically need different menu bars, so that when the view changes, so should the menu bar. In UIQ, menu bar switching is done in the view's `ViewActivatedL()` function using code like this:

```
MEikAppUiFactory* factory = iEikonEnv->AppUiFactory();
factory->MenuBar()->ChangeMenuBarL(0, R_NEW_MENUBAR, EFalse);
```

The second parameter of `CEikMenuBar::ChangeMenuBarL()` is the resource ID of the new view's menu bar. The other two parameters are not used in UIQ.

In Series 60, if you need to dynamically change the menu bar you can do this using code like:

```
MEikAppUiFactory* factory = iEikonEnv->AppUiFactory();
factory->MenuBar()->SetMenuTitleResourceId(R_NEW_MENUBAR);
```

Unlike UIQ, Series 60 does not implement `CEikMenuBar::Change-MenuBarL()`; you must use `SetMenuTitleResourceId()` instead.

Normally, however, Series 60 uses a different scheme for switching between view-specific menu bars. Series 60 views are defined by `AVKON_VIEW` resources and the menu bar and softkeys associated with the view are specified in the resource:

```
STRUCT AVKON_VIEW
    {
    LLINK hotkeys=0;
    LLINK menubar=0;
    LLINK cba=0;
    }
```

In the view's `ConstructL()`, you call `CAknView::BaseConstructL()`, passing it the ID of an `AVKON_VIEW` resource. When the view is activated, its menu bar and softkeys are automatically used.

In Series 60, dialogs can define their own menu bar. The Series 60 dialog base class, `CAknDialog`, owns a menu bar, and the menu bar's resource ID is specified when you construct the dialog. The softkeys used by the dialog are defined in the `DIALOG` resource's `buttons` field. When the dialog is launched, its softkey labels overwrite any labels previously displayed and its menu bar is added to the control stack so that it receives key events before any existing menu bar. Note that Series 60 dialogs that do not need an **Options** menu pane can be derived from `CEikDialog` instead.

A form (`CAknForm`) is a type of dialog derived from `CAknDialog`. Forms provide their own default menu bar and menu pane. List query dialogs (`CAknListQueryDialog`) are also derived from `CAknDialog`.

They are defined by an AVKON_LIST_QUERY resource, which uses a field called softkeys rather than buttons to define its softkeys.

Dialogs and forms are described later in this chapter.

2.4.2 Menu Panes

The contents of menu panes may also need to change according to the state of the application. The function to make menu items available or unavailable is CEikMenuPane::SetItemDimmed(). In UIQ, calling this function with ETrue causes the item to remain in the menu, but it appears dimmed. In Series 60, to save screen space and make menus quicker to navigate, the menu item is removed. Calling the function again with EFalse undims it or causes it to reappear. CEikMenu-Pane::DeleteMenuItem(), on the other hand, removes a menu item altogether, and CEikMenuPane::AddMenuItemL() adds one. Both of these functions have variants that allow multiple menu items to be added or deleted at once. Note that UIQ deprecates dynamically adding and removing menu items because this can confuse users; you should dim and undim them instead. If the user selects a dimmed menu item, an infoprint explaining why the item is unavailable may optionally be used.

Menu items are added, deleted, dimmed and undimmed in the app UI's implementation of DynInitMenuPaneL(). This function is called by the framework just before the menu pane is displayed. For Series 60's menu bar-owning dialogs, DynInitMenuPaneL() is implemented by the dialog rather than the app UI. CAknDialog defines a DynInit-MenuPaneL() function which is empty, but CAknForm overrides it to dim any items that are not required.

Related items can be grouped in a menu pane using separators. This is done in the menu pane's resource definition by specifying

```
flags=EEikMenuItemSeparatorAfter;
```

for the MENU_ITEM preceding the separator. Both Series 60 and UIQ support this flag. EEikMenuItemSeparatorAfter is defined with other menu item flags in uikon.hrh.

2.4.2.1 Cascading Menu Panes

These are panes within another pane. Figure 2.4 shows an example.

A cascading menu pane is defined by a MENU_ITEM's cascade field. As they are considered to add complexity, UIQ deprecates them.

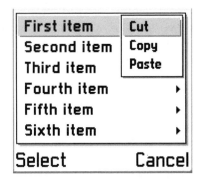

Figure 2.4   Cascading menu pane in Series 60

2.4.3   Dialogs

In both UIs, dialogs appear at the bottom of the screen. They always fill the width of the screen, and have variable height, although in UIQ they never cover the application picker or the status bar. Unlike UIQ, Series 60 does not use dialog buttons; the softkeys are used instead.

All dialogs are ultimately derived from class CEikDialog and are usually defined by a DIALOG resource, or sometimes by a FORM resource in Series 60. The most important fields in a DIALOG resource are the flags, the buttons and the items. The items in a dialog (or a Series 60 form) are defined by DLG_LINE resources. Each dialog line can contain one of a variety of controls. The available controls vary between the two UIs.

The pattern of use for a dialog is the same in both UIs. You construct it using a C++ constructor, call the dialog's ExecuteLD() function, specifying the ID of the DIALOG resource that defines it, then test ExecuteLD()'s return value, which indicates which button or softkey was pressed to dismiss it. For convenience, some dialogs wrap up construction and execution in a static function called RunDl-gLD().

2.4.3.1   Title Bar

Dialogs in UIQ, unlike those in Series 60, have a title bar which serves several purposes:

- It holds the dialog's title (defined by DIALOG struct's title field).

- It can be dragged, so the user can see what is underneath the dialog.

Figure 2.5 A UIQ dialog with a help button, showing a drop-down edit menu

- If the dialog contains a text editor, UIQ adds a button that when pressed activates a drop-down menu pane providing cut, copy and paste functions (see Figure 2.5).

- It can optionally contain a help button which, when pressed, launches a context-specific help topic.

The help button is added if the dialog provides a help context. It can do this either by calling `CEikDialog::SetHelpContext()` or by overriding `CEikDialog::GetHelpContext()`. The UIQ SDK documentation provides more information on the context-sensitive help system.

2.4.3.2 Dialog Flags

Uikon's dialog flags have names beginning with `EEikDialogFlag` and are defined in `uikon.hrh`. They control the behavior and layout of dialogs, but note that not all of them are relevant to both UIs. A useful flag for Series 60 is `EEikDialogFlagFillAppClientRect` which causes the dialog to fill the main pane. In UIQ, the `EEikDialogFlagWait` flag makes the dialog modal; in other words, pen taps outside the dialog are ignored.

Series 60 defines standard sets of flags suitable for Series 60-specific dialogs, in `avkon.hrh`. For instance `EAknErrorNoteFlags` is used for error notes.

2.4.3.3 Dialog Buttons

UIQ supports multiple buttons inside a dialog. They are usually defined using a `DLG_BUTTONS` resource, which is an array of `DLG_BUTTON` resources, or you can use a standard UIQ button resource, for instance `R_EIK_BUTTONS_CANCEL_OK`, defined in `eikcore.rsg` for **Cancel** and **Done** buttons.

Dialog buttons in UIQ are arranged by default in a single horizontal row, although it is possible to use two rows or to arrange the buttons vertically. One button should be set as the default. This is the button that is activated by the **Confirm** hardware key and is drawn with a highlight. It is normally the button labelled **Done**, or equivalent. A button is set as the default by specifying the `EEikLabeledButtonIsDefault` flag in its `DLG_BUTTON` resource definition.

Series 60 does not use `DLG_BUTTON` resources and has no default button. The softkeys are used instead.

2.4.3.4 *Dialog Lines*

In both UIs, the main body of a dialog is defined by an array of `DLG_LINE` resources. The most important fields in a `DLG_LINE` resource are the `type`, the `id`, the `prompt` and the `control`.

The type of the control in the dialog line must always be specified, because both UIs have a control factory that uses it to construct the correct C++ class.

- Common control types have an `EEikCt` prefix and are defined in the `TEikStockControls` enumeration in `uikon.hrh`. Note that not all of these controls are supported by both UIs.

- Series 60-specific control types have an `EAknCt` prefix and are defined in `avkon.hrh`.

- UIQ-specific control types have an `EQikCt` prefix and are defined in the `TQikStockControls` enumeration in `qikstockcontrols.hrh`.

The `id` uniquely identifies the control in the program code and also needs to be specified. If you need to get pointers to the dialog's controls in your application code, you can do this using `CEikDialog::Control()` which takes the control's id as its parameter. You need to cast the return value to the correct type. The `id` should be defined in the `.hrh` file so that it can be used in both the C++ code and the resource file.

Controls in dialogs can have a prompt, which is a caption displayed beside the control. If the prompt and the control combined are too wide for the dialog, UIQ breaks the caption into multiple lines, while Series 60 truncates it. In UIQ, you can alternatively place the caption above the control by specifying `EQikDlgItemCaptionAboveControl` in the dialog line's `itemflags` field.

UIQ controls can optionally have a trailer (using the `DLG_LINE`'s `trailer` field). This is an additional caption in bold, displayed after the control, which is sometimes used to hold units of measurement. This is not supported in Series 60.

Series 60 has support for bitmaps in dialog lines. Use the DLG_LINEs' bmpfile, bmpid and bmpmask fields to specify the mbm file, bitmap and mask. If specified, the bitmap is displayed beside the prompt.

2.4.3.5 Dialog Pages

Both UIs support multi-page dialogs. In Series 60, tabs in the navigation pane identify the current page, and the navigation controller is used to switch between them. Series 60 always displays two tabs for multi-page dialogs (Figure 2.6), even if the dialog has more than two pages.

Note that in Series 60, tabs in the navigation pane are also used for view switching in applications with multiple views. In this case, the navigation pane can contain up to four tabs. Because dialog pages are navigated in the same way as application views, when porting a multiple-view application from UIQ to Series 60 it may be possible to redesign it to use a dialog-based architecture, where each dialog page represents a view.

In UIQ, the dialog page tabs are displayed at the bottom of the page, just above the dialog buttons. UIQ displays as many tabs as will fit on the screen. As in Series 60, left and right navigation arrows are automatically added to the ends if not all the tabs will fit (see Figure 2.7).

Dialog pages are defined using the DIALOG struct's pages field. This is assigned an array of PAGE resources. PAGE resource structs exist in both UIQ and Series 60, but in Series 60 the page tab can contain a bitmap instead of text and the page can contain a form.

Figure 2.6 Navigation tabs for a multi-page dialog in Series 60

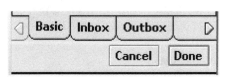

Figure 2.7 Navigation tabs for a multi-page dialog in UIQ

2.4.3.6 UIQ-specific Dialogs

UIQ implements a set of standard dialogs that can be reused in third-party applications. For convenience all of these dialogs define a static function called RunDlgLD() that wraps up construction and execution.

Name	Class	Link against
Add/delete/edit folders	`CQikEditCategoriesDialog`	`qikdlg.lib`
Find text	`CEikEdwinFindDialog`	`eikcdlg.lib`
Set zoom level	`CQikZoomDialog`	`qikdlg.lib`
Send as	`CQikSendAsDialog`	`qikdlg.lib`
Set password	`CEikSetPasswordDialog`	`eikcdlg.lib`
Verify password	`CEikVerifyPasswordDialog`	`eikcdlg.lib`
Set location	`CEikTimeDialogSetCity`	`eikcdlg.lib`
Set date and time	`CEikTimeDialogSetTime`	`eikcdlg.lib`
Set summer times	`CEikTimeDialogSetDst`	`eikcdlg.lib`
Set date and time formats	`CEikTimeDialogOptionFormat`	`eikcdlg.lib`
Set workdays	`CEikTimeDialogOptionWorkday`	`eikcdlg.lib`
Info window	Use `iEikonEnv->InfoWinL()`	`eikcore.lib`
Alert window	Use `iEikonEnv->AlertWin()`	`eikcore.lib`
Query window	Use `iEikonEnv->QueryWinL()`	`eikcore.lib`

2.4.3.7   Series 60-specific Dialogs

Series 60 extends the Uikon dialog hierarchy with its own dialog types, notably `CAknDialog`, which adds support for a menu bar, and its derived classes. These include:

- query dialogs (`CAknQueryDialog` and several derived classes)
- selection and markable list dialogs (`CAknSelectionListDialog` and `CAknMarkableListDialog`)
- forms (`CAknForm`).

Series 60 also implements some dialogs that do not need a menu bar, so are derived from `CEikDialog` instead:

- note dialogs (`CAknNoteDialog` and derived classes) and associated wrappers.

Query Dialogs

Query dialogs are used to request confirmation or other input from the user. There are various types, all derived from `CAknQueryDialog`. The main ones are:

- Confirmation queries. These request user confirmation, so often the dialog uses **Yes** and **No** softkeys.
- Data queries. These use an editor, for instance an `EDWIN` for text input, or a `NUMBER_EDITOR` for numeric input. There are several

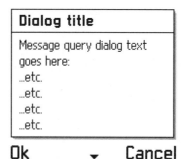

Figure 2.8 A message query dialog

subtypes of data query. They often use **Ok** and **Cancel** buttons, and have their own edit indicator to show what kind of input is accepted.

- List queries. These display a popup single or multiple selection list, and support a wide range of list box types, for instance single or double line, with or without icons.

- Message queries. These display a message and optionally have a title. Figure 2.8 shows an example.

This table lists the classes and resources associated with query dialogs.

Name	Class	Resource struct
Confirmation query	CAknQueryDialog	AVKON_CONFIRMATION_QUERY
Text query	CAknTextQueryDialog	AVKON_DATA_QUERY
Integer query	CAknNumberQueryDialog	AVKON_DATA_QUERY
Date or time query	CAknTimeQueryDialog	AVKON_DATA_QUERY
Duration query	CAknDurationQueryDialog	AVKON_DATA_QUERY
Floating point query	CAknFloatingPointQueryDialog	AVKON_DATA_QUERY
Multiple line query	CAknMultiLineDataQueryDialog	AVKON_DATA_QUERY
Single selection list query	CAknListQueryDialog	AVKON_LIST_QUERY
Multiple selection list query	CAknListQueryDialog	AVKON_MULTISELECTION_LIST_QUERY
Message query	CAknMessageQueryDialog	AVKON_MESSAGE_QUERY

Notes:

- The control type for all query dialogs is EAknCtQuery.

- The confirmation and data query resource structs have a layout field. Its possible values are defined in avkon.hrh, for instance ENumber-Layout for integer queries and EConfirmationQueryLayout for confirmation queries.

- Single selection list query dialogs are constructed with an integer pointer. If the dialog's `ExecuteLD()` returns true, this pointer holds the index of the selected item. Multiple selection list query dialogs use a pointer to an integer array instead.

- Query dialogs can play a tone when they execute. The tone type is specified when the dialog is constructed.

- `CAknQueryDialog::SetEmergencyCallSupport()` enables emergency telephone calls to be made even when a numeric query dialog is executing. By default, this is disabled.

- Text query dialogs may use predictive text entry by calling `CAkn-QueryDialog::SetPredictiveTextInputPermitted()`. The default is not to use it.

List query dialogs support the following types of list box:

Name	Class	Control type
Single line list box	`CAknSinglePopupMenuStyleListBox`	`EAknCtSinglePopupMenuListBox`
Single line list box with icons	`CAknSingleGraphicPopupMenuStyleListBox`	`EAknCtSingleGraphicPopupMenuListBox`
Single line list box with titles	`CAknSingleHeadingPopupMenuStyleListBox`	`EAknCtSingleHeadingPopupMenuListBox`
Single line list box with icons and titles	`CAknSingleGraphicHeadingPopupMenuStyleListBox`	`EAknCtSingleGraphicHeadingPopupMenuListBox`
Double line list box	`CAknDoublePopupMenuStyleListBox`	`EAknCtMenuDoublePopupMenuListBox`
Double line list box with icons	`CAknDoubleLargeGraphicPopupMenuStyleListBox`	`EAknCtDoubleLargeGraphicPopupMenuListBox`

Selection and Markable List Dialogs

`CAknSelectionListDialog` and a derived class, `CAknMarkable-ListDialog`, are convenient list box dialogs with an optional search box (permanent or popup) and a predefined menu bar. They occupy the whole of the main pane (see Figure 2.9).

They are defined using a `DIALOG` resource with a `LISTBOX` control in the first dialog line and, optionally, a search box in a second line. The

Figure 2.9 Selection list dialog with search box

LISTBOX can be of any single or double list box type: the possible values are defined in avkon.hrh, starting with EAknCtSingleListBox.

The dialog needs to define appropriate softkeys and a menu pane. For instance, the markable list dialog uses an **Options** rather than **Ok** softkey, with **Mark** and **Mark all** menu options. The resource ID of the menu bar is specified when the dialog is constructed. Standard menu panes are:

- R_AVKON_MENUPANE_SELECTION_LIST

- R_AVKON_MENUPANE_SELECTION_LIST_WITH_FIND_POPUP

- R_AVKON_MENUPANE_MARKABLE_LIST

- R_AVKON_MENUPANE_MARKABLE_LIST_WITH_FIND_POPUP.

The comments in header file AknSelectionList.h describe how to define and use these dialogs.

Note Dialogs and Wrappers

Note dialogs display a warning, a question or a progress indicator. They do not need a menu bar and most are timed so do not even need softkeys. Because of this, the note dialog base class, CAknNoteDialog, is derived from CEikDialog rather than CAknDialog.

You can define your own notes using an AVKON_NOTE resource and assigning it to the control field in the dialog's DLG_LINE, but because note dialogs are so commonly used, note wrappers, which use predefined resources, are provided.

Wrappers make note dialogs very easy to use by providing a default resource definition with values for the duration, tone, animation and softkeys. The confirmation note wrapper provides a default label (**Done** or equivalent), but for the others you specify the label when calling the dialog's ExecuteLD().

The following table lists all of the note dialogs and their associated wrapper classes.

Name	Dialog class	Wrapper class
Confirmation note	CAknNoteDialog	CAknConfirmationNote
Information note	CAknNoteDialog	CAknInformationNote
Warning note	CAknNoteDialog	CAknWarningNote
Error note	CAknNoteDialog	CAknErrorNote
Progress note	CAknProgressDialog	n/a
Wait note	CAknWaitDialog	CAknWaitNoteWrapper
Permanent note	CAknStaticNoteDialog	n/a

Notes:

- Series 60 defines standard sets of flags for note dialogs in `avkon.hrh`, for instance `EAknErrorNoteFlags`.

- All note dialogs should contain an `AVKON_NOTE` resource, whose type is `EAknCtNote`.

- The layout of note dialogs varies, so you need to specify the right value in the `AVKON_NOTE`'s layout field, for instance `EProgressLayout` for a progress note.

- Like query dialogs, note dialogs can play a tone when they execute.

- `AVKON_NOTE`s can contain a label (in singular and plural versions), an image, an icon and an animation. You can select whether to use the singular or plural version of the text label by calling `CAknNoteDialog::SetTextPluralityL()` and you can set the number in the label dynamically using `CAknNoteDialog::SetTextNumberL()`.

Global Notes and Queries

Global note and query dialogs are similar to standard note and query dialogs except that they are displayed even if the application that launched them is switched away from or closed down. They can be used by UI or engine components, and are not defined by resources. They are used only in unusual circumstances, for example to display the error that caused an application to shut down.

A global note is a `CAknGlobalNote`. After initialization, call `ShowNoteL()`, specifying the type of note (enumerated in `TAknGlobalNoteType`). There are three types of global queries: global confirmation queries (`CAknGlobalConfirmationQuery`), global list queries

(CAknGlobalListQuery) and global message queries (CAknGlob-alMsgQuery).

Forms

A form is a type of dialog specific to Series 60 that looks like a list box, but which allows list items to be edited. Forms support two modes of operation – view mode and edit mode – which users can switch between. Forms can alternatively support edit mode only.

Forms are defined in resource files using a FORM resource which can be assigned to the form field in a DIALOG or to the form field in a PAGE. In either case, the softkeys are defined in the dialog.

FORM flags are defined in uikon.hrh:

- EEikFormShowEmptyFields – shows lines without a value (this is the default).

- EEikFormHideEmptyFields – hides lines without a value.

- EEikFormShowBitmaps – displays bitmaps in dialog lines.

- EEikFormEditModeOnly – the form supports edit mode only.

- EEikFormUseDoubleSpacedFormat – the control is displayed beneath the line containing the bitmap and label (Figure 2.10).

Forms have a default menu bar and pane, which are used if you don't specify a menu bar resource when calling CAknForm::ConstructL(). The default menu pane includes standard options including **Save**, **Add field** and **Delete field**.

A form should minimally implement SaveFormDataL(), which is called when the user chooses **Save**, and DoNotSaveFormDataL(),

Figure 2.10 A double-spaced form in edit mode with bitmaps and prompts

which is called when the user chooses not to save their changes – this may be implemented to reset the form to contain default values.

Popup fields can only be used inside forms: see Section 2.5.1.2.

2.4.4   List Boxes

List boxes display information in columns. All list boxes in UIQ and Series 60 are ultimately derived from `CEikListBox`. The main types of list boxes defined by Uikon are:

- text list boxes – `CEikTextListBox`

- column list boxes – `CEikColumnListBox`

- hierarchical list boxes – `CEikHierarchicalListBox`.

The Series 60 UI is designed with an emphasis on viewing rather than creating data and lists are the main way in which data is displayed. Series 60, unlike UIQ, defines many of its own specialized list box types, described below.

In UIQ, you may need to create your own list boxes. A `LISTBOX` resource is used to define a list box. You can customize it using flags defined in `uikon.hrh`, for instance `EEikListBoxMultipleSelection`. Its contents can be set either in the resource, or in C++, through the list box model (`CTextListBoxModel::SetItemTextArray()`). You may also need to set a list box observer so that you are notified when an item is selected.

In Series 60, if the list box is in a dialog, you could use a list query dialog, or a selection or markable list box dialog instead. In UIQ, **choice lists** are often used in dialogs. These are described in Section 2.5.2. UIQ does not define any standard list box dialogs.

2.4.4.1   Series 60-specific List Boxes

Series 60 defines the following list boxes:

Name	Class	Control type
Single list box	`CAknSingleStyleListBox`	`EAknCtSingleListBox`
Single numbered list box	`CAknSingleNumberStyleListBox`	`EAknCtSingleNumberListBox`
Single list box with titles (titles are displayed to the left of the item text, in a different font)	`CAknSingleHeadingStyleListBox`	`EAknCtSingleHeadingListBox`

Name	Class	Control type
Single list box with graphics	`CAknSingleGraphicStyleListBox`	`EAknCtSingleGraphicListBox`
Single list box with titles and graphics	`CAknSingleGraphicHeading StyleListBox`	`EAknCtSingleGraphicHeading ListBox`
Single numbered list box with titles	`CAknSingleNumberHeading StyleListBox`	`EAknCtSingleNumberHeading ListBox`
Single list box with increased spacing between list items	`CAknSingleLargeStyleListBox`	`EAknCtSingleLargeListBox`
Double list box	`CAknDoubleStyleListBox`	`EAknCtDoubleListBox`
Double numbered list box	`CAknDoubleNumberStyleListBox`	`EAknCtDoubleNumberListBox`
Double list box where items can contain the time, including optional am/pm text	`CAknDoubleTimeStyleListBox`	`EAknCtDoubleTimeListBox`
Double list box with increased spacing between list items	`CAknDoubleLargeStyleListBox`	`EAknCtDoubleLargeListBox`
Double list box with graphics	`CAknDoubleGraphicStyleListBox`	`EAknCtDoubleGraphicListBox`
Settings list box (the user can change the value of each item)	`CAknSettingStyleListBox`	`EAknCtSettingListBox`
Numbered settings list box	`CAknSettingNumberStyleListBox`	`EAknCtSettingNumberListBox`

The contents of each list box column are of the same type, so for instance there are separate columns for numbers, titles, icons and the main list item text. List box contents are defined in the resource file by list item strings that use tab characters as column delimiters. All of the list boxes in the table above are declared in the header file `aknlists.h`, and comments in `aknlists.h` describe the list item string format for each.

Icons are identified in list item strings by their index into the list box's icon array, so before you can use a list box with graphics, the

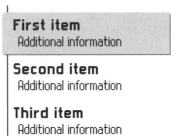

Figure 2.11 A double list box

list box's icon array must have been set using CColumnListBox-Data::SetIconArray(). This applies also to list query dialogs and selection list dialogs, which provide a SetIconArrayL() function.

The difference between single and double list boxes is that double list boxes can have two lines of text per item in the main list item text column, while single list boxes have one. Figure 2.11 shows an example of a double list box.

All the list boxes in the table above except for Settings list boxes can support multiple selection, depending on flags specified in the LISTBOX resource, for instance EAknListBoxMarkableList and EAknListBoxMultiselectionList.

2.4.4.2   Grids

Grids are a type of two-dimensional list box with columns and rows. They are more commonly used in Series 60 than in UIQ, so Series 60 provides special support for them.

CAknGrid is the base class for Series 60 grids. Because it is derived from CEikListBox and its model (CAknGridM) is derived from CTextListBoxModel, grids are used in a similar way to lists.

A grid is defined by a GRID resource and constructed using CAknGrid::ConstructFromResourceL(), or in C++ using CAknGrid::ConstructL().

You can customize the grid's layout either by using a GRID_STYLE resource definition or by calling CAknGrid::SetLayoutL(). Layout variables are:

- the orientation (horizontal or vertical)

- the scroll type, for instance whether to loop back to the first item in the row or column when the last item has been reached, or to stop scrolling, or to move to the start of the next row

- the number of rows and columns to display

- the size of gaps between rows and columns
- the size of grid items.

Cells in the grid can contain text, graphics or both. If the grid contains graphics, you need to set its icon array, in a similar way to a list box, using `CFormattedCellListBoxData::SetIconArrayL()`. To set and get the currently selected item(s) use `CAknGrid::SetCurrentData-Index()` and `CAknGrid::CurrentDataIndex()`.

Popup and multiple selection grids are possible. Popup grids are created using a `CAknPopupList`, which provides a title and softkeys. For a multiple selection grid, specify the `EAknListBoxMarkableGrid` flag on construction.

2.4.5 Editors

UIQ and Series 60 define some standard editors that can be reused in third-party applications, including inside dialogs. They are listed in the table below. Common resource structs are defined in `uikon.rh` or `eikon.rh`, UIQ's in `Qikon.rh`, and Series 60's in `avkon.rh`.

Name	UIQ class name, control type, resource struct	Series 60 class name, control type, resource struct
Integer editor	CQikNumberEditor EQikCtNumberEditor QIK_NUMBER_EDITOR	CEikNumberEditor EEikCtNumberEditor NUMBER_EDITOR
Floating point editor	CQikFloatingPointEditor EQikCtFloatingPointEditor QIK_FLOATING_POINT_EDITOR	CEikFloatingPointEditor EEikCtFlPtEd FLPTED
Time editor	CQikTimeEditor EQikCtTimeEditor QIK_TIME_EDITOR	CEikTimeEditor EEikCtTimeEditor TIME_EDITOR
Popup calendar	CEikCalendar n/a R_EIK_ONE_MONTH_CALENDAR	n/a
Date editor	CQikDateEditor EQikCtDateEditor QIK_DATE_EDITOR	CEikDateEditor EEikCtDateEditor DATE_EDITOR
Time and date editor	CQikTimeAndDateEditor EQikCtTimeAndDateEditor QIK_TIME_AND_DATE_EDITOR	CEikTimeAndDateEditor EEikCtTimeAndDateEditor TIME_AND_DATE_EDITOR
Duration editor	CQikDurationEditor EQikCtDurationEditor QIK_DURATION_EDITOR	CEikDurationEditor EEikCtDurationEditor DURATION_EDITOR

(continued overleaf)

Name	UIQ class name, control type, resource struct	Series 60 class name, control type, resource struct
Time offset editor	n/a	CEikTimeOffsetEditor EEikCtTimeOffsetEditor TIME_OFFSET_EDITOR
Plain text editor	CEikEdwin EEikCtEdwin EDWIN	CEikEdwin EEikCtEdwin EDWIN
Global text editor	CEikGlobalTextEditor EEikCtGlobalTextEditor GTXTED	CEikGlobalTextEditor EEikCtGlobalTextEditor GTXTED
Rich text editor	CEikRichTextEditor EEikCtRichTextEditor RTXTED	CEikRichTextEditor EEikCtRichTextEditor RTXTED
Secret editor	CEikSecretEditor EEikCtSecretEd SECRETED	CEikSecretEditor EEikCtSecretEd SECRETED
PIN editor	n/a	CAknNumericSecretEditor EAknCtNumericSecretEditor NUMSECRETED
Color selector	CQikColorSelector EQikCtColorSelector QIK_COLOR_SEL	n/a
Sound selector	CQikSoundSelector EQikCtSoundSelector n/a	n/a
IP editor	CQikIpEditor n/a n/a	CAknIpFieldEditor EAknCtIpFieldEditor IP_FIELD_EDITOR
Slider	CQikSlider EQikCtSlider QIK_SLIDER	CAknSlider EAknCtSlider SLIDER
Phone number editor	n/a	CAknPhoneNumberEditor EAknPhoneNumberEditor PHONE_NUMBER_EDITOR

Notes:

- The UIQ integer editor uses spinner arrows to change the number.

- The UIQ time, date and duration editors use popup time pickers or calendars.

- The secret editor is alphanumeric in UIQ, and alphabetic in Series 60. For numeric passwords in Series 60, use the PIN editor.

- UIQ supports horizontal and vertical sliders; Series 60 supports horizontal only. Series 60 supports different slider layouts and the slider can display a value, a percentage or a fraction.

2.4.6 Progress Bars

In both UIs, a progress bar can be defined in a resource file using a `PROGRESSINFO` resource struct. This has a control type of `EEikCtProgInfo`, which corresponds to `CEikProgressInfo` in C++. The progress bar can contain text showing the progress as a fraction or a percentage (Figure 2.12), or it can be divided up by vertical lines.

To update and redraw the progress bar, call `CEikProgressInfo::IncrementAndDraw()` or `CEikProgressInfo::SetAndDraw()`.

In Series 60, progress bars are usually displayed using the progress note dialog (`CAknProgressDialog`).

Figure 2.12   UIQ progress bar

2.4.7 Navigation Tabs

In Series 60, tabs in the navigation pane can be used to switch between application views. Figure 2.13 shows an example.

UIQ also supports navigation tabs which are located at the bottom of the screen, in the area occupied by the toolbar (see Figure 2.14). They are not used to switch view in UIQ, but to switch between pages in the same view. A page may be a scrollable container, for instance.

Series 60 navigation pane tabs can be confused with multi-page dialog tabs, which are also displayed in the navigation pane and are used for switching between dialog pages. These were described earlier in this chapter.

Series 60 navigation pane tabs are defined by TABs in a TAB_GROUP resource, each of which has an ID and can contain text or a bitmap or

Figure 2.13   A tab group in the Series 60 Profiles application

Figure 2.14   Two navigation tabs in the UIQ Agenda

both. The tab group consumes left and right key events and the application should change the view depending on which tab is active.

You can customize the number of tabs visible in the pane, with a maximum of four. If not all tabs are visible, navigation arrows are displayed (see Figure 2.13).

To set the navigation pane's tab group you first need access to the status pane (both the app UI and the view provide a StatusPane() function), and through this you can get a handle to the navigation pane. The following code initializes the navigation pane with a TAB_GROUP called R_MY_TAB_GROUP.

```
CAknNavigationControlContainer* naviPane =
STATIC_CAST(CAknNavigationControlContainer*, StatusPane()->
    ControlL(TUid::Uid(EEikStatusPaneUidNavi)));
TResourceReader reader;
iCoeEnv->CreateResourceReaderLC(reader, R_MY_TAB_GROUP);
CAknNavigationDecorator* naviTabGroup = naviPane->CreateTabGroupL(reader);
naviPane->PushL(*naviTabGroup); // set the tab group to be visible in the
                                // navigation pane
```

You can set the application's initial navigation pane by defining a STATUS_PANE_APP_MODEL resource that contains a NAVI_DECORATOR pane. This can be set via the status_pane field in the application's EIK_APP_INFO resource.

The equivalent of navigation pane tabs in UIQ is the tab screen. This is constructed either in code (CQikTabScreen) or from a QIK_TABSCREEN resource. As in Series 60, tabs in the tab screen can contain text or a bitmap or both. Each tab is associated with a page and tab/page pairs are added to the tab screen using CQikTabScreen::AddTabPageL(). The tab screen can contain an additional control that is not a tab, for instance a **Done** button.

2.4.8 Messages and Notifications

Every control in a GUI application has a member called iEikonEnv, which is a pointer to the GUI environment, CEikonEnv. CEikonEnv is common to both UIs and provides many useful functions, including access to information and busy messages.

CAknEnv is a Series 60 extension to CEikonEnv. A CAknEnv instance is created at the same time as CEikonEnv and is owned by CEikonEnv. All applications have access to it through a global pointer called iAvkonEnv, although iEikonEnv is much more useful to third-party applications.

The following messages and notifications are common to both UIs.

- Info messages. These are concise messages displayed by default in the top right-hand corner of the screen for three seconds. They are used to notify the user of a minor error, or that something without visual impact has happened. They are generated using `iEikonEnv->InfoMsg()`.

- Busy messages. These are flashing text messages that notify the user that the application is busy. They are generated using `iEikonEnv->BusyMsgL()`. They do not have a fixed duration; they end when `iEikonEnv->BusyMsgCancel()` is called.

- Info windows. These are used to notify the user of more serious errors, for instance that a file is corrupt. An info window is a dialog with a title, a description and a button labeled **Continue** (UIQ), or **Back** (Series 60), as shown in Figure 2.15. They are generated using `iEikonEnv->InfoWinL()`.

- Alert windows. These are similar to info windows except that resources are guaranteed to be available for them. Also, in UIQ, the title contains the word **Information**. They are generated using `iEikonEnv->AlertWin()`.

- Query windows. These are similar to info windows, except they have buttons labelled **Yes** and **No** and they return a value that indicates which button was pressed (see Figure 2.16). They are generated using `iEikonEnv->QueryWinL()`.

Note that in Series 60, note and query dialogs can alternatively be used for notification.

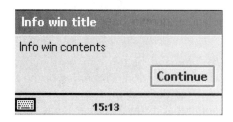

Figure 2.15 A UIQ info window

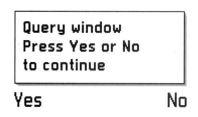

Figure 2.16 A Series 60 query window

2.5 UI-specific Components

2.5.1 Buttons

UIQ's touch-sensitive screen allows it to use buttons extensively. UIQ uses three types of button: command buttons, option buttons and check boxes. Series 60 does not support any of these, although similar behavior can be achieved using other controls.

2.5.1.1 Command Buttons

Command buttons are often used in UIQ's toolbar and in dialogs. Different types are available:

- Standard command buttons (class `CEikCommandButton`/resource struct `CMBUT`) contain text, or an icon, or both. Their control type is `EEikCtCommandButton`.

- Two picture command buttons (`CEikTwoPictureCommandButton`/`PICMBUT`) contain text, or an icon that changes when the button is pressed, or both. Their type is `EEikCtTwoPictureCommandButton`.

- Bitmap buttons (`CEikBitmapButton`/`BMPBUT`) contain no text, just one or two bitmaps. Their type is `EEikCtBitmapButton`.

- Text buttons (`CEikTextButton`/`TXTBUT`) contain text only. Their type is `EEikCtTextButton`.

- Menu buttons (`CEikMenuButton`/`MNBUT`) contain text or an icon, or both, and launch a popout menu pane. Their type is `EEikCtMenuButton`.

A command button can be dimmed if unavailable using its `Set-Dimmed()` function. Properties of the button and its layout can be set using flags defined in `uikon.hrh`. For instance, a button can latch, meaning that it stays set after it has been pressed and released, or it can stay set, regardless of whether it is pressed.

Standard command IDs issued by buttons are defined in `uikon.hrh`, for instance `EEikBidCancel` and `EEikBidOk`.

In Series 60, the softkeys are normally used instead of command buttons.

2.5.1.2 Option Buttons

Option buttons, sometimes referred to as radio buttons, are used to select one from a list of options. In UIQ they are usually held in either a vertical or a horizontal list:

	Vertical option button list	**Horizontal option button list**
Class name	`CQikVertOptionButtonList`	`CEikHorOptionButtonList`
Resource struct	`QIK_VERTOPBUT`	`HOROPBUT`
Control type	`EQikCtVertOptionButtonList`	`EEikCtHorOptionButList`

Option buttons are defined by `OPBUT` resources which have an `id` and a `label`. The option button list classes' `SetButtonById()` function is used to change the selected button.

Option buttons can be used in menus as well as dialogs. This is done using the `MENU_ITEM`'s flags field, specifying one of the flags `EEikMenuItemRadioStart`, `EEikMenuItemRadioMiddle` or `EEikMenuItemRadioEnd` (defined in `uikon.hrh`) for each of the menu options to be included in the option button list. An option button in a menu item is set or unset using `CEikMenuPane::SetItemButtonState()`.

Radio button behavior is available in Series 60 using a popup field (`CAknPopupField`). This is a control with two possible modes: label mode and selection list mode. In label mode, the control displays a text label. When it receives an **Ok** or **Confirm key** event, it switches to selection list mode and displays a popup selection list containing radio buttons, causing the form to redraw its contents around the list. Figure 2.17 shows an example. Note that popup fields can be used only in forms, and the form must be in edit mode before the popup selection list can be displayed. When the field is in label mode and it is selected, an alternative to pressing **Confirm** is to use the left and right navigation keys to cycle through the available options.

Optionally, the popup field can allow users to enter new values, using a button in the popup list usually labeled **Other** ... (or its equivalent).

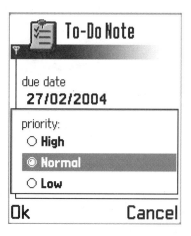

Figure 2.17   Popup selection list from Series 60 To-Do application

This makes them similar to combo boxes (see Section 2.5.2). Selecting this button invokes an editor, allowing the user to add an entry to the list. The `EAknPopupFieldFlagAllowsUserDefinedEntry` flag should be specified in the resource to allow this, or you can call `CAknPopupField::SetAllowsUserDefinedEntry()`.

Popup fields support different data types, for instance text, integers, dates and times, so an appropriate editor must be used to create new entries. The editor, and the initial array of values to display in the popup selection list, are specified by calling `CAknPopupField::SetQueryValueL()`, with an object that implements the `MAknQueryValue` interface.

An exception to this is a text-only popup field, `CAknPopupFieldText` (for example, the selection list shown in Figure 2.17). Its initial values are specified in its resource definition (a `POPUP_FIELD_TEXT` struct, with a control type of `EAknCtPopupFieldText`), which removes the need to call `SetQueryValueL()` and significantly simplifies its construction.

2.5.1.3 Check Boxes

If multiple list options can be selected, then check boxes should be used instead of option buttons.

Figure 2.18 shows a dialog from UIQ's Agenda application that uses seven dialog lines, each of which has a prompt that is the day of the week and a control of type `EikCtCheckBox`, which in C++ corresponds to a `CEikCheckBox`.

Series 60 supports multiple selection in markable list dialogs and list boxes.

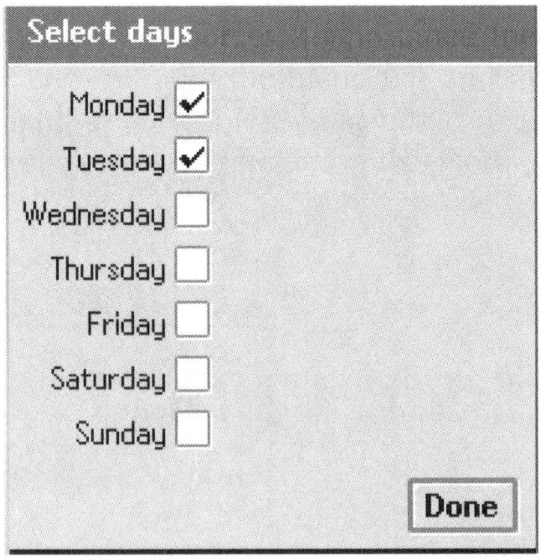

Figure 2.18   Check boxes in UIQ

2.5.2 Choice Lists and Combo Boxes

Choice lists are a type of single selection text list box that display a single list item only, but can use a popout box to display the whole list. In UIQ they are often used where space is limited, for instance in dialogs. Choice lists are not supported in Series 60; the custom list dialogs provide similar behavior.

A choice list is defined using a CHOICELIST resource. This has a control type of EEikCtChoiceList and corresponds to a CEik-ChoiceList in C++. You can specify its width using the maxdisplay-char field, otherwise it defaults to the width of the widest item. The list's contents are set either in the resource file using an ARRAY resource, or in C++ using CEikChoiceList::SetArrayL().

You can customize the list through CEikChoiceList, including its alignment, font and borders, and you can disallow the list popout, using CEikChoiceList::AllowPopout(). SetCurrentItem() and CurrentItem() set and get the selected item.

Combo boxes are a combination of choice list and text editor. They allow new items to be added and existing items to be edited. The COMBOBOX resource defines a combo box. It has a control type of EEikCtComboBox, which corresponds to a CEikComboBox in C++. A COMBOBOX defines the width of the edit box in characters, the maximum number of characters that can be entered by the user and the maximum number of entries in the array.

A CEikComboBox is used in a similar way to a CEikChoiceList. The array can be initialized from a resource ARRAY, or from a C++ array (by calling CEikComboBox::SetArrayL()), and the selected item is set and retrieved using SetTextL() and GetText().

Series 60 does not support combo boxes; its closest equivalent is popup fields in forms.

2.5.3 Scroll Bars

In UIQ, users can scroll either using the hardware navigation keys, or using scroll bars. UIQ supports both vertical and horizontal scroll bars. Scroll bars allow different types of scrolling; for instance, the scroll bar thumb can be dragged, or the scroll arrows can be pressed. As an alternative to using scroll bars, scroll arrows alone can be used. These save screen space by avoiding the need for a horizontal or vertical scroll bar (see Figure 2.19).

Series 60 does not support scroll bars, but instead displays scroll indicators in the control pane, between the softkey labels. Their color tone changes to indicate how much information remains to be scrolled. Series 60 does not support horizontal scrolling.

Figure 2.19   'Floating' scroll arrows in UIQ

In either UI, if you want scroll indicators for a list box or text editor, you must create them yourself and you must call the list box's or editor's `UpdateScrollBarsL()` function when the scroll bar needs updating, for instance when items are added to or removed from the list.

Scroll bars and arrows can be created as follows:

```
listbox->CreateScrollBarFrameL();
// Set the horizontal scroll bars to be invisible and vertical scroll bars
//  to be visible, when needed
listbox->ScrollBarFrame()->SetScrollBarVisibilityL
     (CEikScrollBarFrame::EOff, CEikScrollBarFrame::EAuto);
// Make the scrollbar floating (removing the scroll bars)
listbox->ScrollBarFrame()->SetScrollBarManagement
     (CEikScrollBar::EVertical,CEikScrollBarFrame::EFloating);
// Give the floating scrollbar standard scroll arrows
listbox->ScrollBarFrame()->SetScrollBarControlType
     (CEikScrollBar::EVertical,EQikCtArrowHeadPageScrollBar);
```

UIQ provides a control container called `CQikScrollableContainer` that is useful for managing scroll bars for a list of controls that do not all fit on the screen.

2.6   Skins

Skins (also called themes) are used to customize a phone UI's appearance at runtime. Both UIs support them, but they are implemented differently. Support for skins was introduced in Series 60 v2.0 and in UIQ 2.1.

In both UIs, skins specify a color scheme and optionally a bitmap to display in the background. In UIQ the background bitmap, if specified, is displayed only in the application launcher while in Series 60 it is displayed in all applications. UIQ skins can customize the system sounds, for instance the sounds made when an email or SMS is received. Series 60 skins cannot do this, but can customize the icons used throughout the UI, including application icons.

2.6.1   Skins in UIQ

Skin-aware applications in UIQ must do all drawing using the logical colors enumerated in `TLogicalColor`, rather than physical colors, and controls should implement `CCoeControl::HandleResource-Change()`. This should respond to messages of type `KEikMessage-ColorSchemeChange`, which indicates a color scheme change, by updating any colors to use the new values, for instance:

```
void CMyAppView::HandleResourceChange(TInt aType)
    {
    if (aType == KEikMessageColorSchemeChange)
        {
        Window().SetBackgroundColor(iEikonEnv->
        ControlColor(EColorControlBackground, *this));
        }
    CCoeControl::HandleResourceChange(aType);
    }
```

2.6.2   Skins in Series 60

In Series 60, skins have a greater impact on applications than in UIQ. Skins are automatically applied to menus, note and query dialogs, the control and status panes, the fast swap window and, if present, the Chinese FEP. Other UI components may support skins, depending on whether the application is skin-aware, and the control is skin-enabled.

To make an application skin-aware, call:

```
BaseConstructL(CAknAppUi::EAknEnableSkin);
```

in the app UI's `ConstructL()`. In a skin-aware application, skins are applied to all standard Series 60 controls, for instance search boxes, list boxes and grids. If `BaseConstructL()` is not called with this parameter, only the controls mentioned in the first paragraph are drawn using skins.

To draw non-standard, custom controls and container controls using skins, you need to use the skin-drawing utility class `AknsDraw-Utils` (the `Akns` prefix stands for Avkon Skins). Before you can do this, you first need to get the currently active skin by calling `Akns-Utils::SkinInstance()`, and non-container controls also need to call `AknsDrawUtils::ControlContext()` to get the skin control context, described below.

If the skin does not provide a resource for a particular UI component, then it should be drawn using a graphics context as in Series 60 1.x, instead of using `AknsDrawUtils`. This is the case for all controls, even in skin-aware applications, if the user selects the default Series 60 skin. The return values from `AknsDrawUtils` draw methods indicate whether a skin was used for drawing.

As in UIQ, applications may need to check for skin change events in their `HandleResourceChange()` function. In Series 60, these are identified by `KAknsMessageSkinChange` messages.

2.6.2.1 Skin Control Context

The skin control context provides the ID of the background image to use from the active skin, the rectangle to draw it in, and image attributes, for instance whether to draw the bitmap tiled or stretched. The skin instance and skin control context are needed by all `AknsDrawUtils` draw functions.

It is up to the container control to provide a skin control context to its contained controls. Skin control contexts are objects derived from `MAknsControlContext`, for instance:

- `CAknsBasicBackgroundControlContext`

- `CAknsListBoxBackgroundControlContext`

- `CAknsFrameBackgroundControlContext`.

The container owns the context and returns a pointer to it in its implementation of `MopSupplyObject()`. For example,

```
TTypeUid::Ptr CMyAppContainer::MopSupplyObject(TTypeUid aId)
    {
    if(aId.iUid == MAknsControlContext::ETypeId)
        {
        return MAknsControlContext::SupplyMopObject(aId,
            iSkinControlContext);
        }
    return CCoeControl::MopSupplyObject(aId);
    }
```

An object provider chain must exist from the container to all the controls that need the control context; the chain is set up by calling `CCoeControl::SetMopParent()` for all controls in the chain.

When constructing a skin control context, the ID of the background bitmap must be specified. These IDs are defined in `aknsconstants.h`. For instance, `KAknsIIDQsnBgAreaMain` identifies the bitmap to use in the main pane.

2.7 Handling User Input

The main types of user input that an application needs to handle are commands, key events and, in UIQ, pointer events.

2.7.1 Commands

When a command is generated, for instance by a user selecting a menu item, or pressing a Series 60 softkey or a UIQ toolbar button, the framework passes the ID of the command to the app UI of the foreground application, by calling its `HandleCommandL()` function. The app UI can handle the command itself, or can pass it to a view or other control for handling.

Most command IDs are application-specific, but both UIs provide some standard ones. In `avkon.hrh`, Series 60 defines the command IDs generated by the standard softkey resource definitions, for instance `EAknSoftkeyOptions` and `EAknSoftkeyExit`, which are issued by the `R_AVKON_SOFTKEYS_OPTIONS_EXIT` CBA.

UIQ defines the command IDs generated by standard toolbar and dialog buttons in `uikon.hrh`, for instance `EEikBidCancel`. It also defines some command IDs in `Qikon.hrh`, including those generated by the set zoom dialog.

`EEikCmdExit` is defined in Symbian OS in `uikon.hrh` and is generated by the system when it needs to close down applications. It must be handled by app UIs in both Series 60 and UIQ.

2.7.2 Key Events

A keypress generates three types of key event: a key down (with an event code of `EEventKeyDown`), one or more standard key events (`EEventKey`) and a key up (`EEventKeyUp`). On receiving any of these types of key event, the UI framework calls `CCoeControl::OfferKeyEventL()` for all controls in the foreground application's control stack, starting with the control at the top of the stack, until a control consumes it. Applications conventionally ignore up and down key events and consume only standard events.

The parameters of `OfferKeyEventL()` are the key event and the event code. The key event contains the key code (`TKeyEvent::iCode`) and any modifier keys (`TKeyEvent::iModifiers`), for instance Shift, that were held down, and specifies whether the key was held down long enough to repeat (`TKeyEvent::iRepeats`).

In both UIs, it is rare for applications to need to handle modifier keys. The only modifier key supported in Series 60 is Shift (`EModifierShift`), which is generated by pressing the Edit key. In UIQ, modifier keys may not be supported, depending on the active FEP, so it is not recommended for applications to rely on them.

In Series 60, some applications treat long and short key presses differently. For instance, the Series 60 Calculator clears all user input if the Clear key is pressed and held down, in other words if its `TKeyEvent::iRepeats` value is greater than zero.

Note that in Series 60, although the left and right softkeys generate key events (`EKeyCBA1` and `EKeyCBA2`), applications should in general handle the generated commands instead. The mapping between softkeys and command IDs is defined in the application's resource file.

As well as the standard alphanumeric keypad keys, applications may need to handle some of the special Symbian OS key codes defined in e32keys.h and some UI-specific ones. Series 60's are defined in uikon.hrh and UIQ's in quartzkeys.h. The following table lists the most common key codes that applications need to handle.

Series 60	UIQ	Purpose
EKeyLeftArrow, EKeyRightArrow	EQuartzKeyFourWayLeft, EQuartzKeyFourWayRight	Used to navigate horizontally, for instance between navigation tabs.
EKeyUpArrow, EKeyDownArrow	EQuartzKeyTwoWayUp or EQuartzKeyFourWayUp, EQuartzKeyTwoWayDown or EQuartzKeyFourWayDown	Used to navigate vertically, for instance in lists and menu panes. EQuartzKeyTwoWayUp and EQuartzKeyTwoWayDown may sometimes be used to implement horizontal navigation.
EKeyOK	EQuartzKeyConfirm	Generated by pressing the Confirm/Selection key to open or activate an item. In Series 60, in some circumstances, it may be used to activate a short context-sensitive popup menu instead. EKeyEnter is not generated in Series 60 or UIQ, except on the emulator.
EKeyBackspace	EKeyBackspace	In Series 60, this is generated by the Clear key and is used for deletion.

2.7.2.1 FEPs

A FEP (short for front end processor) is a DLL that allows users to input characters that are not directly available on the phone's keypad. In both UIs, the FEP communicates directly with editors, so no action is required by third-party developers to use a FEP.

In UIQ, the FEP is the main source of text input. UIQ supports multiple FEPs on a phone, which the user can switch between, but only one can be enabled at any time. Most UIQ phones provide at least a handwriting recognition FEP and an on-screen virtual keyboard.

Series 60 phones support a single FEP only that is always loaded and available to applications. It supports multitap text entry, where one or more numeric key presses outputs a single alphanumeric character, and predictive text entry, where each key press causes a single character to be predicted and output by consulting a multiple language dictionary.

The Series 60 FEP distinguishes between short and long key presses. For instance, in numeric input mode, a short press of the # key inputs the hash character, but a long press changes the FEP to alphabetic input mode. It is also aware of the input capabilities of the target editor, so for instance will automatically change to numeric input mode when a phone number editor has focus. In the Asia Pacific region, it has separate modes for inputting text in different languages. Sometimes a device may require both. For example, in Hong Kong, both Chinese input and English input may be needed on the same phone.

2.7.3   Pointer Events

As with key events, the UI framework ensures that pointer events are sent to the right control for handling. `CCoeControl::HandlePointer-EventL()` takes a `TPointerEvent` parameter that packages information about the event. This includes the event's type, for instance pen down and pen up, any keyboard modifiers, and the event's position on the screen.

In UIQ, items are selected and activated by a single pen tap, so pen down events (of type `TPointerEvent::EButton1Down`) and drag events (`TPointerEvent::EDrag`) generally change the selection, and pen up events (`TPointerEvent::EButton1Up`) activate the selection.

Some controls need to handle pointer drag events, for example sliders (`CQikSlider`). To do this you must first enable them, by calling `CCoeControl::EnableDragEvents()`. Repeat pointer events (`TPointerEvent::EButtonRepeat`) may also need to be handled, for instance by spinner arrows in a numeric editor or the hours and minutes controls in a popout time picker. In order to receive repeating pointer events, the control must first call `RequestPointerRepeatEvent()` on its window, specifying the initial delay before the first repeat event is generated. Then it should be called repeatedly, specifying the delay before the next repeat event.

As in key event handling, modifiers are usually ignored.

2.8   Summary

The aim of this chapter was to show that despite the obvious differences to a user, the Series 60 and UIQ application frameworks have a lot in common.

- Both are built on top of Uikon, the common UI framework, which means applications have the same structure and have common base classes.

- In both UIs, menus are generally defined using standard Uikon resources and C++ classes.

- Dialogs are widely used in both UIs. Although Series 60 defines many specialized dialog types, all of them are ultimately derived from Uikon's dialog base class `CEikDialog`, so follow a standard pattern of use.

- Most editors either are common to both UIs or have a near equivalent in the other UI.

- Buttons are widely used in UIQ, but not in Series 60, although equivalent behavior can usually be achieved using other means.

- Command and event handling is similar in both UIs, although some of the command IDs and key event codes differ, and UIQ applications need to handle pointer as well as key events.

3

A Running Application

In this chapter we walk through an application from startup to closedown. In particular, we look at how system calls work during the launch and termination of an application.

3.1 Introduction

On a Symbian OS phone, when an application's icon is selected from the shell, a special application framework executable is started, called `Apprun.exe`.

All programs running in Symbian OS are processes; therefore, the OS creates a new process for `Apprun.exe`.

In addition, as part of the process initialization, the OS creates a default thread of execution within this process, initially called 'Main'. It is within the context of this thread that the application will eventually run.

The shell passes the full path and filename (of the application DLL that is to be launched), as well as any document file that the application must open, as command-line parameters to the executable, which the OS in turn makes available to the process via an API.

Within this new process, a Uikon environment (`CEikonEnv`) is created that, in brief:

- loads the application DLL, and checks its type, i.e., that its second UID is `0x100039ce`, which defines an application type polymorphic DLL (a `.app`)

- calls its ordinal one function, `NewApplication()`, to create an application object, i.e., an object of a class derived from `CApaApplication`

- calls `CreateDocumentL()`, which is overridden by the application to create the document

Symbian OS C++ for Mobile Phones, Volume 2.   Edited by Richard Harrison
© 2004 Symbian Software Ltd   ISBN: 0-470-87108-3

- calls `CreateAppUiL()`, which creates the app UI for the document (which then goes on to create the app view)

- starts the CONE event loop

- when the loop terminates, destroys itself.

3.1.1   Application Structure

If you implement anything less than the following four classes, you don't have a Symbian OS GUI application. The classes are as follows.

- **An application**. The application class serves to define the properties of the application, and also to manufacture a new blank document. In the simplest case the only property that you have to define is the application's unique identifier, or UID.

- **A document**. A document represents the data model for the application. If the application is file-based, the document is responsible for storing and restoring the application's data. Even if the application is not file-based, it must have a document class, even though that class doesn't do much apart from creating the application user interface (app UI).

- **An app UI**. The app UI is entirely invisible. It creates an application view (app view) to handle drawing and screen-based interaction. In addition, it provides the means for processing commands that may be generated, for example, by menu items.

- **An app view**. This is, in fact, a concrete control, whose purpose is to display the application data on screen and allow you to interact with it. In the simplest case, an app view provides only the means of drawing to the screen, but most application views will also provide functions for handling input events.

These four classes are shown in Figure 3.1. Three of the four classes in a Series 60 application are derived from base classes in Avkon which themselves are derived from Uikon classes. In applications written for other target machines, these three classes may be derived directly from the Uikon classes or from an alternative layer that replaces Avkon.

In turn, Uikon is based on two important frameworks:

- CONE, the **control environment**, is the framework for graphical interaction.

- APPARC, the **application architecture**, is the framework for applications and application data.

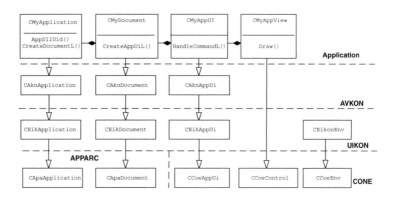

Figure 3.1 The four classes for implementing a (Series 60) application

3.2 System Calls

The code example on pages 100–101 lists the major system calls from application startup to closedown. Those underlined concern application calls involving the classes described in Section 3.1.1. The arrows in the left margin show the main functional responsibilities of the significant blocks of code.

3.2.1 Application Startup

Being an executable, `Apprun.exe` has a single main entry point, `E32Main()`. When `Apprun.exe` is launched, the system creates a new process and within it a new thread, initially called 'Main'.

We now discuss each system call shown in the code example.

E32Main()

- E32Main() is defined in `app-framework/uikon/apprun/apprun.cpp` and takes no arguments.

The entry point for processes. Called by the operating system to start the program.

Gets the command line arguments of `Apprun.exe`, such as the name of the application. At this point the thread is simply called 'Main'.

Calls the following function.

EikDll::RunAppInsideThread()

- EikDll::RunAppInsideThread() is a public EXPORTed static function declared in `app-framework/uikon/coreinc/eikdll.h`.

- called by E32Main() in apprun.cpp.

```
Apprun's E32Main()
    EikDll::RunAppInsideThread()
        CEikonEnv* coe=new CEikonEnv
        coe->CEikonEnv::ConstructAppFromCommandLineL()
            CEikonEnv::ConstructL()
                CCoeEnv::ConstructL()
                    CCoeEnv::CreateActiveSchedulerL()
                    CActiveScheduler::Add(this)
                    CCoeEnv::ConnectToFileServerL()
                    CCoeEnv::ConnectToWindowServerL()
                CEikonEnv::InitSystemFontsL()
                    CEikonEnv::InitSystemResourceFileL()
                        iSystemResourceFileOffset=LafEnv::LoadCoreResFileL()
                    CEikonEnv::InitPrivateResourceFileL()
                        iPrivateResourceFileOffset=LafEnv::LoadPrivResFileL()
                    LafEnv::CreateSystemFontsL()
                    iNormalFont=LafEnv::MatchFont()
                iLafEnv = new(ELeave) CEikLafEnv
                CEikonEnv::LoadLibrariesL()
                    for each CEikLibrary* library
                    {
                        library->CEikLibrary::InitializeL()
                        // one of the above will call CEikonEnv::SetAppUiFactoryL(),
                        // which calls iAppUiFactoryArray = new(ELeave) CArrayFixFlat<TEikAppUiFactory>
                        library->CEikLibrary::ControlFactoryArrayL()
                        for each control factory
                        {
                            // Store the control-factory function pointer in CEikonEnv's iControlFactoryFuncArray
                            // for later use by EikControlFactory::CreateByTypeL()
                        }
                        library->CEikLibrary::ResourceFileArrayL()
                        // virtual for each resource file
                        {
                            BaflUtils::NearestLanguageFile()
                            CCoeEnv::AddResourceFileL()
                        }
                    }
                CEikonEnv::SetUpFepL()
```

Framework
initialization

```
(CApaProcess*) iProcess=CEikProcess::NewL()
CEikDocument* doc=iProcess ->CApaProcess::AddNewDocumentL()
    CApaProcess::CreateApplicationL()
        CApaDll::CreateApplicationL()
            RLibrary::Load()
            functionPointer=RLibrary::Lookup()
            (CApaApplication*) iApplication=(call functionPointer, i.e. the application's NewApplication()}
    iApplication->CEikApplication::PreDocConstructL()
        CEikApplication::ResourceFileName()
            CEikApplication::ParseWithAppName()
            BaflUtils::NearestLanguageFile()
        CCoeEnv::AddResourceFileL()
        CEikApplication::OpenAppInfoFileL()
    CApaProcess::CreateDocL()
        CApaDocument* doc=iApplication->CEikApplication::CreateDocumentL()
doc->CEikDocument::PrepareToEditL()
    (CEikAppUi*) iAppUi=CEikDocument::CreateAppUiL()
    iAppUi->CEikAppUi::SetDocument()
    iAppUi->CEikAppUi::ConstructL()
        CEikAppUi::BaseConstructL()
            CCoeEnv::SetAppUi(this)
doc->CEikDocument::OpenFileL()
    CApaDocument::RestoreL()

coe->CCoeEnv::ExecuteD()
    CActiveScheduler::Start() // where the normal running of the application takes place
        // one of the first active-objects to RunL() is the CCoeFepLoader active object,
        // which loads the FEP that was set up by the SetUpFepL() call

        // ... time passes .....

coe->CEikonEnv::DestroyEnvironment()
    CCoeEnv::DeleteResourceFile(iSystemResourceFileOffset)
    CEikonEnv::CloseLibraries()
    DeleteArrayOfResourceFilesRemovingEach()
    CCoeEnv::DestroyEnvironment()
        CCoeEnvExtra::DestroyFep()
        delete iAppUi
        delete this
        CEikonEnv::~CEikonEnv()
            CCoeEnv::DeleteResourceFile(iPrivateResourceFileOffset)
            delete iAppUiFactoryArray
            delete iLafEnv
            iClockDll->RAnimDll::Destroy()
        CCoeEnv::~CCoeEnv()
            delete CActiveScheduler::Current()
            iWsSession.Close()
            iFsSession.Close()
```

Loading app DLL

Creating app Ui and restoring document data

Event loop

Closing down

Responsible for the initial application framework preparation, application wait loop, and high-level error management.

Runs the application inside its own process. It is self-sufficient in that it does not assume anything about its environment, or rather, it assumes there is nothing in its environment, for example no `CCoeEnv`, no trap handler, no active scheduler, and so on.

Does not return until the application has exited.

Note that rather than using `EikDll`, `RApaLsSession` can be used to launch applications from the APPARC server thread.

Calls the following major functions:

- `CEikonEnv* coe = new CEikonEnv`

- `coe->CEikonEnv::ConstructAppFromCommandLineL()`

- `coe->CCoeEnv::ExecuteD()`.

CEikonEnv coe = new CEikonEnv*

- `CEikonEnv::CEikonEnv()` is a public EXPORTed function declared in `app-framework/uikon/coreinc/eikenv.h`.

- called by `EikDll::RunAppInsideThread()` in `eikdll.cpp`.

Creates a pointer to a new `CEikonEnv` object.

`CEikonEnv` provides an environment for creating Uikon controls and utility functions for manipulating those controls. `CEikonEnv` is a derivative of `CCoeEnv`, the CONE environment object.

Applications do not need to create a new `CEikonEnv` object. An instance is always available to the application through `CEikApplication::iCoeEnv`, `CEikonEnv::Static()`, or the `iEikonEnv` macro of `CEikApplication`.

coe->CEikonEnv::ConstructAppFromCommandLineL()

- `CEikonEnv::ConstructAppFromCommandLineL()` is a public EXPORTed function declared in `app-framework/uikon/core-inc/eikenv.h`.

- called by `EikDll::RunAppInsideThread()` in `eikdll.cpp`.

Initializes the application.

Constructs the new application consisting of a `CEikAppUi`, a `CEikApplication`, and a `CEikDocument` object.

The function is responsible for:

- extracting the command-line parameters that were passed to `Apprun.exe` (i.e., the application DLL filename, document name, and so on)

- clearing the screen, if the application was launched in the foreground
- updating the window server with task and window group information (renames the thread from 'Main' to the application's name)
- calling the following major functions:
    - `CEikonEnv::ConstructL()`
    - `(CApaProcess*) iProcess = CEikProcess::NewL()`
    - `CEikDocument* doc = iProcess->CApaProcess::Add-NewDocumentL()`
    - `doc->CEikDocument::PrepareToEditL()`
    - `doc->CEikDocument::OpenFileL()`.

CEikonEnv::ConstructL()

- `CEikonEnv::ConstructL()` is a public EXPORTed function declared in `app-framework/uikon/coreinc/eikenv.h`.
- called by `CEikonEnv::ConstructAppFromCommandLineL()` in `eikenv.cpp`.

Offers second-phase construction of the `CEikonEnv` object.

The main function for application-framework initialization, responsible for ensuring that all the relevant UI framework objects are initialized (e.g. colors, parsers, bitmaps, application window group, FEP, and so on).

Calls the following major functions:

- `CCoeEnv::ConstructL()`
- `iLafEnv = new (ELeave) CEikLafEnv`
- `CEikonEnv::LoadLibrariesL()`
- `CEikonEnv::SetUpFepL()`.

CCoeEnv::ConstructL()

- `CCoeEnv::ConstructL()` is a public EXPORTed function declared in `app-framework/cone/inc/coemain.h`.
- called in `eikenv.cpp` by `CEikonEnv::ConstructL()` with the `aInitialFocusState` parameter.

Completes the construction of the `CCoeEnv` object.

`CCoeEnv` is the Control environment (alias CONE), which is an active environment for creating controls. It implements active objects and an active scheduler, which provide access to the window server, simplifying the API for application programs. It also provides utility functions that are useful to many applications.

CCoeEnv is the parent class of CEikonEnv, and furthermore all of CCoeEnv's functionality is directly available to applications via CEikonEnv.

CCoeEnv::ConstructL() creates the active scheduler, and adds active objects to it, for standard and redraw events. However, the active scheduler is not started, and no events are received, until ExecuteD() is called.

CCoeEnv contains a list of open resource files (an array of RResourceFile objects, which read resource files) and manages the mapping of resource ID to resource file.

The resource compiler and the RResourceFile class provide two mechanisms to support the use of an arbitrary number of resource files: the NAME statement, and the resource file signature (both defined within the .rss resource file).

The NAME statement (which is the first non-comment statement in the .rss file, between one and four characters long) uniquely identifies a resource file, and hence all the resources in it. The name is converted into a 20-bit value referred to as the offset, and from this value, resource IDs are allocated in sequence.

The resource file signature (which is the first resource in the .rss file, and consequently the first resource ID in the sequence) is used in conjunction with the NAME statement. It allows a developer to query what the offset is for a particular resource file, and may be used by the RResourceFile class to identify the offset for resource IDs.

Multiple resource files are supported to enable resource files for UI libraries to be loaded into the application space. This is so that applications can freely use any resource ID defined by these shared libraries, in addition to any application-specific resources.

ConstructL() also does the following:

- creates a connection to the window server

- creates a connection to the file server so that the control environment can access resource files

- creates a graphics context (the system GC), a standard font, a screen device, and the application's window group.

Calls the following major functions:

- CCoeEnv::CreateActiveSchedulerL()

- CActiveScheduler::Add(this)

- `CCoeEnv::ConnectToFileServerL()`
- `CCoeEnv::ConnectToWindowServerL()`
- `CEikonEnv::InitSystemFontsL()`.

CCoeEnv::CreateActiveSchedulerL()

- `CCoeEnv::CreateActiveSchedulerL()` is a private function declared in `app-framework/cone/inc/coemain.h`.
- called from `CCoeEnv::ConstructL()` in `coemain.cpp`.

Installs the specified active scheduler as the current active scheduler.

The installed active scheduler now handles events for this thread, controlling the handling of asynchronous requests as represented by active objects.

CActiveScheduler::Add(this)

- `CActiveScheduler::Add()` is a public EXPORTed function declared in `base/e32/include/e32base.h`.
- called from `CCoeEnv::ConstructL()` in `coemain.cpp`.

Adds the specified active object to the current active scheduler.

CCoeEnv::ConnectToFileServerL()

- `CCoeEnv::ConnectToFileServerL()` is a private function declared in `app-framework/cone/inc/coemain.h`.
- called from `CCoeEnv::ConstructL()` in `coemain.cpp`.

Connects the client process to the file server.

See also `iFsSession.Close()` in Section 3.2.2, 'Application close-down', below.

CCoeEnv::ConnectToWindowServerL()

- `CCoeEnv::ConnectToWindowServerL()` is a private function declared in `app-framework/cone/inc/coemain.h`.
- called from `CCoeEnv::ConstructL()` in `coemain.cpp`.

Connects the client session to the window server, creating a corresponding session object in the server, by calling `RWsSession::Connect()`.

See also `iWsSession.Close()` in Section 3.2.2, 'Application close-down', below.

`CEikonEnv::InitSystemFontsL()`

- `CEikonEnv::InitSystemFontsL()` is a private EXPORTed function declared in `app-framework/uikon/coreinc/eikenv.h`.
- called, during the construction of a CCoeEnv object, by `CCoeEnv::ConstructL()` in `coemain.cpp`.

Initializes system fonts.

Overrides the default implementation provided by `CCoeEnv::InitSystemFontsL()` declared in `coemain.h`.

Calls the following major functions:

- `CEikonEnv::InitSystemResourceFileL()`
- `CEikonEnv::InitPrivateResourceFileL()`
- `LafEnv::CreateSystemFontsL()`
- `iNormalFont = LafEnv::MatchFont()`.

`CEikonEnv::InitSystemResourceFileL()`

- `CEikonEnv::InitSystemResourceFileL()` is a private function declared in `app-framework/uikon/coreinc/eikenv.h`.
- called by `CEikonEnv::InitSystemFontsL()` in `eikenv.cpp`.

Adds the Uikon system resource file to the underlying cone environment to allow applications to make use of standard localized strings or constants.

Implements the following statement.

`iSystemResourceFileOffset = LafEnv::LoadCoreResFileL(*this);`

- `LafEnv::LoadCoreResFileL()` is a public EXPORTed static function declared in `app-framework/uiklafgt/inc/lafenv.h`.
- called by `CEikonEnv::InitSystemResourceFileL()` in `eikenv.cpp`.

Loads the system resource file.

`CEikonEnv::InitPrivateResourceFileL()`

- `CEikonEnv::InitPrivateResourceFileL()` is a private function declared in `app-framework/uikon/coreinc/eikenv.h`.
- called by `CEikonEnv::InitSystemFontsL()` in `eikenv.cpp`.

Implements the following statement only.

iPrivateResourceFileOffset = LafEnv::LoadPrivResFileL()

- LafEnv::LoadPrivResFileL() is a public EXPORTed static function declared in app-framework/uiklafgt/inc/lafenv.h.

- called by CEikonEnv::InitPrivateResourceFileL() in eikenv.cpp.

Loads the private resource file.

The private resource file defines resources required by the system environment but not intended for application use.

LafEnv::CreateSystemFontsL()

- LafEnv::CreateSystemFontsL() is a public EXPORTed static function declared in app-framework/uiklafgt/inc/lafenv.h.

- called by CEikonEnv::InitSystemFontsL() in eikenv.cpp.

Populates an array with a set of system fonts.

CEikonEnv calls this to get system fonts, and uses the array in subsequent calls to MatchFont().

iNormalFont = LafEnv::MatchFont()

- LafEnv::MatchFont() is a public EXPORTed static function declared in app-framework/uiklafgt/inc/lafenv.h.

- called by CEikonEnv::InitSystemFontsL() in eikenv.cpp.

Gets the nearest match in the specified fonts for a specified logical system font.

iLafEnv = new (ELeave) CEikLafEnv

- CEikLafEnv::CEikLafEnv() is a public constructor declared in app-framework/uikon/coresrc/eikenv.cpp.

- called by CEikonEnv::ConstructL() in eikenv.cpp.

Creates a new CEikLafEnv object.

The product Look And Feel (LAF) library sets the appearance, such as color and size, of Uikon controls.

CEikonEnv::LoadLibrariesL()

- CEikonEnv::LoadLibrariesL() is a private function defined in app-framework/uikon/coreinc/eikenv.h.

- called by CEikonEnv::ConstructL() in eikenv.cpp.

Loads all plugin DLLs (using RLibrary) in z:\system\libs\uikon_ init with a middle UID of 0x10004CC1; this yields a list of CEikLibrary objects.

CEikLibrary is an abstract base class specifying the interface for a dynamically loaded DLL, which goes on to configure other Uikon statically loaded DLLs. CEikLibrary allows CEikonEnv to define types of control and resources.

RLibrary is a class to load named DLLs, allowing late binding, at runtime.

Calls the following major functions for each CEikLibrary* library:

- library->CEikLibrary::InitializeL()

- library->CEikLibrary::ControlFactoryArrayL()

- library->CEikLibrary::ResourceFileArrayL()

- BaflUtils::NearestLanguageFile() (and for each resource file)

- CCoeEnv::AddResourceFileL() (and for each resource file).

library->CEikLibrary::InitializeL()

- CEikLibrary::InitializeL() is a pure virtual function declared in app-framework/uikon/coreinc/eiklibry.h.

- called for each library by CEikonEnv::LoadLibrariesL() in eikenv.cpp.

When CEikonEnv dynamically loads a DLL, it calls its InitializeL() function to set up any variables that CEikonEnv requires.

One of these calls to InitializeL() will call the public EXPORTed function CEikonEnv::SetAppUiFactoryL() (declared in eikenv.h), which in turn implements the statement iAppUiFactoryArray = new(ELeave) CArrayFixFlat<TEikAppUiFactory>.

A TEikAppUiFactory object consists of a CEikAppUi object and a MEikAppUiFactory object. The former handles application-wide user interface details. The latter is an app UI factory interface, which defines a mixin interface to break the dependency of the Uikon Application User Interface on controls.

library->CEikLibrary::ControlFactoryArrayL()

- CEikLibrary::ControlFactoryArrayL() is a pure virtual function declared in app-framework/uikon/coreinc/eik-libry.h.

- called for each library by CEikonEnv::LoadLibrariesL() in eikenv.cpp.

Gives CEikonEnv access to the library's Control Factories and returns an array of TCreateByTypeFunction function pointers.

The Control Factory is a plugin to CEikonEnv that offers an interface between the platform (e.g., Series 60) and the application. It is an extensible mechanism for adding further mappings to control types.

For each Control Factory, the Control Factory function pointer is stored in CEikonEnv's iControlFactoryFuncArray for later use by EikControlFactory::CreateByTypeL(), for example for creating dialogs containing controls of different types (these control types being specified by numeric identifiers in the particular dialog's definition in its resource file).

EikControlFactory is an abstract factory that creates controls by type. The class provides a unified way to create controls according to the specified control integer ID.

EikControlFactory::CreateByTypeL() creates the specified type of control by going through the dynamically loaded DLLs' Control Factories until one is found that successfully creates the desired control.

library->CEikLibrary::ResourceFileArrayL()

- CEikLibrary::ResourceFileArrayL() is a pure virtual function declared in app-framework/uikon/coreinc/eiklibry.h.

- called for each library by CEikonEnv::LoadLibrariesL() in eikenv.cpp.

Gives CEikonEnv access to the library's resources.

BaflUtils::NearestLanguageFile()

- BaflUtils::NearestLanguageFile() is a public static function declared in syslibs/bafl/inc/bautils.h.

- called for each resource file within each library by CEikonEnv::LoadLibrariesL() in eikenv.cpp.

Searches for the file with the correct language extension for the language of the current locale or, failing this, the best matching file.

For example, if the current language was `ELangPolish` (enumerated by `TLanguage` in `e32std.h`), the function would look for a localized resource file with the suffix `.r27`. If this file is found, it will be used in preference to the default `.rsc` resource file.

CCoeEnv::AddResourceFileL()

- `CCoeEnv::AddResourceFileL()` is a public EXPORTed function declared in `app-framework/cone/inc/coemain.h`.
- called for each resource file within each `library` by `CEiko-nEnv::LoadLibrariesL()` in `eikenv.cpp`.

Adds the specified resource file to the list maintained by `CCoeEnv`, giving applications easy access to the resources. This enables resources to be used without having to explicitly state which resource file they come from, though this only works if each resource file specifies a unique NAME.

As mentioned previously, each resource file is identified by a resource file offset. This identifier can be used to unload a resource file, i.e., a resource file offset must be passed as a parameter to the `Delete-ResourceFile()` function. `DeleteResourceFile()` functions are described in Section 3.2.2 below.

CEikonEnv::SetUpFepL()

- `CEikonEnv::SetUpFepL()` is a private function declared in `app-framework/uikon/coreinc/eikenv.h`.
- called by `CEikonEnv::ConstructL()` in `eikenv.cpp`.

Sets up the Front End Processor (FEP); reads resources from resource files and sets its resource reader.

The FEP provides a framework for components that enable the input of characters that are not directly supported by the keyboard (if present). It forms an interface between the application and the likes of non-trivial keyboards (as found on mobile phones), handwriting and voice input, and so on.

(CApaProcess) iProcess = CEikProcess::NewL()*

- `CEikProcess::NewL()` is a public static function declared in `app-framework/uikon/coreinc/eikproc.h`.
- called by `CEikonEnv::ConstructAppFromCommandLineL()` in `eikenv.cpp`.

Creates a Uikon process to manage the application's files. Note that it does not actually create a new process.

`CApaProcess`, which is the base class for `CEikProcess`, maintains a list of documents and all the potentially shared resources used by documents (including DLL files).

CEikDocument* doc =
iProcess->CApaProcess::AddNewDocumentL()

- `CApaProcess::AddNewDocumentL()` is a public EXPORTed function declared in `app-framework/apparc/inc/apparc.h`.

- called by `CEikonEnv::ConstructAppFromCommandLineL()` in `eikenv.cpp`.

Creates and adds the new (possibly embedded) document of the specified application. If this is the first document of that type to be added to the application, it creates a new `CApaApplication` object.

Calls the following major functions:

- `CApaProcess::AddAppDllL()`

- `CApaProcess::CreateDocL()`.

CApaProcess::AddAppDllL()

- `CApaProcess::AddAppDllL()` is a private function declared in `app-framework/apparc/inc/apparc.h`.

- called by `CApaProcess::AddNewDocumentL()` in `apparc.cpp`.

Searches storage media for a DLL with the correct name and UID, creates one and adds it to the array. If the full path is specified, the function loads the DLL; if a UID is specified, it will locate the DLL by UID; otherwise, it locates the DLL by name. (The DLL in question here is the 'real' application DLL object.)

Calls the following major functions:

- `CApaDll::CreateApplicationL()`

- `iApplication->CEikApplication::PreDocConstructL()`.

CApaDll::CreateApplicationL()

- `CApaDll::CreateApplicationL()` is a public function declared in `app-framework/apparc/apparc/apadll.h`.

- called by `CApaProcess::AddAppDllL()` in `apparc.cpp`.

Opens the DLL, checks its type and calls the ordinal one function to create an instance of the CApaApplication object.

If the DLL is of an incorrect type, the function leaves KErrNot-Supported. If the first exported DLL function is not a constructor, the function leaves KErrBadEntryPoint.

Calls the following major functions:

- RLibrary::Load()

- functionPointer = RLibrary::Lookup()

- (CApaApplication*) iApplication = {call functionPoin-ter}.

RLibrary::Load()

- RLibrary::Load() is a public EXPORTed function declared in base/e32/include/e32std.h.

- called by CApaDll::CreateApplicationL() in apparc.cpp.

Loads the named DLL that matches the specified UID type. If successful, the function increments the usage count by one.

functionPointer = RLibrary::Lookup()

- RLibrary::Lookup() is a public EXPORTed function declared in base/e32/include/e32std.h.

- called by CApaDll::CreateApplicationL() in apparc.cpp.

Returns a pointer to a given function, identified by ordinal number.

(CApaApplication*) iApplication = { call functionPointer}

- NewApplication() is defined by the application.

- called by CApaDll::CreateApplicationL() in apparc.cpp.

Calls functionPointer (see above), i.e., the application's single EXPORTed function, NewApplication(); and creates the CApaApplication object.

A GUI application EXPORTs the NewApplication() factory function which, when called, causes a new instance of the class derived from the base application class to be constructed.

iApplication->CEikApplication::PreDocConstructL()

- CEikApplication::PreDocConstructL() is a protected EX-PORTed function declared in app-framework/uikon/coreinc/ eikapp.h.

- called by CApaProcess::AddAppDllL() in apparc.cpp.

Completes construction of, and initializes, the new application object.

After this function has been called, CEikApplication can create document objects. If there is a default resource file for this application, then it is added to the control environment.

PreDocConstructL() is also responsible for obtaining the application's capabilities and caption (from the application information file, .aif).

CApaApplication::PreDocConstructL() is a pure virtual function declared in apparc.h, but is implemented by CEikApplication (although it may be overridden by 'CMyApplication'). An implementation of this function is supplied by the UI framework.

Calls the following major functions:

- CEikApplication::ResourceFileName()

- CCoeEnv::AddResourceFileL()

- CEikApplication::OpenAppInfoFileL().

CEikApplication::ResourceFileName()

- CEikApplication::ResourceFileName() is a public EX-PORTed virtual function declared in app-framework/uikon/ coreinc/eikapp.h.

- called by CEikApplication::PreDocConstructL() in eikapp.cpp.

Gets the name of the resource file used by this application.

By default, this file has an extension of .rsc, uses the same basename as the application .app file, and is located in the same directory. Language variants are supported through BaflUtils::NearestLanguageFile().

If the application has no resource file, ENoAppResourceFile should be given as an argument to CEikAppUi::BaseConstructL(), which is called later.

Calls the following major functions:

- CEikApplication::ParseWithAppName()

- BaflUtils::NearestLanguageFile().

CEikApplication::ParseWithAppName()

- `CEikApplication::ParseWithAppName()` is a private function declared in `app-framework/uikon/coreinc/eikapp.h`.

- called by `CEikApplication::ResourceFileName()` in `eik-app.cpp`.

Utility function that seeds the `TParse` derived parameter with the application filename and path. Used by the framework when loading bitmaps and resource files.

BaflUtils::NearestLanguageFile()

- `BaflUtils::NearestLanguageFile()` is a public EXPORTed static function declared in `syslibs/bafl/inc/bautils.h`.

- called by `CEikApplication::ResourceFileName()` in `eik-app.cpp`.

Gets the correct localized version of the resource file, if there is one; otherwise, gets the `.rsc` file.

See the description for `BaflUtils::NearestLanguageFile()` earlier.

CCoeEnv::AddResourceFileL()

- `CCoeEnv::AddResourceFileL()` is a public EXPORTed function declared in `app-framework/cone/inc/coemain.h`.

- called by `CEikApplication::PreDocConstructL()` in `eik-app.cpp`.

Adds the specified resource file to the list maintained by `CCoeEnv`.

See the description for `CCoeEnv::AddResourceFileL()` earlier.

CEikApplication::OpenAppInfoFileL()

- `CEikApplication::OpenAppInfoFileL()` is a public EXPORTed function declared in `app-framework/uikon/coreinc/eikapp.h`.

- called by `CEikApplication::PreDocConstructL()` in `eik-app.cpp`.

Opens the application information file (`.aif`) associated with the application, constructs the associated `CApaAppInfoFileReader` object and returns a pointer to it.

Overrides the pure virtual `CApaApplication::OpenAppInfo-FileL()` function declared in `apparc.h`.

CApaProcess::CreateDocL()

- `CApaProcess::CreateDocL()` is a private function declared in `app-framework/apparc/inc/apparc.h`.

- called by `CApaProcess::AddNewDocumentL()` in `apparc.cpp`.

High-level function that creates a new document associated with the application, adds it to the available document list for this application, and returns a pointer to the new document to the caller.

Calls the following major function.

CApaDocument doc = iApplication->CEikApplication::CreateDocumentL()*

- `CEikApplication::CreateDocumentL()` is a protected EX-PORTed function declared in `app-framework/uikon/coreinc/eikapp.h`.

- called by `CApaProcess::CreateDocL()` in `apparc.cpp`.

Firstly, this parameterized `CreateDocumentL()` function takes a pointer to the underlying `CApaProcess` object and stores it as a data member. Secondly, it calls the virtual unparameterized application-framework factory function, `CEikApplication::CreateDocumentL()`, which must be implemented by applications, and returns a new concrete application document object.

doc->CEikDocument::PrepareToEditL()

- `CEikDocument::PrepareToEditL()` is a public EXPORTed function declared in `app-framework/uikon/coreinc/eikdoc.h`.

- called by `CEikonEnv::ConstructAppFromCommandLineL()` in `eikenv.cpp`.

Prepares the document for editing.

A class derived from `CEikDocument` is used in each GUI application. In file-based applications, the document provides an intermediate layer between the user interface, the model (also known as the engine), and the file the model will be stored in. Whether they are file based or not, most well-behaved Symbian applications use a model; the document is usually the owner of the model. Non-file-based applications use a document to create the application UI; the derived class must, at least, implement the `CreateAppUiL()` function.

Calls the following major functions:

- `(CEikAppUi*) iAppUi = CEikDocument::CreateAppUiL()`
- `iAppUi->SetDocument()`
- `iAppUi->ConstructL()`.

(CEikAppUi) iAppUi = CEikDocument::CreateAppUiL();*

- `CEikDocument::CreateAppUiL()` is a pure virtual function declared in `app-framework/uikon/coreinc/eikdoc.h`.
- called by `CEikDocument::PrepareToEditL()` in `eikdoc.cpp`.

Creates an instance of an application UI, a `CEikAppUi` object, for this document and its associated application.

The application must implement this function.

IAppUi->CEikAppUi::SetDocument()

- `CEikAppUi::SetDocument()` is a public EXPORTed function declared in `app-framework/uikon/coreinc/eikappui.h`.
- called by `CEikDocument::PrepareToEditL()` in `eikdoc.cpp`.

Sets the application UI's document pointer from the supplied `CEikDocument*` parameter.

IAppUi->CEikAppUi::ConstructL()

- `CEikAppUi::ConstructL()` is a virtual EXPORTed function declared in `app-framework/uikon/coreinc/eikappui.h`.
- called by `CEikDocument::PrepareToEditL()` in `eikdoc.cpp`.

The implementation of `CEikAppUi::ConstructL()` simply calls the following function, `CEikAppUi::BaseConstructL()`.

CEikAppUi::BaseConstructL()

- `CEikAppUi::BaseConstructL()` is a protected EXPORTed function declared in `app-framework/uikon/coreinc/eikappui.h`.
- called by `CEikAppUi::ConstructL()` in `eikappui.cpp`.

Initializes the application UI with standard values. The application's standard resource file will be read unless either the `EnoAppResource-File` flag or the `EnonStandardResourceFile` flag is specified.

Calls the following function.

CCoeEnv::SetAppUi (this)

- `CCoeEnv::SetAppUi()` is a public EXPORTed function declared in `app-framework/cone/inc/coemain.h`.
- called by `CEikAppUi::BaseConstructL()` in `eikappui.cpp`.

Causes the `CCoeEnv/CEikonEnv` object to take ownership of the `CEikAppUi` object.

doc->CEikDocument::OpenFileL()

- `CEikDocument::OpenFileL()` is a public EXPORTed virtual function declared in `app-framework/uikon/coreinc/eikdoc.h`.
- called by `CEikonEnv::ConstructAppFromCommandLineL()` in `eikenv.cpp`.

Opens, or creates then opens, the specified file, depending on the supplied parameters, returning a file store object.

Unlike in Eikon, Series 60 applications have to specifically enable document and `.ini` file processing (`.ini` or initialization files store application preferences). Only if document files are enabled will the system load the document file (which is specified in the application's `.rss` resource file) automatically at start time.

In Series 60, `CAknDocument` is the normal document base class; it is this that disables document files. To enable documents, the application must override its document's `OpenFileL()` function to call `CEik-Document::OpenFileL()`. This does provide a functional default implementation, though it may need to be overridden by the application.

Similarly, `CAknApplication::OpenIniFileLC()` removes `.ini` file opening functionality from the application. To enable the use of an `.ini` file, the application's `OpenIniFileLC()` function must be overridden to call `CEikApplication::OpenIniFileLC()`.

The system does not save data to file automatically; the application needs to provide an implementation of the `StoreL()` function derived from `CEikDocument`.

Calls the following function.

CApaDocument::RestoreL()

- `CApaDocument::RestoreL()` is a pure virtual function declared in `app-framework/apparc/inc/apparc.h`.

- called by `CEikDocument::OpenFileL()` in `eikdoc.cpp`.

Restores the document's content and state from data persisted in the specified store.

The Uikon implementation of this function, `CEikDocument::RestoreL()`, is empty. Applications that wish to load any persisted data must provide their own version.

coe->CCoeEnv::ExecuteD()

- `CCoeEnv::ExecuteD()` is a public EXPORTed function declared in `app-framework/cone/inc/coemain.h`.

- called by `EikDll::RunAppInsideThread()` in `eikdll.cpp`.

Starts the active scheduler wait loop (owned by the `CCoeEnv`), enabling asynchronous requests to be processed (for example, events from the window server to be handled such as redraws, key presses, and so on). It forms the outer loop of all Control Environment applications.

The environment is destroyed by the time the `ExecuteD()` function returns (as the trailing D signifies).

Calls the following major functions:

- `CActiveScheduler::Start()`

- `CEikonEnv::DestroyEnvironment()`.

CActiveScheduler::Start()

- `CActiveScheduler::Start()` is a public EXPORTed function declared in `base/e32/include/e32base.h`.

- called by `CCoeEnv::ExecuteD()` in `coemain.cpp`.

Starts a new wait loop under the control of the current active scheduler.

The wait loop (which, for instance, waits for UI input) is responsible for dispatching the processing of all asynchronous requests. It continues until one of the active objects' `RunL()` functions requests termination using `CActiveScheduler::Stop()`.

This is where the normal running of the application takes place. The active scheduler, which encapsulates the wait loop, runs for practically the whole lifetime of the application.

`CActiveScheduler` controls the handling of asynchronous requests as represented by active objects. An active scheduler is used to schedule the sequence in which active object request completion events are handled by a single event-handling thread.

When standard events occur, the active scheduler calls `CCoeEnv::RunL()`. When non-standard events occur, such as redraw events, it calls `CCoeRedrawer::RunL()`; priority key events are accessed using the Window Server API directly.

Active objects send requests to service providers, such as the window server, and wait for the requests to complete. A client application connects to the window server by creating a window server session and calling `RWsSession::Connect()`. Events generated by the window server are then delivered to the client via the session, and the client must handle these events appropriately. The window server generates events of three different classes:

- a general event, `TWsEvent`, which should be handled by an active object of standard priority, and represents user input events such as pointer and key events

- a redraw event, `TWsRedrawEvent`, which should be handled by an active object of lower priority

- a priority key event, `TWsPriorityKeyEvent`, which should be handled by an active object of higher priority. Note that active objects that have very high priorities might 'slow' the appearance of an application, that is, they might well run before the window server CONE active objects that are responsible for initiating application redraw.

One of the first active objects to `RunL()` is the `CCoeFepLoader` active object, which loads the FEP that was set up by the `SetUpFepL()` call earlier. As the FEP takes events from the user as input, it needs to exist before the first event from the user is handled; hence the need for the `CCoeFepLoader` active object's high priority.

3.2.2 Application Closedown

When the active scheduler's wait loop terminates, the following functions are called.

coe->CEikonEnv::DestroyEnvironment()

- `CEikonEnv::DestroyEnvironment()` is a public EXPORTed function declared in `app-framework/uikon/coreinc/eik-env.h`.

- called by `CCoeEnv::ExecuteD()` in `coemain.cpp` when the active scheduler's wait loop terminates.

Deletes resources owned by the `CEikonEnv` object.

Calls the following major functions:

- `CCoeEnv::DeleteResourceFile()`
- `CEikonEnv::CloseLibraries()`
- `CCoeEnv::DestroyEnvironment()`.

CCoeEnv::DeleteResourceFile(iSystemResource-FileOffset)

- `CCoeEnv::DeleteResourceFile()` is a public EXPORTed function declared in `app-framework/cone/inc/coemain.h`.
- called by `CEikonEnv::DestroyEnvironment()` in `eikenv.cpp`.

Deletes the specified resource file from the list maintained by `CCoeEnv`.
  The resource file is identified by its offset value, which the `CCoeEnv::AddResourceFileL()` function returned earlier.

CEikonEnv::CloseLibraries()

- `CEikonEnv::CloseLibraries()` is a private function declared in `app-framework/uikon/coreinc/eikenv.h`.
- called by `CEikon::DestroyEnvironment()` in `eikenv.cpp`.

Deletes all the 'library' objects loaded by the `CEikLibrary` framework, i.e., deletes all DLL-owned data and closes the DLL `RLibrary`s.
  Calls the following `static` function.

DeleteArrayOfResourceFilesRemovingEach()

- `DeleteArrayOfResourceFilesRemovingEach()` is a LOCAL_C function defined in `app-framework/uikon/coresrc/eikenv.cpp`.
- called by `CEikonEnv::DestroyEnvironment()` in `eikenv.cpp`.

Deletes the dynamic DLL-owned resource files.

CCoeEnv::DestroyEnvironment()

- `CCoeEnv::DestroyEnvironment()` is a public EXPORTed virtual function declared in `app-framework/cone/inc/coemain.h`.

- called by `CEikonEnv::DestroyEnvironment()` in `eikenv.cpp`.

Deletes several resources owned by the `CCoeEnv` object.

This includes the application UI, the system GC (graphics context), and the active object that receives redraw events. Also, it closes the window group, the connection to the window server, and the connection to the file server.

Calls the following major functions:

- `CCoeEnvExtra::DestroyFep()`

- `delete iAppUi`

- `delete this`.

CCoeEnvExtra::DestroyFep()

- `CCoeEnvExtra::DestroyFep()` is a public static function declared in `app-framework/cone/src/coemain.cpp`.

- called from `CCoeEnv::DestroyEnvironment()` in `coemain.cpp`.

Destroys the `CCoeStatic` objects with a destruction priority >0.

delete iAppUi

- called from `CCoeEnv::DestroyEnvironment()` in `coemain.cpp`.

Destroys the `CCoeStatic` objects with a destruction priority <0.

delete this

- called from `CCoeEnv::DestroyEnvironment()` in `coemain.cpp`.

Calls the following destructors:

- `CEikonEnv::~CEikonEnv()`

- `CCoeEnv::~CCoeEnv()`.

CEikonEnv::~CEikonEnv

- `CEikonEnv::~CEikonEnv()` is a public EXPORTed function declared in `app-framework/uikon/coreinc/eikenv.h`.

- called by `delete this` within `CCoeEnv::DestroyEnviron-ment()` in `coemain.cpp`.

Frees any resources acquired by the `CEikonEnv` during its construction. Calls the following major functions:

- `CCoeEnv::DeleteResourceFile()`
- `delete iAppUiFactoryArray`
- `delete iLafEnv`
- `iClockDll->Destroy().`

CCoeEnv::DeleteResourceFile(iPrivateResource-FileOffset)

- `CCoeEnv::DeleteResourceFile()` is a public EXPORTed function declared in `app-framework/cone/inc/coemain.h`.
- called by `CEikonEnv::~CEikonEnv()` in `eikenv.cpp`.

Deletes the specified resource file from the list maintained by the CCoeEnv object.

delete iAppUiFactoryArray

- called by `CEikonEnv::~CEikonEnv()` in `eikenv.cpp`.

Deletes the array of `TEikAppUiFactory` objects.

delete iLafEnv

- called by `CEikonEnv::~CEikonEnv()` in `eikenv.cpp`.

Deletes the `CEikLafEnv` object.

iClockDll->RAnimDll::Destroy()

- `RAnimDll::Destroy()` is a public EXPORTed function declared in `graphics/wserv/inc/w32std.h`.
- called by `CEikonEnv::~CEikonEnv()` in `eikenv.cpp`.

Closes and deletes a previously loaded polymorphic DLL.
    Equivalent to calling `this->Close()` followed by `delete this`, in accordance with the `RLibrary` mechanism provided for managing polymorphic DLLs.

RAnimDll is a client-side interface to the server-side animation DLL.

CCoeEnv::~CCoeEnv()

- CCoeEnv::~CCoeEnv() is a public EXPORTed function declared in app-framework/cone/inc/coemain.h.
- called by delete this within CCoeEnv::DestroyEnvironment() in coemain.cpp.

Deletes any resources owned by the CCoeEnv object that were not deleted by the CCoeEnv::DestroyEnvironment() function.
  Calls the following major functions:

- delete CActiveScheduler::Current()
- iWsSession.Close()
- iFsSession.Close().

delete CActiveScheduler::Current()

- CActiveScheduler::Current() is a public EXPORTed function declared in base/e32/include/e32base.h.
- called by CCoeEnv::~CCoeEnv() in coemain.cpp.

Deletes the currently installed active scheduler.

iWsSession.Close()

- RWsSession::Close() is a public EXPORTed function declared in graphics/wserv/inc/w32std.h.
- called by CCoeEnv::~CCoeEnv() in coemain.cpp.

Closes the window server session.
  Cleans up all resources in the RWsSession and disconnects it from the server. Prior to disconnecting from the window server, the client-side window server buffer is destroyed without being flushed.

iFsSession.Close()

- RHandleBase::Close() is a public EXPORTed function declared in base/e32/include/e32std.h.
- called by CCoeEnv::~CCoeEnv() in coemain.cpp.

Closes the handle to a file server session.

3.3   Summary

In this walkthrough of the life of an application from startup to closedown, we have seen:

- what the system does for you
- what the system expects you to do
- how the resource files, icons and other resources are accessed
- how events are processed.

4

Using Controls and Dialogs

This chapter explains the basic principles of writing and using controls and dialogs. Although the examples are written to run on a phone using the Series 60 UI, they have deliberately been kept general, in order to make the explanations apply as far as possible to all UIs. You will need to consult the Style Guide and other documentation in the appropriate SDK for information on topics such as the preferred layout and color schemes that are specific to a particular phone.

4.1 What is a Control?

In Symbian OS, controls provide the principal means of interaction between an application and the user. Applications make extensive use of controls: each of an application's views is a control, and controls form the basis of all dialogs and menu panes.

A control occupies a rectangular area on the screen and, in addition to responding to user-, application- and system-generated events, may display any combination of text and image. Depending upon the particular user interface, the user-generated events may include:

- keypresses, either alphanumeric or from device-specific buttons
- pointer events, generated by the user touching the screen with a pen.

Drawing a control's content can be initiated by the application itself, for example when the control's displayable data changes. Alternatively, the system might initiate drawing, for example when all or part of the control is exposed by the disappearance of an overlying control – which may belong to the same, or another, application.

Remember that Symbian OS is a full multitasking system in which multiple applications may run concurrently, and the screen is a single resource that must be shared among all these applications. Symbian

Symbian OS C++ for Mobile Phones, Volume 2.   Edited by Richard Harrison
© 2004 Symbian Software Ltd   ISBN: 0-470-87108-3

OS implements this sharing by associating one or more **windows** with each application, to handle the interaction between the controls and the screen. The windows are managed by the **window server**, which ensures that the correct window or windows are displayed, managing overlaps and exposing and hiding windows as necessary.

In order to gain access to the screen, each control must be associated with a window, but there is not necessarily a separate window for each control. Some controls, known as **window-owning controls**, use an entire window, but many others – known as **non-window-owning controls** or (more concisely) **lodger controls** – simply share the window that is owned by another control.

You can test whether a control owns a window by checking whether `OwnsWindow()` returns `ETrue` or `EFalse`, although its main use is within `CCoeControl`'s framework code. This test is made, for example, in `CCoeControl`'s destructor to determine whether `CloseWindow()` should be called.

For a system such as Symbian OS, which runs on devices with limited power and resources, windows are expensive. Each window obviously uses resources in the application that uses it, but it also needs corresponding resources in the window server. Furthermore, each window results in additional client-server communication between the application and the window server. It therefore makes sense to use lodger controls as much as possible. Fortunately, as we shall see, there isn't a great deal of difference in programming terms between the two types.

Any normal application requires some form of access to the screen, and so must contain at least one window-owning control. Other controls might need to own a window if they require the properties of a window, such as the ability to overlap another window.

All controls are ultimately derived from the `CCoeControl` class, defined in `coecntrl.h`. The API for this class contains almost 90 separate member functions; we'll look at the more significant ones in this chapter.

One very common use for a class derived from `CCoeControl` is as an application view and we'll use that to illustrate the workings of controls. Chapter 5, 'Views and the View Architecture', looks at this use in more detail, especially for cases where the application has more than one view, and where one or more of the application's views may need to be accessed from other applications. In this chapter we'll restrict the discussion to exclude these possibilities.

4.2   Simple Controls

In addition to whether or not they own a window, controls may be either simple or compound. A compound control is one that contains one or more component controls; a simple control does not.

Perhaps the simplest of simple controls that it is possible to write is one that draws a blank rectangle and does not respond to user input. (It is technically possible to create an instance of the `CCoeControl` base class, but its usage is obscure as the resulting control cannot draw itself.) As an example, we'll look at the application view used in `HelloBlank`, a 'Hello World' application that uses such a blank control for its view. The class definition for the application view is simply:

```
class CBlankAppView : public CCoeControl
    {
public:
    void ConstructL(const TRect& aRect);
    };
```

We don't need the class constructor or destructor to do anything special, so we're simply relying on the default implementations. For this control, the only member function we need to write is `ConstructL()`, the second-stage constructor, whose implementation is just:

```
void CBlankAppView::ConstructL(const TRect& aRect)
    {
    CreateWindowL();   // Create a window for this control
    SetRect(aRect);    // Set the control's size
    SetBlank();        // Make it a Blank control
    ActivateL();       // Activate the control, which makes it ready to be
                       // drawn
    }
```

The four functions that `ConstructL()` calls are all, in one way or another, associated with enabling the control to display itself.

Since all controls must be associated with a window, and this is the only control in the application, it follows that it has to be a window-owning control. The `CreateWindowL()` function both creates the window and sets the control as its owner.

Unsurprisingly, `SetRect()` sets the size and position, in pixel units, of the control's rectangle on the screen, relative to its associated window. Since this control owns a window, the window's size is also adjusted to that of the control. As we'll see later, a control is responsible for ensuring that it is capable of drawing every pixel within the specified area and that its content is correctly positioned within this rectangle.

The call to `SetBlank()` is what makes this a blank control. All it does is enable the control subsequently to fill its rectangle with a plain background color. You'll see how this happens a little later, when we discuss the `Draw()` function.

Without the final call to `ActivateL()`, drawing of the control remains disabled. For drawing to be enabled, the control must be both visible and activated. Controls are visible by default, but need explicit activation by means of a call to `ActivateL()`. This function is required because controls are not always in a fit state to draw their content immediately after construction. For example, the data that a control is to display may not be available at the time of its construction, and this may also affect the control's required size and position. In such a case, the call to `ActivateL()` would not be made from within `ConstructL()`, but would be called from elsewhere, once the necessary additional initialization is complete.

The blank control class is instantiated as a Series 60 application view from the application's app UI class in a typical way:

```
void CMyApplicationAppUi::ConstructL()
    {
    BaseConstructL(EAknEnableSkin);
    iAppView = new(ELeave) CBlankAppView;
    iAppView->ConstructL(ClientRect());
    }
```

4.3   Compound Controls

To illustrate some of the differences between simple and compound controls, I'll use the Console Application Launcher (ConsLauncher) example.

The purpose of this application is to provide a simple means of running text-based, or console, applications on target phones. By convention, console application files have a `.exe` filename extension and reside in a `\system\programs` directory.

Most current Symbian OS phones do not expose the filing system to the user, so once you have downloaded a console application, there is no obvious means of selecting or running it. There might be a file manager application available, either in the relevant SDK or from other sources for your particular target machine. If there is, you can use it to select and launch console applications. Alternatively, you can use the Console Application Launcher described here.

The requirements of this application are to:

- display a list of console applications
- allow the user to select a specific application from those in the list
- launch the selected application.

The appearance of the application is shown in Figure 4.1. The implementation uses a standard list box as a component control in the application view. The list box supplies all the mechanisms for displaying its content and selecting one of its items, using either the cursor keys or the pointer. In consequence, we don't have to concern ourselves too

Figure 4.1 The ConsLauncher application

much with these aspects, and can concentrate on just those features that are necessary to allow a control to include component controls.

The list box is, itself, a compound control of some complexity, with separate controls to display each item in the list, optional vertical and horizontal scroll bars, and so on. Fortunately, other than the knowledge required to correctly initialize the list box and the data it displays, you do not need to understand or manage this internal complexity.

We'll start by looking at the app UI class definition:

```
class CConsLauncherAppUi : public CAknAppUi
    {
public:
    CConsLauncherAppUi();
    ~CConsLauncherAppUi();
    void ConstructL();

public: // from CAknAppUi
    void HandleCommandL(TInt aCommand);

private:
    CConsLauncherAppView* iAppView;
    HBufC* iDirFspec;
    CDir* iFileList;
    };
```

Apart from the additional data members, there are no significant differences from other simple app UI definitions. By and large, the second-stage constructor and destructor simply illustrate that the class owns the view, a text buffer (iDirFspec) specifying where to look for the console application files, and a list (iFileList) of the names of all found applications.

```
void CConsLauncherAppUi::ConstructL()
    {
    BaseConstructL(EAknEnableSkin);

    iDirFspec     =
        iEikonEnv->AllocReadResourceL(R_CONSLAUNCHER_DIR_FSPEC);
    RFs& fs = iCoeEnv->FsSession();
    CDir *DirList = NULL;
    fs.GetDir(*iDirFspec,KEntryAttNormal,ESortNone,iFileList,DirList);
    delete DirList;
    iAppView = new(ELeave) CConsLauncherAppView;
    iAppView->ConstructL(ClientRect(),iFileList);

    AddToStackL(iAppView);
    }
```

```
CConsLauncherAppUi::~CConsLauncherAppUi()
    {
    if (iAppView)
        {
        iEikonEnv->RemoveFromStack(iAppView);
        delete iAppView;
        iAppView = NULL;
        }
    delete iDirFspec;
    delete iFileList;
    }
```

The line that is of particular interest in `ConstructL()` is

```
AddToStackL(iAppView);
```

which adds the view to the application's **control stack**, thereby ensuring that the view will have the opportunity to process keypresses. We'll see the consequences later, during the discussion of the view itself.

> *The control stack maintains a prioritized list of all controls that have an interest in processing key events and ensures that key events are offered to the controls in priority order. In addition to controls from GUI applications, the list contains controls associated with any front-end processor, with the processing of debug keys (Ctrl+Alt+Shift+key combinations) and active dialogs.*

If any view has been placed on the control stack, the destructor must ensure their removal, using code of the form

```
iEikonEnv->RemoveFromStack(iAppView);
```

For completeness, here is the implementation of the `HandleCom-mandL()` function, although the only point that is of any direct relevance to the discussion of compound controls is the call to the view's `GetCur-rentItemIndex()` function to obtain the index of the currently selected application's name.

```
void CConsLauncherAppUi::HandleCommandL(TInt aCommand)
    {
    switch(aCommand)
        {
    case EEikCmdExit:
    case EAknSoftkeyExit:
        Exit();
        break;
    case EConsLauncherCmdLaunch:
        // Run the selected .exe file
        TInt i = iAppView->GetCurrentItemIndex();
        if (iFileList && (i<iFileList->Count()))
            {
            TPtrC name = (*iFileList)[i].iName;
            RFs& fs = iCoeEnv->FsSession();
            TParse fp;
            fs.Parse(name, *iDirFspec,fp);
            const TDesC& runExeDesC = fp.FullName();
            RProcess consProcess;
            TInt retval = consProcess.Create(runExeDesC, runExeDesC);
            if (retval == KErrNone)
                {
                consProcess.Resume();
                consProcess.Close();
                break;
                }
            }
        TBuf<40> text;
        iEikonEnv->ReadResource(text, R_CONSLAUNCHER_TEXT_RUNFAILED);
        CAknInformationNote* informationNote =
            new (ELeave) CAknInformationNote;
        informationNote->ExecuteLD(text);
        break;
    default:
        Panic(EConsLauncherUi);
        break;
        }
    }
```

The `Launch` command first gets the current item from the view and, after checking that it is a valid selection, uses the filing system's parsing service to construct the full path name of the selected file. It then attempts to create and launch the corresponding process. If the selection is not valid, or if the process cannot be created, an information note informs the user.

Getting back to the discussion of controls, the application's view class is defined as:

```
class CConsLauncherAppView : public CCoeControl
    {
public:
```

```
    CConsLauncherAppView();
    ~CConsLauncherAppView();
    void ConstructL(const TRect& /* aRect */);
private: // from CCoeControl
    CCoeControl* ComponentControl(TInt aIndex) const;
    TInt CountComponentControls() const;
    TKeyResponse OfferKeyEventL(const TKeyEvent& aKeyEvent,TEventCode
        aType);
public:
    TInt GetCurrentItem();
private:
    void AppendItemsL(const CDir* aFileList);
private:
    CEikTextListBox* iTextListBox; // Text list box to display file names
    CDesCArray* iNamesArray;       // Ref to array owned by text list box
    };
```

Support for the component list box control is supplied by implementations of two CCoeControl functions: CountComponentControls() and ComponentControl(). Since the view needs to respond to keypresses, we also need an implementation of CoeControl's OfferKeyEvent() function. We'll come back to these a little later, but first we'll look at the second stage constructor and the destructor:

```
void CConsLauncherAppView::ConstructL(const TRect& aRect,  CDir* aCDir)
    {
    CreateWindowL();
    SetRect(aRect);
    SetBlank();
    iTextListBox = new(ELeave) CEikTextListBox;
    iTextListBox->ConstructL(this,0);
    iTextListBox->CreateScrollBarFrameL();
    iTextListBox->ScrollBarFrame()->
      SetScrollBarVisibilityL(CEikScrollBarFrame::EOff,
          CEikScrollBarFrame::EAuto);
    iTextListBox->SetBorder(TGulBorder::EDeepSunken);
    TRect lbRect = Rect();
    lbRect.Shrink(10,10);
    iTextListBox->SetRect(lbRect);
    iNamesArray =
        static_cast<CDesCArray*>(iTextListBox->Model()->ItemTextArray());
    AppendItemsL(aCDir);
    ActivateL();
    }
```

The highlighted code in ConstructL() shows that the view itself is just a blank control, exactly as was used in the earlier HelloBlank example. The remainder of the code is entirely concerned with creating and initializing the instance of CEikTextListBox. The list box is set to have no horizontal scroll bar, and a vertical scroll bar that appears only if it is needed, and the border style is set to EDeepSunken.

The list box is surrounded by a 10-pixel-wide outer border, which is drawn (blanked) by the containing view. Admittedly, the efficiency of drawing like this is not ideal, since the whole control will be blanked before the list box draws itself, leaving the control susceptible to flicker. It would be better if the container control drew only the outer blank border. We'll deal with this when we discuss drawing later in the chapter.

The code for the destructor does little more than confirm that the list box is owned by the view – and that the name array is not:

```
CConsLauncherAppView::~CConsLauncherAppView()
    {
    delete iTextListBox;
    }
```

The `GetCurrentItem()` function has no direct relevance to the understanding of controls, but it's very short so we'll include it here:

```
TInt CConsLauncherAppView::GetCurrentItem()
    {
    return iTextListBox->CurrentItemIndex();
    }
```

This function does, however, illustrate a point about function naming conventions made in Chapter 1. In the early days of Symbian OS, we used the prefix `Set` *on any function that set the value of an item of member data, and the prefix* `Get` *on any function that retrieved the value. Nowadays we still use the prefix* `Set` *in these circumstances, but prefer to reserve* `Get` *for use in functions that do more than just retrieve the value of an item of member data.*

The implementation of the `AppendItemsL()` function is mainly concerned with supplying the displayable content for the list box, and so only the final line, which sets the list box to have keyboard focus, is relevant to our discussion of controls.

```
void CConsLauncherAppView::AppendItemsL(const CDir* aFileList)
    {
    TInt n = 0;
    if (aFileList)
        {
        n = aFileList->Count();
        }
    if (n>0)
        {
        for (TInt i = 0; i < n; i++)
            {
            TPtrC ThisName = (*aFileList)[i].iName;
            iNamesArray->AppendL(ThisName);
            }
        }
    else
```

```
    {
    HBufC* msg=iCoeEnv->
        AllocReadResourceLC(R_CONSLAUNCHER_TEXT_NOTFOUND);
    iNamesArray->AppendL(*msg);
    CleanupStack::PopAndDestroy(msg);
    }
iTextListBox->HandleItemAdditionL();
iTextListBox->SetCurrentItemIndex(0);
iTextListBox->SetFocus(ETrue);
}
```

Setting a component control to have keyboard focus will normally have an effect on the appearance of the component. It will usually highlight itself in some way and, if the component accepts text, may cause a text cursor to become visible. Setting and unsetting keyboard focus also has other implications, which we'll explain later in this chapter.

Supplying implementations for CountComponentControls() and ComponentControl() is mandatory for any control that contains component controls. The two functions work as a pair and should always be coded to be consistent with each other. The control framework calls both functions in order to access the component controls, using code of the form:

```
for (TInt i=0; i<CountComponentControls(); i++)
    {
    CCoeControl* component = ComponentControl(i);
    ...
    }
```

The default implementation of CountComponentControls() returns zero, as you would expect. In the current example, there is only one component and we've hard-coded the implementation as:

```
TInt CConsLauncherAppView::CountComponentControls() const
    {
    return 1;
    }
```

If a control contains a fixed number of components, a convenient Symbian OS idiom is to supply an enumeration, such as:

```
enum
    {
    EFirstControl,
    ESecondControl,
    ...
    ELastControl,
    ENumberOfControls
    };
```

Your `CountComponentControls()` function can then simply return `ENumberOfControls`. If a control contains a variable number of controls then the return value will obviously have to be evaluated dynamically.

The default implementation of `ComponentControl()` panics, with the panic code `ECoePanicControlIsNotContainer`, and clearly should never be called. In the ConsLauncher example, `ComponentControl()` is, like `CountComponentControls()`, hard coded – this time to return a pointer to the list box:

```
CCoeControl* CConsLauncherAppView::ComponentControl(TInt /*aIndex*/) const
    {
    return iTextListBox;
    }
```

If the control contains more than one component, a simple implementation might be as a switch statement, with each case hard coded to the address of the appropriate component control. If your control contains a variable number of components, then one possibility would be to store pointers to them in an array and let `ComponentControl()` return the content of the appropriate array element.

4.3.1 The Noughts and Crosses Application

Before looking at the behavior of controls in more detail, we'll introduce the Noughts and Crosses example application, OandX. Figure 4.2 illustrates the appearance of the application, running on a phone using the Series 60 user interface.

This application runs a game of noughts and crosses between two players, who take turns to make their moves. We've kept the logic

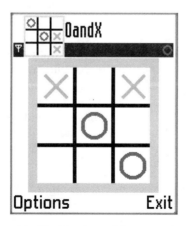

Figure 4.2   The Noughts and Crosses application

relating to the game itself very simple so that the bulk of the code illustrates issues of a general nature, relevant to most applications.

Since the game control logic is so simple, we've implemented it within the app UI class, rather than defining a separate controller class. The relevant parts of `oandxappui.h` are:

```
class MOandXGameControl
    {
public:
    virtual TBool HitSquareL(TInt aIndex)=0;
    virtual TInt SquareStatus(TInt aIndex)=0;
    };

class COandXAppUi : public CAknAppUi, public MOandXGameControl
    {
    ...
public: // New functions
    TBool IsCrossTurn();
    TInt GameWonBy();

private:
    ...
    void SwitchTurnL();

private: // From MOandXGameControl
    TBool HitSquareL(TInt aIndex);
    TInt SquareStatus(TInt aIndex);

private:
    COandXAppView* iAppView; // AppUi owns the application view
    TInt iGameStatus;
    TBool iCrossTurn;
    RArray<TInt> iTileState;
    RArray<TInt> iWinLines;
    ...
    };
```

Following a pattern that is commonly used in Symbian OS code – briefly described in Chapter 1 – we've implemented the two functions `HitSquareL()` and `SquareStatus()`, used by the view to communicate with the control logic, in an `MOandXGameControl` interface class. Note that the two member functions are declared as private in the app UI class, but are public in the interface class, so that they are accessible only as explicit members of the interface.

In addition, we've implemented the application in a way that makes it easy to change the board layout, so that the example could form the basis of other grid-based games. The layout is specified, in `oandxdefs.h`, by the three defined constants

```
#define KTilesPerRow 3
#define KTilesPerCol 3
#define KNumberOfTiles (KTilesPerRow*KTilesPerCol)
```

which are used in both the app UI and the view. You can alter the board layout simply by changing the values of KTilesPerRow and KTilesPerCol and recompiling. The example code should produce satisfactory results for any layout up to an 8 × 8 grid. The remaining content of oandxdefs.h is:

```
#define KTilesPerWinLine 3
#define KCountWinLines 8

enum TOandXTileStates
    {
    ETileBlank  = 0,
    ETileNought = 1,
    ETileCross  = 4
    }; // Values chosen so that:
        // KTilesPerRow*ETileNought < ETileCross
        // KTilesPerCol*ETileNought < ETileCross

enum TOandXGameStatus
    {
    EGameNew,
    EGamePlaying,
    EGameEnded
    };
```

The logic of the remainder of the game control is closely tied to the noughts and crosses game, and would obviously need rewriting to support a different game.

Since the two integer array members, iTileState and iWinLines, are small and fixed in size, we could have used standard C-style arrays as follows:

```
TInt iTileState[KNumberOfTiles];
TInt iWinLines[KCountWinLines][KTilesPerWinLine];
```

Instead, we've chosen to base them on the templated RArray class, which provides a simple and efficient way to create one-dimensional arrays of fixed size objects. An RArray's content is set dynamically so that, unlike for C-style arrays, there is no need to specify its size in the declaration. The initialization of RArrays is illustrated in the app UI's ConstructL():

```
void COandXAppUi::ConstructL()
    {
    BaseConstructL(EAknEnableSkin);
    iAppView = new(ELeave) COandXAppView();
    iAppView->ConstructL(ClientRect(), this);
    AddToStackL(iAppView); // Enable keypresses to the view
    for (TInt i=0; i<KNumberOfTiles; i++)
        {
        User::LeaveIfError(iTileState.Append(ETileBlank));
        }
```

```
// Set the eight possible winning lines
SetWinLineL(0,1,2);
SetWinLineL(3,4,5);
SetWinLineL(6,7,8);
SetWinLineL(0,3,6);
SetWinLineL(1,4,7);
SetWinLineL(2,5,8);
SetWinLineL(0,4,8);
SetWinLineL(2,4,6);

CEikStatusPane* sp = StatusPane();
iNaviPane=(CAknNavigationControlContainer*)sp->
                    ControlL(TUid::Uid(EEikStatusPaneUidNavi)));
SetNavigationPaneL(iCrossTurn);
}

void COandXAppUi::SetWinLineL(TInt aTile0, TInt aTile1, TInt aTile2)
{
User::LeaveIfError(iWinLines.Append(aTile0));
User::LeaveIfError(iWinLines.Append(aTile1));
User::LeaveIfError(iWinLines.Append(aTile2));
}
```

The tile state array is simply initialized to contain KNumberOfTiles elements, each initially equal to ETileBlank. Conceptually, iWin-Lines is a two-dimensional array that lists all possible winning lines. For the game of noughts and crosses there are eight distinct winning lines, each consisting of three tiles. For simplicity we've implemented iWin-Lines as an RArray, but we've arranged its initialization to emphasize the two-dimensional nature of the data.

The check for a possible winner takes place in the GameWonBy() function, which again emphasizes the two-dimensional nature of the data:

```
TInt COandXAppUi::GameWonBy()
{
TInt sum = 0;
for (TInt i=0; i<KCountWinLines; i++)
    {
    for (TInt j=0; j<KTilesPerWinLine; j++)
        {
        sum += iTileState[iWinLines[i*KTilesPerWinLine + j]];
        }
    if ((sum == KTilesPerWinLine * ETileNought) ||
                        (sum == KTilesPerWinLine * ETileCross))
        { // all noughts or all crosses
        sum /= KTilesPerWinLine; // equal to ETileNought or ETileCross
        break;
        }
    else
        {
        sum = 0; // no winner
        }
    }
return sum;
}
```

The function returns `ETileNought` or `ETileCross` to indicate a winner, or zero if neither player has yet won.

The application's view is a compound control, owning all nine component tile controls. Pointers to the tiles are held in an `RPointerArray` array, declared in the view's member data, in `oandxappview.h`, as:

```
private:
    ...
    RPointerArray<COandXTile> iTiles; // View owns the tiles
```

The tiles are indexed, using the standard array indexing convention, so that the indices 0, 1 and 2 refer to the tiles on the first row. The implementations of the view's `CountComponentControls()` and `ComponentControl()` are therefore particularly simple:

```
TInt COandXAppView::CountComponentControls() const
    {
    return KNumberOfTiles;
    }
```

```
CCoeControl* COandXAppView::ComponentControl(TInt aIndex) const
    {
    return iTiles[aIndex];
    }
```

4.4 Control Layout

A control is responsible for the layout of its content. If it is a compound control, it is also responsible for setting the position and size of its component controls. If a control's size is set when it is created, and does not subsequently change, it can usually set the position and size of any components from within its `ConstructL()` function. If a control's size is likely to change during its lifetime, then this approach will not work and the control will need to revise the layout of its content accordingly. The control framework provides a `SizeChanged()` function specifically to meet this need. Its default implementation is empty.

Although the view of the noughts and crosses application does not change size after its creation, we've used `SizeChanged()` to calculate and set the size and position of the border surrounding the playing area and the component tiles. We've coded it to cope with different screen sizes and, as mentioned earlier, alternative board layouts, containing different numbers of tiles.

```
void COandXAppView::SizeChanged()
    {
```

```
    __ASSERT_DEBUG(iTiles[KNumberOfTiles-1], Panic(EOandXNoTiles));
    // all component tiles must already exist

TSize controlSize = Rect().Size();
    TSize tileSize;
    tileSize.iWidth=2*((controlSize.iWidth-2*KBorderWidth
                     -(KTilesPerRow-1)*KLineWidth)/(2*KTilesPerRow));
    tileSize.iHeight=2*((controlSize.iHeight-2*KBorderWidth
                     -(KTilesPerCol-1)*KLineWidth)/(2*KTilesPerCol));
    iTileSide = tileSize.iWidth < tileSize.iHeight ?
                                   tileSize.iWidth :tileSize.iHeight;
    TSize boardSize;
    boardSize.iWidth = KTilesPerRow*iTileSide +
        (KTilesPerRow-1)*KLineWidth;
    boardSize.iHeight = KTilesPerCol*iTileSide +
        (KTilesPerCol-1)*KLineWidth;
    iBoardRect.iTl.iX = (controlSize.iWidth - boardSize.iWidth)/2;
    iBoardRect.iTl.iY = (controlSize.iHeight - boardSize.iHeight)/2;
    iBoardRect.iBr.iX = iBoardRect.iTl.iX + boardSize.iWidth;
    iBoardRect.iBr.iY = iBoardRect.iTl.iY + boardSize.iHeight;
    iBorderRect = iBoardRect;
    iBorderRect.Grow(KBorderWidth,KBorderWidth);

    for (TInt i=0; i<CountComponentControls(); i++)
        {
        TInt row = i / KTilesPerRow;
        TInt col = i % KTilesPerRow;
        TRect tileRect;
        tileRect.iTl.iX = iBoardRect.iTl.iX + col * (iTileSide +
            KLineWidth);
        tileRect.iTl.iY = iBoardRect.iTl.iY + row * (iTileSide +
            KLineWidth);
        tileRect.iBr.iX = tileRect.iTl.iX + iTileSide;
        tileRect.iBr.iY = tileRect.iTl.iY + iTileSide;
        ComponentControl(i)->SetRect(tileRect);
        }
    }
```

We first perform independent calculations of the tile width and height that will enable the required number of tiles – and their borders – to fit in the available area, making sure that the tile dimensions are always an even number of pixels. We then take the smaller of the two dimensions and use it for the side of the tile squares.

Once we've done that, it is a fairly simple matter to calculate overall board and border positions, which are stored in member data for later use when drawing the board. We finally calculate the tile positions and set them by calling each tile's SetRect() function.

SizeChanged() is called whenever one of the control's size-changing functions

- SetExtent()

- SetSize()

- SetRect()

- `SetCornerAndSize()`

- `SetExtentToWholeScreen()`

is called on the control.

In noughts and crosses, the call is triggered by the call to `SetRect()` in the view's `ConstructL()` function:

```
void COandXAppView::ConstructL(const TRect& aRect, COandXAppUi* aAppUi)
    {
    CreateWindowL();
    iAppUi = aAppUi;
    for (TInt i = 0; i < CountComponentControls(); i++)
        {
        iTiles[i] = CreateTileL();
        }
    ComponentControl(0)->SetFocus(ETrue);
    SetRect(aRect);
    ActivateL();
    }
```

All we have to do is to ensure that `SetRect()` is not called until after the tiles have been created.

There is no system-wide mechanism for notifying controls of size changes, and `SizeChanged()` is called only as a result of calling one of the five functions listed above. In consequence, if resizing one control affects the size of other controls, it is the application's responsibility to ensure that it handles the resizing of all affected controls.

Other related functions are:

`SetSizeWithoutNotification()`	Sets a control's size without calling `SizeChanged()`.
`SetPosition()`	Sets the pixel position of the top-left corner of the control. If the control owns its containing window, the function achieves this by setting the position of the window. Otherwise the position of the control is set relative to its containing window. The positions of the control's components are adjusted accordingly and `PositionChanged()` is called.
`PositionChanged()`	Responds to changes in the position of a control. It has an empty default implementation, which may be overridden by a control. This function is called whenever the application calls `SetPosition()` on the control.

4.5 Handling Key and Pointer Events

Whether an application needs to handle both types of event depends in part on the UI of the target phone. One of the design considerations for the Series 60 UI is that it should support one-handed operation. In consequence, Series 60 phones do not have touch-sensitive screens, so applications written specifically for this UI do not need to handle pointer events. Applications running on pointer-based phone UIs, such as UIQ, will generally be expected to handle both types of event. Such phones generally have some keys, and also use front-end processors (FEPs) associated with, say, handwriting recognition, or a virtual (on-screen) keyboard, to generate key events.

Whether you choose to handle one or both types of event may also depend on the nature of your application. In general, if there are no specific reasons to the contrary, handling both types of event will make it easier to convert your application to run on a phone that uses a different UI.

The two types of event have some fundamental differences. A pointer event occurs at a specific position on the screen, which usually corresponds to one particular application's window and, more precisely, one particular control associated with that window. A key event has no such intrinsic connection to a particular control. It is the internal logic of an application that determines this connection, and it is often the case that different types of key event may be processed by different controls within the application.

This fundamental difference is reflected in the way that Symbian OS handles the two types of event.

4.5.1 Key Events

A key event is represented by an instance of a `TKeyEvent` and a key event type, which may be one of `EEventKeyDown`, `EEventKey` or `EEventKeyUp`. Unless your application is particularly interested in detecting when a key is pressed or released, you can safely ignore key events of types other than `EEventKey`. The `TKeyEvent` class has four data members: `iCode`, `iModifiers`, `iRepeats` and `iScanCode`. Of these, `iCode`, containing the key's character code, is usually of most significance to an application. The other data members have the following meanings:

`iModifiers`	The state of any modifier keys, such as Control or Shift, as defined in the `TEventModifier` class.
`iRepeats`	A value of 0 means an event without repeats and a value of 1 or more represents the number of auto repeat events. It is normal to ignore this value.
`iScanCode`	The scan code of the key that caused the event. Standard scan codes are defined in `TStdScanCode`. In general, only game applications for which the position of the key on the keyboard is more important than the character it represents would take note of this value.

As we saw earlier in this chapter, a control should register its interest in processing key events by being added to the control stack by means of a call to the app UI's `AddToStackL()` function.

All key events that may potentially be handled by controls, whether they are generated from a real keypress or from a front-end processor (FEP), are processed by the control stack's `OfferKeyL()`. This function calls the `OfferKeyEventL()` function of each object on the control stack until either:

- all the controls have been offered the key event and have indicated, by returning `EKeyWasNotConsumed`, that they can not process the event, or

- a control can process the key event, and indicates that it has done so by returning the value `EKeyWasConsumed`.

It follows that a control's implementation of `OfferKeyEventL()` *must* ensure that the function returns `EKeyWasNotConsumed` if it does not do anything in response to a key event, otherwise other controls or dialogs may be prevented from receiving the key event. The default action of `CCoeControl`'s `OfferKeyEventL()` function is just to return `EKeyWasNotConsumed`.

A compound control may process (and consume) keys and/or may offer key events to its component controls. A typical pattern is for the container control to process cursor key events, using them to navigate between its component controls. Other key events may be passed to one or other of the components.

In the Console launcher example, which only has one component control, the view itself does not process any keys, so all it does is pass on the offer to its component list box:

```
TKeyResponse CConsLauncherAppView::OfferKeyEventL(const TKeyEvent&
    aKeyEvent, TEventCode aType)
    {
    return iTextListBox->OfferKeyEventL(aKeyEvent,aType);
    }
```

The noughts and crosses application demonstrates the more typical behavior to key events mentioned above. If we look first at the view's `OfferKeyEventL()` function, you'll see that it consumes only the up, down, left and right cursor key events, using them to navigate between the tiles:

```
TKeyResponse COandXAppView::OfferKeyEventL(const TKeyEvent& aKeyEvent,
                                              TEventCode aType)
    {
    TKeyResponse keyResponse = EKeyWasNotConsumed;
```

```
if (aType!=EEventKey)
    {
    return keyResponse;
    }
TInt index = IdOfFocusControl();
switch (aKeyEvent.iCode)
    {
case EKeyLeftArrow: // check not in first column
    if (index % KTilesPerRow)
        {
        MoveFocusTo(index-1);
        keyResponse = EKeyWasConsumed;
        }
    break;
case EKeyRightArrow: // check not in last column
    if ((index % KTilesPerRow) < KTilesPerRow - 1)
        {
        MoveFocusTo(index+1);
        keyResponse = EKeyWasConsumed;
        }
    break;
case EKeyUpArrow: // check not on top row
    if (index >= KTilesPerRow)
        {
        MoveFocusTo(index-KTilesPerRow);
        keyResponse = EKeyWasConsumed;
        }
    break;
case EKeyDownArrow: // check not in bottom row
    if (index < KNumberOfTiles - KTilesPerRow)
        {
        MoveFocusTo(index+KTilesPerRow);
        keyResponse = EKeyWasConsumed;
        }
    break;
default:
    keyResponse =
        ComponentControl(index)->OfferKeyEventL(aKeyEvent,aType);
    break;
    }
return keyResponse;
}
```

Firstly, events other than `EEventKey` are ignored. The control therefore ignores – and does not consume – events such as the `EEventKey-Down` and `EEventKeyUp` events that bracket every `EEventKey` event.

We find the 'current' tile with a call to `IdOfControlWithFocus()`, which is coded as:

```
TInt COandXAppView::IdOfControlWithFocus()
    {
    TInt ret = -1;
    for (TInt i=0; i<CountComponentControls(); i++)
        {
        if (ComponentControl(i)->IsFocused())
            {
```

```
            ret = i;
            break;
            }
        }
    __ASSERT_ALWAYS(ret>=0, Panic(EOandXNoCurrentTile));
    return ret;
    }
```

If the requested move is allowed (for example, an `EKeyUp` event is okay only if the current tile is not in the top row) we determine the required change in the tile index and transfer focus, by means of calls to `SetFocus()` to the new current tile.

If the key event is not a cursor key, we just pass it to the current tile by calling its `OfferKeyEventL()`.

```
TKeyResponse COandXTile::OfferKeyEventL(const TKeyEvent& aKeyEvent,
                                        TEventCode aType)
    {
    if (aType!=EEventKey)
        {
        return EKeyWasNotConsumed;
        }
    TKeyResponse keyResponse;
    switch (aKeyEvent.iCode)
        {
    case EKeyOK:
        TryHitL();
        keyResponse = EKeyWasConsumed;
        break;
    default:
        keyResponse = EKeyWasNotConsumed;
        break;
        }
    return keyResponse;
    }
```

The tile is only interested in the key event with key code `EKeyOK`, corresponding to an 'enter' press on a Series 60 phone's joystick control. The call to `TryHit()` does nothing if the current tile already contains a nought or a cross; otherwise it sets the tile to contain a nought or a cross, depending on whose turn it is.

In general, a control that supplies its own `OfferKeyEventL()` should also provide a matching implementation of `InputCapabilities()`. Front-end processors (FEPs) call this function, via the app UI and the control framework, to determine what classes of key event are appropriate to send to the currently focused control. In the case of a compound control, the FEP receives an `ored` combination of the input capabilities of the control and all focused component controls. In consequence, is not always necessary for a container control to override

CCoeControl's default implementation (which returns TCoeInput-Capabilities::ENone). In the noughts and crosses example, only the COandXTile class overrides this function:

```
TCoeInputCapabilities COandXTile::InputCapabilities() const
    {
    return TCoeInputCapabilities::ENavigation;
    }
```

This information is supplied to the FEP to help it to optimize its performance. The nature of the optimization is left to the FEP itself, so you should not assume that a control will never receive key events other than those specified in its own InputCapabilities() function and those of its component controls.

4.5.1.1 Focus

The basic concept of **focus** – sometimes referred to as **keyboard focus** – is fairly straightforward. In the previous discussion I equated the 'current' tile with the one that had 'focus', and that is essentially the case. In any compound control there is generally one component control that is the current center of interest and is the one with which the user is interacting. In the noughts and crosses application, a player might be adding a nought or a cross into a particular tile, but other examples are:

- typing text into an edit box
- selecting one item from those in a list box
- changing a date in a date editor.

You might be tempted to define the control with focus as being the one to which key events are offered, but that is not entirely true. Certainly, *some* key events are directed to the control, but others are not. You only have to look at the noughts and crosses application to see that, in that case, the control with focus is never offered cursor key events. Key events might be offered to, and consumed by, other types of control before there is any chance of their being offered to the control with focus. Such potential consumers of keys include FEPs, dialogs, and the application's menu bar.

Some FEPs consume key events as well as generate them. An example is one that consumes a key event sequence to generate the key event for a single Chinese character. Such FEPs own a control that is placed on the control stack at a higher priority than visible controls, giving it first refusal of all key events.

As in the noughts and crosses example, a compound control normally treats the control with focus as the default key-processing component, offering it all key events that are not explicitly consumed by the parent control.

By default, all controls are capable of having focus. You can set a control to be unable to receive focus by calling its `SetFocusing()` function with an `EFalse` parameter, and can check whether a control is in this state by calling `IsNonFocusing()`. Once a control has been set to this state, you can change it back by calling `SetFocusing(ETrue)`.

Only one component control should have focus at any one time and it is the application's responsibility to ensure that this is obeyed by its views. In the noughts and crosses example code above, we've seen that a container can find out whether a control has focus by calling its `IsFocused()` function, and can set or remove focus on one of its component controls with that control's `SetFocus()` function. `SetFocus()` is defined in `coecntrl.h` as:

```
IMPORT_C void SetFocus(TBool aFocus,TDrawNow aDrawNow=ENoDrawNow);
```

In addition to setting or removing focus on a control, `SetFocus()` also calls the control's `FocusChanged()` function, passing the value of `aDrawNow` – unless the control either is invisible or has not yet been activated, in which case `FocusChanged()` is always called with a parameter value of `ENoDrawNow`. The purpose of this function is to give the control an opportunity to change its appearance in response to the change in focus. The default implementation is empty, so controls should implement it as and when necessary. The `COandXTile` class implements it simply as:

```
void COandXTile::FocusChanged(TDrawNow aDrawNow)
    {
    if (aDrawNow == EDrawNow)
        {
        DrawNow();
        }
    }
```

There are two other focus-related functions: `PrepareForFocus-GainL()` and `PrepareForFocusLossL()`. The control framework does not call them and their default implementations are empty. We'll return to them in the discussion of dialogs, later in this chapter.

4.5.2 Pointer Events

Running the ConsLauncher application in the emulator shows that it responds to pointer events, even though the application source has no explicit code to handle them.

To verify this, you need to make sure that there are two or more .exe files in the emulated `c:\system\programs` directory (`\epoc32\ <target>\c\system\programs`, where `<target>` is one of `wins`, `winsb` or `winscw`, depending on which compiler and IDE you are using). You will then find that you can select any listed item by clicking on it.

Series 60-specific controls have no need to respond to pointer events, but all the general-purpose controls available in Symbian OS – and all UIQ-specific controls – respond to pointer as well as key events. Unlike for key events, framework code directs pointer events to the appropriate control without any explicit assistance from the application itself.

Pointer events originate in the digitizer driver, which passes them to the window server. Normally, the window server associates a pointer event with the foremost window whose rectangle encloses the event's position (but there are exceptions, which we'll describe later). The window server sends the event to the application that owns the window group containing that particular window.

Within the application, the event is passed to the app UI's `Handle-WsEventL()` function which, as we saw earlier, is the same function that handles key events. This function recognizes the event as a pointer event associated with a particular window and calls the `ProcessPoint-erEventL()` of the control that owns the window. This, in turn, calls the control's `HandlePointerEventL()` function. If the control has components, the default implementation of `HandlePointerEventL()` scans the visible non-window-owning components to locate one that contains the event's position. If one is found, `HandlePointerEvent()` calls its `ProcessPointerEventL()` function.

So, to customize the response to pointer events in a simple control, you should override its `HandlePointerEventL()` function. You wouldn't normally override this function in a compound control, but if you do, make sure that you don't prevent pointer events from being passed to a component control.

In the noughts and crosses example, we implement a tile's `Handle-PointerEventL()` function as:

```
void COandXTile::HandlePointerEventL(const TPointerEvent& aPointerEvent)
    {
    if (aPointerEvent.iType == TPointerEvent::EButton1Down)
        {
        TryHitL();
        }
    }
```

The `TryHitL()` function uses controller code to attempt to add a nought or a cross to the relevant tile and, if successful, redraws the tile.

In addition to calling `HandlePointerEventL()`, the `Process-PointerEventL()` function also performs some useful preprocessing of pointer events. This includes:

- discarding any pointer down events on invisible controls

- implementing pointer grab

- ignoring pointer events until the next pointer up event (which is initiated by calling the control's `IgnoreEventsUntilNextPointerUp()` function, normally after handling a pointer down event)

- reporting, but then discarding, any pointer down events on dimmed controls

- reporting any pointer down event on an unfocused control that is capable of taking focus

- reporting, if necessary, and *after* calling `HandlePointerEventL()`, the need to transfer focus to this control.

We'll discuss the three possible reporting options below, in the next section, on observing a control.

Pointer grab is one of the exceptions that we mentioned early in this section. It is used to ensure that the control that received a pointer down event receives all following pointer events until the next pointer up event, even if the pointer is dragged so that these later events occur outside the original control. Implementing pointer grab needs cooperation between the window server, which has to ensure that the later events are directed to the same window, and the control framework, which has to ensure that they go to the same control within that window.

The other exception is **pointer capture**, which is initiated and terminated by calling a control's `SetPointerCapture()` function. While set, pointer capture prevents any other control from receiving pointer down events. The most common uses of pointer capture are:

- in dialogs, to throw away pointer events that occur outside the dialog

- in menus, to dismiss the menu in response to a pointer down event outside its window.

4.6 Observing a Control

It is frequently useful for a control to be able to notify some other class of significant events, and the standard mechanism for doing this is to use an observer interface. The control framework defines the `MCoeControlObserver` interface class with a single member function,

`HandleControlEventL()`, to provide the mechanism for receiving and processing notification of a range of common control events. Any class that derives from this interface is known as a control observer. A control's observer is frequently, but not necessarily, the control's container.

A control's observer can be set using its `SetObserver()` function and subsequently referenced via the `Observer()` function. A control wishing to report an event to its observer should call `ReportEventL()`, passing one of the enumerated event types listed below. `ReportEventL()` will only report the event if the control's observer has previously been set by means of a call to `SetObserver()`.

There are three general-purpose events that your control may report. These are:

`EEventStateChanged`	A control should use this event to report that some significant piece of internal data has changed. A dialog's container control responds to a report of this event from one of its component controls by calling the dialog's `HandleControlStateChangeL()`. You should implement this function to make the appropriate changes to any other controls that are affected by the change of state. The event is not used elsewhere.
`EEventRequestExit`	The intended use of this event is to indicate the successful completion of an operation, for example, selecting an item from a choice list. The control framework does not use this event.
`EEventRequestCancel`	This event should be used to indicate that an operation has been canceled, for example, backing out of a choice list, leaving the original choice unchanged. The control framework does not use this event.

The control framework uses a further three events to handle the three reports made by `ProcessPointerEventL()`. They are:

`EEventInteractionRefused`	The control framework notifies this event when a pointer down event occurs on a dimmed control. A dialog responds to this event by calling `HandleInteractionRefused()`, whose default behavior is to display a message informing the user that the control is not available.

`EventPrepareFocusTransition`	The control framework notifies this event when a pointer down event occurs on a control that does not yet have, but could get, focus. An appropriate response, which is used in dialogs, would be to call the currently focused control's `PrepareForFocusLossL()` function. This gives the control an opportunity to ensure that it is in a suitable state to lose focus. If the control is not in a suitable state, and cannot change itself into such a state, its implementation of `PrepareForFocusLossL()` should leave, thereby aborting the attempted change in focus. In some cases, it might also be appropriate to call `PrepareForFocusGainL()` in the control that is about to gain focus. This function should also leave if the control is not prepared to gain focus.
`EEventRequestFocus`	The control framework also notifies this event when a pointer down event occurs on a control that does not yet have, but could get, focus. The event is notified after notification of the `EEventPrepareFocusTransition` event, and after calling the control's `HandlePointerEventL()` function. The appropriate response is to transfer focus to the control in which the pointer down event occurred.

In the noughts and crosses application, the tiles have the application's view as their observer. The view implements `HandleControlEventL()` as follows:

```
void COandXAppView::HandleControlEventL(CCoeControl* aControl,
                                        TCoeEvent aEventType)
    {
    switch (aEventType)
        {
    case EEventRequestFocus:
        TInt index;
        index = IdOfFocusControl();
        ComponentControl(index)->SetFocus(EFalse, EDrawNow);
        aControl->SetFocus(ETrue, EDrawNow);
        break;
    default:
```

```
        break;
        }
    }
```

None of the tiles is ever dimmed, and the tiles are never in a state that would prevent them losing focus. In consequence, of the three pointer-related events that it could receive, it needs to handle only EEventRequestFocus.

4.7   Drawing a Control

The HelloBlank and ConsLauncher example applications both used blank controls for their views. The noughts and crosses application uses custom controls for both its view and the board's tiles, and both controls need to draw their own specific content.

The standard way of customizing how a control draws its content is to implement a Draw() function, so the class definition of our earlier simple blank control would need to change to:

```
class CBlankAppView : public CCoeControl
    {
public:
    CBlankAppView();
    ~CBlankAppView();
    void ConstructL(const TRect& aRect);
private:
    void Draw(const TRect& aRect) const;
    };
```

Obviously, the view's ConstructL() function should not call Set-Blank() and becomes simply:

```
void CHelloBlankAppView::ConstructL(const TRect& aRect)
    {
    CreateWindowL();    // Create a window for this control
    SetRect(aRect);     // Set the control's size
    ActivateL();        // Activate the control, which makes it ready to be
                        // drawn
    }
```

The Draw() function must be capable of drawing to every pixel within the rectangular area occupied by the control. A simple implementation for a blank window-owning control might be:

```
void CHelloBlankAppView::Draw(const TRect& /*aRect*/) const
    {
    CWindowGc& gc = SystemGc();    // Get the standard graphics context
    gc.Clear();                    // Clear the whole window
    }
```

Drawing requires a graphics context; the appropriate one (a CWin-dowGc) for drawing to a window is found by calling CCoeControl::SystemGc().

As in this case, a Draw() function is free to ignore the rectangular region passed as a parameter, provided it does not draw outside its own rectangle. An alternative version of Draw() that respects the passed parameter would use a different overload of the graphics context's Clear() function:

```
void CHelloBlankAppView::Draw(const TRect& aRect) const
    {
    CWindowGc& gc = SystemGc();   // Get the standard graphics context
    gc.Clear(aRect);              // Clear only the specified rectangle
    }
```

This is effectively the code that the default implementation of Draw() uses to draw a blank window.

Whether or not to ignore the passed rectangle will depend on what the control needs to draw, and considerations such as whether it is more efficient in terms of time and/or memory usage to do one rather than the other. In the current example, as is commonly the case, drawing is simple and fast, so there is no great advantage in restricting the drawing to the passed rectangle.

We'll return to the discussion of the Draw() function, with a less trivial example, later in this section.

4.7.1 Drawing and the Window Server

Before continuing with the practical issues of drawing a control, it will be useful to explain a bit more about how Symbian OS supports drawing. If you wish, you can skip to Section 4.7.7 and read this and the following sections at a later time.

It is virtually axiomatic that all window-owning controls must create the window they own, and this is normally done from the control's ConstructL() function. Most controls will use a standard window, of type RWindow, created with the CreateWindowL() function, as in the earlier example. In addition to creating the client-side RWindow resource, the function also creates a corresponding resource in the window server, to record little more than the window's position and size, together with the region of the window that is visible and any region that needs to be redrawn (the **invalid** region). Once the window is created, the control can access it via the Window() function.

Drawing a control may be initiated either by the control itself if, for example, the application modifies the data displayed by the control, or by the window server when, say, a previously hidden part of a window is exposed. Drawing initiated by the window server is known as a **redraw**

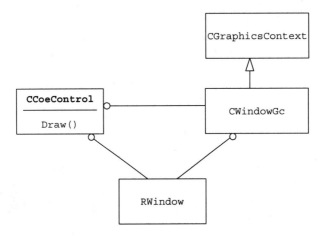

Figure 4.3 A window-owning control (client side)

operation, to distinguish it from application-initiated drawing but, as you'll see, the mechanisms are similar.

To draw to its window, a control uses a window-specific graphics context, CWindowGc, derived from the generic CGraphicsContext. The relationship between the control, window and graphics context is illustrated in Figure 4.3.

The control's drawing code makes calls to CWindowGc functions to set attributes, such as the pen color and line width, or a font style and size, and to draw lines, shapes or text. The graphics context in turn issues a corresponding sequence of commands to a buffer, not shown in Figure 4.3, whose content is subsequently transferred to the window server. Buffering the commands greatly reduces the number of inter-process communications between the application and the window server, thereby improving performance.

As illustrated by Figure 4.4, the window server uses another specialized graphics context, CFbsBitGc, to transform the sequence of commands in its input buffer (again not shown) into a sequence of direct bitmap drawing operations to the screen.

4.7.2 Preparing to Draw

Before such a drawing sequence, the control must perform a little house-keeping to ensure that both the application and the window server are in the correct state for drawing to start. An example of the required code is:

```
TRect theRectToDraw;
...
Window().Invalidate(theRectToDraw);
ActivateGc();
Window().BeginRedraw(theRectToDraw);
```

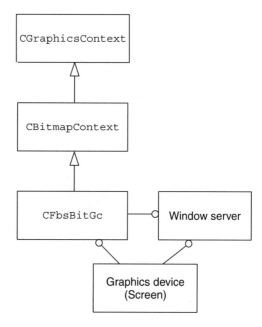

Figure 4.4   Drawing to the screen (client side)

The first call is to the window's `Invalidate()` function, to invalidate the region we are about to draw. The window server keeps track of all invalid regions on each window, and clips drawing to the total invalid region. So, before you do an application-initiated draw, you must invalidate the region you are about to draw, otherwise nothing will appear (unless the region happened to be invalid for some other reason).

The call to the control's `ActivateGc()` activates the standard graphics context owned by `CCoeEnv`, the control environment. Activation prepares the graphics context for drawing to the control's window, and ensures that all its settings are restored to their default values.

The final requirement, before drawing, is to inform the window server of the region that is about to be drawn, by means of a call to the window's `BeginRedraw()` function. In response, the window server:

- creates and sets a clipping region to the intersection of the invalid region, the region specified by `BeginRedraw()`, and the region of the window that is visible on the screen

- marks the region specified by `BeginRedraw()` as valid.

At this point the application's draw code can be called. It must guarantee to cover every pixel of the region specified in the call to `BeginRedraw()`.

4.7.3   When Drawing is Complete

On completion of the application's draw code, a little further housekeeping is required:

```
Window().EndRedraw();
DeactivateGc();
```

The call to the window's `EndRedraw()` function informs the window server that the drawing sequence is complete, enabling it to delete the clipping region that it allocated during `BeginRedraw()` processing.

Calling `DeactivateGc()` deactivates the standard graphics context owned by the UI control framework, freeing it to be used with another window.

4.7.4   Redrawing

When the window server knows that a region of a window is invalid, for example when an overlying window is removed, it sends the bounding rectangle of the invalid region in a redraw event to the window's owning application. The application framework then calls the appropriate window-owning control's `HandleRedrawEvent()` function.

This function, which you should not need to override, calls the `Draw()` function for the control (and, if necessary, the `Draw()` functions of any component controls that intersect with the rectangular region to be redrawn), preceded by calls to `ActivateGc()` and the window's `BeginRedraw()`, and followed by calls to `EndRedraw()` and `DeactivateGc()`. From the application's point of view, the only difference between a redraw and an application-initiated draw is that the code to handle a redraw does not (and should not) call the window's `Invalidate()` function.

4.7.5   Backed Up Windows

In certain, fairly rare, circumstances a window-owning control can alternatively use a backed up window, of class `RBackedUpWindow`. Perhaps the most common case where this might be necessary is when drawing a control is sufficiently complicated, and takes so long, that redraws severely degrade the performance and/or appearance of the application.

A control that uses such a backed up window should create it by using `CreateBackedUpWindowL()` and access it via `BackedUpWindow()` rather than `Window()`. Both window classes are derived from `RDrawableWindow`, so if you are not sure which type of window the control owns, you should access the window via `DrawableWindow()`.

The window server creates a backup bitmap for each such window, and uses application-initiated drawing to update the bitmap as well as to draw to the screen. The advantage of using this type of window is that the window server can handle redraws by drawing from the bitmap, without having to send a redraw event to the application.

An obvious disadvantage is that a backed up window uses much more memory than a standard window, and memory is a scarce resource on mobile phones. If there is insufficient memory for a control to create a backed up window, this will usually have severe consequences for the whole application. It follows that you should only make the decision to use a backed up window if there is no realistic alternative.

4.7.6 Backed Up Behind Windows

Backed up behind windows provide an optimization that is particularly useful for controls that are displayed for only a short time, such as menu panes, pop-ups and dialogs. As with backed up windows, the window server creates a bitmap, but the image that it contains is of the area *behind* the window, not the content of the window itself. When the backed up behind window disappears, the window server can use the bitmap to restore the underlying image, without needing to generate redraw events.

A backed up behind window is not a separate window class. Either an RWindow or an RBackedUpWindow may be converted to a backed up behind window by calling the window's EnableBackup() function. This function is declared (in the class definition of RDrawableWindow in w32std.h) as:

```
IMPORT_C void EnableBackup(TUint aBackupType=EWindowBackupAreaBehind);
```

The two values associated with the TUint parameter, and their meanings, are:

EWindowBackupAreaBehind	Create a backed up behind bitmap to store the image for just the area behind the window
EWindowBackupFullScreen	Create a backed up behind bitmap for the whole underlying screen image

The first of these is frequently used for an overlying window that can be moved around the screen. As the window moves, the exposed underlying area is restored from the bitmap, and the bitmap content is updated to the image behind the window's new position.

The second value is typically used in cases where the underlying screen image is faded when the overlying window appears. In this case, the bitmap stores the unfaded image.

A window-owning control may set its window to be one of these two types by calling, for example:

```
DrawableWindow()->EnableBackup(EWindowBackupAreaBehind);
```

Alternatively, it may set its window to own both types of bitmap simultaneously by calling:

```
DrawableWindow()->
    EnableBackup(EWindowBackupAreaBehind|EWindowBackupFullScreen);
```

This option is useful to display a movable window on a faded background. In this case, the full screen bitmap stores the unfaded image, but the bitmap of the area behind the window stores the faded image.

The bitmap is created at the time the window becomes visible, not when the call to `EnableBackup()` is made. When backing up the whole screen, you need to call `EnableBackup()` before the window is activated, and before you call fade behind on it (`RWindowBase::FadeBehind()`).

We should emphasize that backed up behind windows are supported for purposes of optimization and, as such, any call to `EnableBackup()` is not guaranteed to result in the creation of the requested bitmap. It may, for example, fail if there is insufficient memory available to create the necessary backup bitmap. Unlike in the case of an `RBackedUpWindow`, such a failure is not catastrophic for the application. The only consequence is that the application will be sent redraw events that would not otherwise have been received.

In addition, the usage of such windows is subject to a number of limitations, designed to minimize the amount of memory consumed by this type of window:

- An application may own only one window of each type at any one time. The application may therefore have a maximum of two backed up behind windows, one of each type, or it may have one window set to own both types of bitmap, as described earlier.

- If a window owning control requests a backup of type `EWindowBackupAreaBehind`, any window that has already claimed a backup of this type will lose it.

- If a window owning control requests a backup of type `EWindowBackupFullScreen`, this request will fail if another window has already claimed a backup of this type.

4.7.7 Application-initiated Drawing

As was mentioned earlier, an application does not normally make direct calls to `Draw()`. If the application needs to initiate drawing of a control, it should call either `DrawNow()` or `DrawDeferred()`.

The application may call `DrawNow()` on a control after it has been created and is ready for drawing, or if a change in application data or the control's internal state means that the entire control's appearance is no longer up-to-date. If the control is not ready to be drawn, that is, if it either is invisible or has not yet been activated, `DrawNow()` does nothing. Otherwise, for a control without a backed up window, `DrawNow()` surrounds a call to the control's `Draw()` function with a call to the window's `Invalidate()` function and the other window-server-related housekeeping function calls that we saw earlier. For a compound control, it also calls the `DrawNow()` function of the component controls.

`DrawNow()` is coded to behave correctly if the control is associated with a backed up window.

For a particularly complex control, partial redrawing of the control may sometimes be more appropriate than drawing the entire control. In such a case you should not use `DrawNow()` but will need to write your own customized drawing code.

An application may call a control's `DrawDeferred()` function in the same circumstances as it might call `DrawNow()`. If the control's window does not have a backup bitmap, `DrawDeferred()` simply causes the control's area to be marked as invalid, which will eventually cause the window server to initiate a redraw.

The control framework handles redraw events at a lower priority than user input events, which means that any pending user input events will be processed before the redraw event. `DrawDeferred()` therefore allows a control to do drawing at a lower priority than drawing performed by `DrawNow()`.

An advantage of using `DrawDeferred()` is that if you make multiple calls to `DrawDeferred()` on the same area of a control, the window server will not generate a redraw event to do drawing that has already been superseded. If you make multiple calls to `DrawNow()`, however, all of them get processed, even if they have already been superseded by the time they are processed.

In the noughts and crosses application, a tile's `TryHit()` function calls `DrawDeferred()` following a successful attempt to set the tile to contain a nought or a cross:

```
void COandXTile::TryHitL()
    {
    if (iAppView->TryHitSquareL(this))
        {
        DrawDeferred();
        }
    }
```

If we used `DrawNow()` here, the control would be drawn twice in the case where a pointer event switched focus to this control as well as setting it to contain a nought or a cross.

Calling `DrawDeferred()` on a control associated with a backed up window has effectively the same effect as calling `DrawNow()`.

Both `DrawNow()` and `DrawDeferred()` are implemented by `CCoeControl` and may not be overridden.

4.7.8 The `Draw()` Function

The `Draw()` function needs to be implemented by all non-blank controls, but is intended to be called only from the control's own member functions; hence `Draw()` is declared as a private member function of `CCoeControl`.

The function is passed a reference to a `TRect`, indicating the region of the control that needs to be redrawn, and the function must guarantee to draw to every pixel within the rectangle. It may, subject to considerations of efficiency, ignore the rectangle and draw the whole of the control. Depending on how drawing is implemented, drawing outside the specified rectangle may cause the display to flicker.

The `Draw()` function should not draw outside the control's area. For window-owning controls, drawing is clipped to the window boundary, so drawing outside the control will only have the effect of reducing efficiency. Drawing is not clipped to the boundary of a non-window-owning control, so in this case it is essential that the control does not draw outside its boundary.

You may safely assume that:

* before any `Draw()` function is called, the graphics context has been activated and its properties set to their defaults

* the graphics context is deactivated for you, after the return from `Draw()`.

Your implementation of `Draw()` should gain access to a `CWindowGc` graphics context by means of a call to `SystemGc()` and all drawing should be done using this graphics context.

In the noughts and crosses example, responsibility for drawing is divided between the view and the component tiles. The view draws only the outer regions and the lines between the tiles:

```
void COandXAppView::Draw(const TRect& /*aRect*/) const
    {
    CWindowGc& gc = SystemGc();
    TRect rect = Rect();

    // Draw outside the border
    gc.SetPenStyle(CGraphicsContext::ENullPen);
    gc.SetBrushStyle(CGraphicsContext::ESolidBrush);
    gc.SetBrushColor(KRgbWhite);
```

```
DrawUtils::DrawBetweenRects(gc, rect, iBorderRect);

// Draw a border around the board
gc.SetBrushStyle(CGraphicsContext::ESolidBrush);
gc.SetBrushColor(KRgbGray);
DrawUtils::DrawBetweenRects(gc, iBorderRect, iBoardRect);

//Draw the first vertical line
gc.SetBrushColor(KRgbBlack);
TRect line;
line.iTl.iX = iBoardRect.iTl.iX + iTileSide;
line.iTl.iY = iBoardRect.iTl.iY;
line.iBr.iX = line.iTl.iX + KLineWidth;
line.iBr.iY = iBoardRect.iBr.iY;
gc.DrawRect(line);
TInt i;
// Draw the remaining (KTilesPerRow-2) vertical lines
for (i = 0; i < KTilesPerRow - 2; i++)
    {
    line .iTl.iX += iTileSide + KLineWidth;
    line .iBr.iX += iTileSide + KLineWidth;
    gc.DrawRect(line);
    }
// Draw the first horizontal line
line.iTl.iX = iBoardRect.iTl.iX;
line.iTl.iY = iBoardRect.iTl.iY + iTileSide;
line.iBr.iX = iBoardRect.iBr.iX;
line.iBr.iY = line.iTl.iY + KLineWidth;
gc.DrawRect(line);
// Draw the remaining (KTilesPerCol -2) horizontal lines
for (i = 0; i < KTilesPerCol - 2; i++)
    {
    line .iTl.iY += iTileSide + KLineWidth;
    line .iBr.iY += iTileSide + KLineWidth;
    gc.DrawRect(line);
    }
}
```

The lines are drawn as thin rectangles rather than by using `gc.Draw-Line()`. This ensures that the lines cover the precise region required, without the need for messy calculations involving the line width, and potential problems related to the rounded ends of lines are avoided.

Each tile is responsible for drawing itself:

```
void COandXTile::Draw(const TRect& /*aRect*/) const
    {
    TInt tileType;
    tileType = iAppView->SquareStatus(this);

    CWindowGc& gc = SystemGc();
    TRect rect = Rect();

    if (IsFocused())
        {
        gc.SetBrushColor(KRgbYellow);
        }
```

```
gc.Clear(rect);
if (tileType!=ETileBlank)
    {
    TSize size;
    size.SetSize(rect.iBr.iX- rect.iTl.iX, rect.iBr.iY - rect.iTl.iY);
    rect.Shrink(size.iWidth/6,size.iHeight/6); // Shrink by about 15%
    gc.SetPenStyle(CGraphicsContext::ESolidPen);
    size.iWidth /= 9; // Pen size set to just over 10% of the shape's
                      // size
    size.iHeight /= 9;
    gc.SetPenSize(size);
    if (tileType == ETileNought)
        {
        gc.SetPenColor(KRgbRed);
        gc.SetBrushStyle(CGraphicsContext::ESolidBrush);
        gc.DrawEllipse(rect);
        }
    else // draw a cross
        {
        gc.SetPenColor(KRgbGreen);
        // Cosmetic reduction of cross size by half the line width
        rect.Shrink(size.iWidth/2,size.iHeight/2);
        gc.DrawLine(rect.iTl, rect.iBr);
        TInt temp;
        temp = rect.iTl.iX;
        rect.iTl.iX = rect.iBr.iX;
        rect.iBr.iX = temp;
        gc.DrawLine(rect.iTl, rect.iBr);
        }
    }
}
```

The tile background is drawn first. Note that the background color is set to yellow only for the tile that currently has focus. White is the default brush color and you can safely assume that the graphics context defaults have been set before `Draw()` was called.

> *Highlighting the control with focus isn't always necessary. The noughts and crosses application, running on, say, the Sony Ericsson P900 (using UIQ), doesn't receive key events. You don't need a tile to be highlighted in order to know whether you can tap on it to enter a nought or a cross.*

Having drawn the background, we then find the status of the tile and, if it isn't blank, draw either a red circle or a green cross, as appropriate. This isn't ideal drawing code, as we're drawing over an area that we have previously blanked, which means there is some potential for the display to flicker; ideally, each pixel should be drawn only once. However, the redrawing is confined to a small region and is done immediately so, in practice, the display behavior is acceptable. Note also that, when drawing the cross, we shrink the rectangle by half the line width. This allows for the rounded ends of the lines, without which the cross would look larger than a nought. It's worth paying attention to such details in an application, to create the best possible impression on the user.

These two functions satisfy the basic requirements for the effective drawing of a control:

- The combined drawing code of the view and the tiles covers every pixel of the required area.

- Neither function draws outside the relevant control.

- Apart from the redrawing of the noughts and crosses mentioned earlier, each pixel is drawn only once.

4.7.8.1   *Dimmed and invisible controls*

The control framework provides support for dimming controls or making them invisible, as a means of indicating the controls with which the user is, or is not, allowed to interact.

You can set a control to be visible or invisible by calling `MakeVisible(ETrue)` or `MakeVisible(EFalse)` and you can test the visibility of a control by calling `IsVisible()`. By default, a control's components do not inherit the containing control's visibility, but you can change this behavior by calling the container's `SetComponentsToInheritVisibility()` function, which also takes an `ETrue` or `EFalse` parameter. The effect of this call only extends to the immediate component controls and so you will need to ensure that this function is also called on any components which themselves are compound controls.

A control's `Draw()` function will never be called while it is invisible. It follows that if the control is a component of a compound control, the container control must take responsibility for drawing any area that is occupied by an invisible control.

A dimmed control is conventionally displayed in a subdued palette of colors. You can add or remove dimming by calling `SetDimmed(ETrue)` or `SetDimmed(EFalse)`, and check on the dimmed state by calling `IsDimmed()`. A control should be dimmed if you want to indicate that it is temporarily unavailable but still want the user to be able to see its content.

There is no explicit support for displaying a dimmed control. The control's `Draw()` function is responsible for testing, by means of `IsDimmed()`, the state of the control and then drawing the content in the appropriate colors. The standard way of dimming a control varies from UI to UI, so you need to follow the examples and style guidelines provided in the appropriate SDK.

Clearly, a compound control should never set focus to a dimmed or invisible control.

4.8   Dialogs

A dialog is simply a specialized compound control, so most of what has
been described so far in this chapter applies to dialogs as well. Some of
the particular features of Symbian OS dialogs are:

- Most dialogs are **modal**. Once a modal dialog is running, you, as the
  user, can only interact with the dialog until the dialog is dismissed.
  You are therefore forced to respond, and cannot interact with the
  underlying application until you have done so.

- The component controls that make up a specific dialog are generally defined in the application's resource file. The dialog framework
  code uses the definitions to construct the appropriate controls and
  incorporate them into the dialog.

- The dialog framework provides support for the various possible inter-
  actions between dialog components, and for setting and querying a
  component's data.

- Due in part to both of the previous points, dialogs need relatively
  small amounts of additional code to be written.

All dialog classes are derived, directly or indirectly, from CEikDialog
which, itself, is derived from CCoeControl, as illustrated in Figure 4.5.
   Also, as shown in Figure 4.5, dialogs inherit from a number of interface
classes, including MCoeControlObserver, described earlier in this
chapter. Depending on the particular UI, dialog classes may also inherit
from other interfaces: in Series 60, for example, CEikDialog also inherits

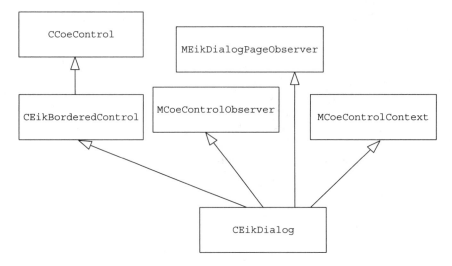

Figure 4.5   Generic derivation of CEikDialog

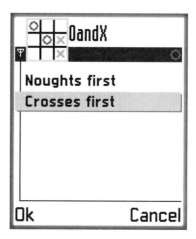

Figure 4.6 Player selection dialog

from a `MAknFadedComponent` interface. We'll describe the interfaces later, but we'll cover only the ones that are common to all UIs.

4.8.1 A Simple Dialog

We'll start by explaining how to modify the noughts and crosses example to use a dialog to select who should play first. Figure 4.6 shows the dialog in action.

The dialog needs to display the two player options to allow the user to select one. We've chosen to use a list box to display the items, since list box controls are available in all Symbian OS UIs. Despite being so simple, this example will explain much of the basic information you need in order to use dialogs. The code is present, but commented out, in the source files of the noughts and crosses example.

First you'll need to add the line:

```
LIBRARY            eikdlg.lib
```

to the libraries section of the application's project definition file, `oandx.mmp`.

The dialog resource is:

```
RESOURCE DIALOG r_oandx_first_move_dialog
    {
    title = "First player";
    flags = EAknDialogSelectionList;
    buttons = R_AVKON_SOFTKEYS_OK_CANCEL;
    items =
        {
        DLG_LINE
```

```
    {
    type = EAknCtSingleListBox;
    id = EOandXPlayerListControl;
    control = LISTBOX
        {
        flags = EAknListBoxSelectionList;
        array_id = r_oandx_player_list;
        };
    }
  };
}
```

The dialog is defined by a DIALOG resource, containing title text, dialog-specific flags, a button resource and an array of dialog line items which, in this case, contains only one member.

The flags value is often, as in this case, UI-specific. The value used here is defined in avkon.hrh to be a combination of the flag bit values EEikDialogFlagNoDrag, EEikDialogFlagNoTitleBar, EEik-DialogFlagFillAppClientRect, EEikDialogFlagCbaButtons and EEikDialogFlagWait. The last of these is the value that ensures the dialog is modal. All normal dialogs will include EEikDialogFlag-Wait, regardless of the UI. In UIQ that is often the only flag value you need. In the present case, the remaining bit values ensure that the appearance of the dialog conforms with Series 60 UI style guide. Note that the dialog has no title bar, so the title item is ignored. We could have left the title entry out of the resource struct, but have left it in to illustrate its use.

The buttons item specifies a resource that defines the buttons to use in the dialog. Phones with touch-sensitive screens, such as those built on UIQ, will often use the resource R_EIK_BUTTONS_CANCEL_OK, which displays two graphical buttons at the bottom of the dialog. The R_AVKON_SOFTKEYS_OK_CANCEL resource used here is commonly used on Series 60 phones. It labels the phone's two softkeys with the labels 'Ok' and 'Cancel'.

Resource structs are defined in resource header files, with a .rh extension, and the DIALOG resource struct is no exception, being defined in eikon.rh. Constants, such as flag values and IDs, that might need to be included in either source code or resource files – or, more commonly, both – are defined in files that have a .hrh extension. If you are not sure about, say, the nature of a resource structure, or the range of flag values available, take a look at these files – they are all in the \epoc32\include directory. In particular, the .rh files are a handy reference to the stock controls available in any particular UI.

The DLG_LINE struct specifies a dialog item. We have chosen a suitable value for the type from the types of list box available in the

Series 60 UI. The value for the id can be anything you like, provided each ID is non-zero and unique within any one dialog. We have chosen to define an ID value in the application's own oandx.hrh.

The dialog item contains a control, which we've declared to be a LISTBOX. The list box flags are, like the item type, chosen from the available Series 60 options. Finally, the two items of text to be displayed in the dialog box are contained in a further resource, which is:

```
RESOURCE ARRAY r_oandx_player_list
    {
    items =
        {
        LBUF { txt= "\tNoughts first"; },
        LBUF { txt= "\tCrosses first"; }
        };
    }
```

Incidentally, the leading tab character that you see in each text item is specific to the Series 60 UI. It is needed because Series 60 list boxes display structured data. The elements of the structure are separated by tabs and may include icons as well as text.

For clarity in the noughts and crosses example, we've included the text strings directly in the application's resource file. In a real application, it's good practice to collect all text strings (and any other language-dependent information) in a separate file. To do this for the R_OANDX_PLAYER_LIST resource, you should define the strings in a separate file – named, say, oandx.loc – as:

```
#define qtn_oandx_noughts_first "\tNoughts first"
#define qtn_oandx_crosses_first "\tCrosses first"
```

and then #include it in the resource file, which is modified to refer to the strings by their symbolic names:

```
RESOURCE ARRAY r_oandx_player_list
    {
    items =
        {
        LBUF { txt= qtn_oandx_noughts_first; },
        LBUF { txt= qtn_oandx_crosses_first; }
        };
    }
```

You'll see the real advantage of doing this if you are intending to translate your application into another language. Using this technique allows the translator to work with just one simple file, which can be annotated with information such as the maximum allowed length of each string and any other context-related information that will help with the task of translation.

A dialog resource essentially defines the content and appearance of the corresponding dialog, but it takes code to specify its behavior. The dialog's class definition is:

```
class COandXFirstPlayDialog : public CEikDialog
    {
public:
    COandXFirstPlayDialog(TBool aCrossTurn, TBool& aFlag);
private: // from CEikDialog
    void PostLayoutDynInitL(); // initialization
    TBool OkToExitL(TInt aKeycode); // termination
private:
    TBool iCrossTurn;
    TBool& iChanged;
    };
```

The dialog is run from the app UI's `HandleCommandL()` function and isn't referenced from anywhere else. For that reason, we've chosen to place both the dialog class definition and source code in the app UI source file, `oandxappui.cpp`. By doing so, we're affirming that the dialog is private to the app UI.

The constructor should take whatever parameters are necessary to enable the dialog to obtain, and modify, the data that it displays. In this case, it is just data related to which player's turn it is.

```
COandXFirstPlayDialog::COandXFirstPlayDialog(TBool aCrossTurn,
    TBool& aFlag)
    : iCrossTurn(aCrossTurn), iChanged(aFlag)
    {
    }
```

The dialog uses the flag reference to indicate whether or not it changed who should play first.

Before continuing with the reminder of the dialog code, we'll look at the code, in the app UI's `HandleCommandL()` function, that launches the dialog and handles its completion:

```
TBool playerChanged=EFalse;
CEikDialog *dialog=
    new(ELeave) COandXFirstPlayDialog(iCrossTurn,playerChanged);
if (dialog->ExecuteLD(R_OANDX_FIRST_MOVE_DIALOG))
    {
    if (playerChanged)
        {
        SwitchTurnL();
        }
    }
```

The highlighted lines illustrate the standard way of constructing and running *any* dialog. The first line C++ constructs the dialog and, as mentioned earlier, provides it with whatever data it needs. The second line does just about everything else. The parameter to `ExecuteLD()` specifies the dialog's resource, which is used by the function to complete the construction of the dialog and all its component controls. Once construction is complete, `ExecuteLD()` runs the dialog and destroys it on completion (as indicated by the `D` suffix). It leaves if the dialog fails because of failures such as insufficient available memory. On successful completion, `ExecuteLD()` returns `ETrue` if the user terminated the dialog by accepting the result (by selecting the 'Ok' option) or `EFalse` if the dialog was canceled. There's no need to involve the cleanup stack since nothing can fail between C++ construction and the call to `ExecuteLD()`, which guarantees the dialog will be destroyed in all circumstances.

`ExecuteLD()` can't handle dialog-specific initialization and processing of the results: these tasks are performed by additional dialog functions. Dynamic initialization is usually handled in one or both of the two virtual functions, `PreLayoutDynInitL()` and `PostLayoutDynInitL()`, that the dialog calls during its second-stage construction. As their names suggest, the first is called before the dialog lays out the dialog, calculating the position and size of each of its component controls, and the second is called after this process is complete. The default implementations of both functions are empty.

You should use `PreLayoutDynInitL()` to perform all initialization that you want to affect the appearance of the dialog. An example might be to set the text in a text editing control, to ensure the dialog has a chance to set the size of the control to show the whole of the text. Use `PostLayoutDynInitL()` for any initialization that you don't want to affect how the dialog is laid out.

In the noughts and crosses player selection dialog, we need only use `PostLayoutDynInit()` to set the initially highlighted item in the list box:

```
void COandXFirstPlayDialog::PostLayoutDynInitL()
    {
    CEikListBox *listBox =
        (CEikListBox *)Control(EOandXPlayerListControl);
    listBox->SetCurrentItemIndex(iCrossTurn ? 0:1);
    };
```

We obtain a pointer to the list box control using the dialog's `Control()` function, passing the ID of the control, as set in the dialog resource. We then call the control's `SetCurrentItemIndex()` function to set the highlighted item. Note that if the current first player is 'Noughts', we highlight 'Crosses' and vice versa, anticipating that the user will only enter the dialog with the intention of changing who plays first.

We need to use `PostLayoutDynInit()` because of the effect of highlighting the second item in the list box. During the layout of a list box, to ensure the highlighted item is visible, it is usually positioned at the top of the list box display area. If we used `PreLayoutDynInitL()`, the net result in this case would be that, at dialog start-up, the list box would show only the second item and we would have to move the cursor up to bring the first item into view. By default, a list box sets the highlight to the first item, so all will be well if we initialize the highlight position after layout is complete. However, if the list contained too many items to have them all visible in the list box, then we would have to set the highlight in `PreLayoutDynInitL()`, otherwise the highlighted item might not be visible.

Dialog-specific processing on termination of the dialog is handled in the `OkToExitL()` function. It should return `ETrue` if the processing concludes successfully; otherwise it should return `EFalse` (or leave). The dialog calls `OkToExitL()` if the user terminates the dialog by pressing any button other than 'Cancel'; the button that was pressed is indicated by the value of the keycode parameter. If there is no processing to be done – say, if the dialog is merely informing the user of something – then you don't need to supply an `OkToExitL()` function, since the default implementation simply returns `ETrue`.

In the current example, we don't need to check the keycode because the dialog has only 'Ok' and 'Cancel' buttons. All we need to do is check whether the dialog has changed who should play first and set a flag appropriately:

```
TBool COandXFirstPlayDialog::OkToExitL(TInt /*aKeycode*/)
    {
    CEikListBox *listBox =
            (CEikListBox *)Control(EOandXPlayerListControl);
    TInt index = listBox->CurrentItemIndex();
    if (((index == 1) && !iCrossTurn) || ((index == 0) && iCrossTurn))
        {
        iChanged = ETrue;
        }
    return ETrue;
    }
```

4.8.2 Series 60 Variants

The Series 60 UI supplies a range of alternatives to using generic dialog boxes. The `CAknDialog` class derives from `CEikDialog` and allows Series 60 dialogs to use a pop-up menu instead of the straightforward 'Ok' softkey. In addition to a range of list-based dialogs, Series 60 uses specialized dialogs known as **forms**. Forms are effectively lists in which all lines are editable but can be displayed in either editable or view-only states. As mentioned earlier, Series 60 list elements may include icons as well as text.

You'll have to refer to the documentation in the relevant Series 60 SDK for more information, but we'll illustrate how you can use one of the Series 60 list dialogs to select the first player in the noughts and crosses application.

We use the `CAknSelectionListDialog`. The dialog resource is similar to the one used above:

```
RESOURCE DIALOG r_oandx_first_move_dialog
    {
    flags = EAknDialogSelectionList;
    buttons = R_AVKON_SOFTKEYS_OK_CANCEL;
    items =
        {
        DLG_LINE
            {
            type = EAknCtSingleListBox;
            id = ESelectionListControl;
            control = LISTBOX
                {
                flags = EAknListBoxSelectionList;
                };
            }
        };
    }
```

The differences are:

- title text removed, since it won't be used
- no longer free to choose the value of the dialog line ID, since it is referenced by existing `CAknSelectionListDialog` code
- listbox content array not specified.

`CAknSelectionListDialog` defines a `NewL()` function, which takes the following parameters:

- a reference to a `TInt`, to take the index of the selected listbox item
- a pointer to a `CDesCArray`, containing the text items to be displayed
- a `TInt` menu resource ID.

Since we don't want a menu, we can continue to use 'Ok' and 'Cancel' buttons in the dialog. To ensure the menu is empty we use a standard Series 60 menu pane resource to define a menu resource containing a single, blank, pane:

```
RESOURCE MENU_BAR r_oandx_first_move_menubar
    {
    titles =
```

```
      {
   MENU_TITLE
      {
      menu_pane = R_AVKON_MENUPANE_SELECTION_LIST;
      }
   };
   }
```

The code to create and run the dialog is highlighted below. As you can see, it follows exactly the same pattern as we saw in the previous dialog example.

```
TInt openedItem = 0;
TInt initialItem = iCrossTurn ? 1:0;
CDesCArrayFlat *array = new(ELeave) CDesCArrayFlat(2);
CleanupStack::PushL(array);
_LIT(KNoughts, "\tNoughts first");
_LIT(KCrosses, "\tCrosses first");
array->AppendL(KNoughts);
array->AppendL(KCrosses);
CAknSelectionListDialog *dialog = CAknSelectionListDialog::NewL
                (openedItem, array, R_OANDX_FIRST_MOVE_MENUBAR);
TInt result = dialog->ExecuteLD(R_OANDX_FIRST_MOVE_DIALOG);
CleanupStack::PopAndDestroy(array);
if (result && (openedItem != initialItem))
   {
   SwitchTurnL();
   }
```

For simplicity, we create the array from literal text. In order to handle `CDesCArray` classes, don't forget to add the line

```
LIBRARY        bafl.lib
```

to the libraries section of the application's project definition file, `oandx.mmp`.

We don't need an `OkToExit()` function since the dialog writes the index of the selected item to `openedItem`. The one thing this version of the code does not do is highlight the appropriate listbox item; on entry to the dialog, it's always the first item that is selected. We could get round this by reordering the text in the array before passing it to the dialog's `NewL()` function. Otherwise we would have to write a derived class to provide the necessary `PostLayoutDynInitL()` function.

4.9   More Complex Dialogs

Unlike the previous examples, most dialogs contain more than one line. Fortunately, that does not necessarily mean that such dialogs are significantly more complicated to write. If the various lines are truly independent of each other, then the only added complexity is the need to initialize the extra components appropriately, and to process the results on completion of the dialog.

When dealing with more complex dialogs, there are issues that did not arise in the earlier example, such as:

- Focus can change from one control to another, and controls may not be in a fit state to lose, or gain, focus.

- Controls may be dependent on each other, so that modifying the state of one requires the state of one or more other controls to change.

4.9.1 Change of Focus

At any time, you can find the ID of the currently focused control by means of the dialog's `IdOfFocusControl()` function, and then obtain a pointer to the control itself by calling `Control()`, which takes a control ID as its parameter.

Whenever focus is about to be lost by a control, the dialog framework first calls its `PrepareForFocusTransitionL()` function. The default action is to identify the control in the currently focused line and call its `PrepareForFocusLossL()` function. The control's response has to be to ensure that it is in a fit state to relinquish focus or, if that is not possible, to leave. If the function leaves, focus will not be removed from the control. You should override this function if you might need to update other controls, depending on the content of the control that is about to lose focus. Your replacement code must still call the focused control's `PrepareForFocusLossL()` function.

Immediately after focus has been transferred to a control, the framework calls the dialog's `LineChangedL()` function. The default implementation is empty. You are less likely to need to override this function but can do so if you need to take any action in response to a control gaining focus.

You can initiate a change of focus by calling `TryChangeFocus-ToL()`, passing the ID of the control to which you wish focus to be transferred. As for any other focus change, the dialog's `PrepareForFo-cusTransitionL()` and `LineChangedL()` functions are called, as described above.

Any attempt to transfer focus to a dimmed dialog item results in a call to the dialog's `HandleInteractionRefused()` function, whose default action is to display an information message. You may override the behavior if you need to take a more specific action. The parameter passed to the function is normally the ID of the relevant control, but may be zero if the user is attempting to transfer to a disabled page in a multi-page dialog.

4.9.2 Change of State

Some controls may notify their observer of significant changes of state by reporting an EEventStateChanged event. A dialog responds to this event by calling its HandleControlStateChangeL() function, passing the ID of the relevant control. The default implementation is empty. You may override the function to perform any required action.

4.9.3 Multi-page Dialogs

If you need to use a dialog that has more lines than can comfortably fit the screen of a Symbian OS phone, you can split the dialog into a number of pages and display it as a multi-page dialog. It obviously makes sense to ensure, if at all possible, that the items on each page are more closely related to each other than they are to the items on other pages.

Each page of a multi-page dialog has a labeled tab, used to navigate from page tp page. The appearance and positioning of the tabs is different in the different Symbian OS UIs. In UIQ, for example, the tabs appear beneath the dialog, but in Series 60 they appear above the dialog, in the navigation pane. Figure 4.7 shows an example of a Series 60 multi-page dialog.

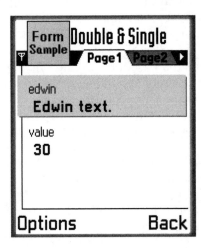

Figure 4.7   A multi-page dialog

Since there isn't much scope for multi-page dialogs in the noughts and crosses application, I've used the 'form' example from the version 2.0 Series 60 SDK. Although this example uses Series 60 forms rather than standard dialog pages, it illustrates the basic techniques.

From a programming point of view, most of the differences between single- and multi-page dialogs lie in the dialog's resource file definitions. Here is the resource used to create the dialog of Figure 4.7:

```
RESOURCE DIALOG r_aknexform_double_and_single_line_form
    {
    flags = EEikDialogFlagNoDrag | EEikDialogFlagFillAppClientRect |
            EEikDialogFlagNoTitleBar | EEikDialogFlagNoBorder |
            EEikDialogFlagCbaButtons;
    buttons = R_AVKON_SOFTKEYS_OPTIONS_BACK;
    pages = r_aknexform_double_and_single_line_form_pages;
    }
```

As you can see, it is a fairly standard, but non-modal (no `EEikDia-logFlagWait` flag) Series 60 `DIALOG` resource. It happens to have an associated menu and so uses 'Options' and 'Back' buttons, rather than 'Ok' and 'Cancel', but that won't affect the description of the dialog itself.

The most important change is that the resource includes a `pages` element rather than the `items` array that is used for single-page dialogs. The `pages` item specifies the identifier of a second resource which in this case is:

```
RESOURCE ARRAY r_aknexform_double_and_single_line_form_pages
    {
    items =
        {
        PAGE
            {
            id = EAknExFormPageCtrlIdPage01;
            text = "Page 1";
            form = r_aknexform_double_line_text_number_field_form;
            },
        PAGE
            {
            id = EAknExFormPageCtrlIdPage02;
            text = "Page 2";
            form = r_aknexform_text_number_field_form;
            },
        PAGE
            {
            id = EAknExFormPageCtrlIdPage03;
            text = "Page 3";
            form = r_aknexform_double_line_icon_form;
            },
        PAGE
            {
            id = EAknExFormPageCtrlIdPage04;
            text = "Page 4";
            form = r_aknexform_double_line_text_number_with_icon_form;
            }
        };
    }
```

As you can see, this is an `ARRAY` resource, whose elements are `PAGE` resource structs. The content of this struct varies with the UI, but the first two elements, a page ID and text to be displayed in the tab for that page, are common to all UIs. Like the IDs for individual dialog lines, page IDs should be non-zero and unique within that dialog. The IDs used in this example are defined, in the application's .hrh file, as follows:

```
enum TAknExFormPageControlIds
    {
    EAknExFormPageCtrlIdPage01 = 1,
    EAknExFormPageCtrlIdPage02,
    EAknExFormPageCtrlIdPage03,
    EAknExFormPageCtrlIdPage04,
    ...
    };
```

All PAGE structs contain at least one item that specifies a further resource. In UIQ this is a lines item and in Series 60 you may specify either a lines or a form item. The two variants of the struct are listed below, for comparison.

```
STRUCT PAGE  // UIQ                  STRUCT PAGE  // Series60
    {                                    {
    WORD id=0;                           WORD id=0;
    LTEXT text;                          LTEXT text;
    WORD controlMinWidth=0;              LTEXT bmpfile = "" ;
    WORD trailerMinWidth=0;              WORD bmpid = 0xffff ;
    LLINK buttons=0;                     WORD bmpmask ;
    LLINK lines=0;                       LLINK lines=0;
    WORD flags=0;                        LLINK form=0 ;
    }                                    WORD flags=0 ;
                                         }
```

UIQ uses the buttons *resource optionally to specify the number and types of the buttons to be included on that page. Series 60 phones have only two softkeys to act as buttons. One of the reasons why Series 60 dialogs can have menus is to provide the flexibility that variable numbers of buttons supply in UIQ.*

Use the lines element if you want to create a standard multi-line dialog. The resource it specifies is simply an array of DLG_LINE items, exactly like the array used to specify the lines of a single-page dialog:

```
RESOURCE ARRAY r_dialog_page_lines_array
    {
    items =
        {
        DLG_LINE
            {
            ...
            },
        DLG_LINE
            {
            ...
            },
```

```
   ...
   DLG_LINE
      {
      ...
      }
   };
}
```

For Series 60 you have the option to use the `form` element to specify that you want the page to be handled as a form. In this case the referenced resource must be a `FORM` struct which, in addition to an array of `DLG_LINE` structs, specifies a form-specific `flags` item. The resource used for the first page of the multi-page dialog in the form example application (slightly truncated to improve clarity) is:

```
RESOURCE FORM r_aknexform_double_line_text_number_field_form
   {
   flags = EEikFormUseDoubleSpacedFormat;
   items =
      {
      DLG_LINE
         {
         type = EEikCtEdwin;
         prompt = qtn_aknexform_form_label_edwin;
         id = EAknExFormDlgCtrlIdEdwin11;
         itemflags = EEikDlgItemTakesEnterKey |
            EEikDlgItemOfferAllHotKeys;
         control = EDWIN
            {
            flags = EEikEdwinNoHorizScrolling | EEikEdwinResizable;
            width = AKNEXFORM_EDWIN_WIDTH;
            lines = AKNEXFORM_EDWIN_LINES;
            maxlength = EAknExFormEdwinMaxLength;
            ...
            };
         tooltip = qtn_aknexform_hint_text_edwin;
         },
      DLG_LINE
         {
         type = EEikCtNumberEditor;
         prompt = qtn_aknexform_form_label_number;
         id = EAknExFormDlgCtrlIdNumber06;
         itemflags = EEikDlgItemTakesEnterKey |
            EEikDlgItemOfferAllHotKeys;
         control = NUMBER_EDITOR
            {
            min = AKNEXFORM_NUMBER_EDITOR_MIN_VALUE01;
            max = AKNEXFORM_NUMBER_EDITOR_MAX_VALUE01;
            };
         tooltip = qtn_aknexform_hint_text_number;
         }
      };
   }
```

The control IDs must be unique across all pages of the dialog and, as always, must be non-zero. The form example application defines them, in its `.hrh` file, in the following way:

```
enum TAknExFormDialogControlIds
    {
    EAknExFormDlgCtrlIdEdwin01 = 0x100,
    ...
    EAknExFormDlgCtrlIdEdwin11,
    ...
    EAknExFormDlgCtrlIdNumber06,
    ...
    };
```

Apart from avoiding the possibility of confusion in the mind of the programmer, there is no particular reason to keep page IDs and control IDs distinct from each other.

Once you have mastered the construction of resources for a multi-page dialog, there is very little else you need to know in order to use them.

There are a few additional dialog functions to be aware of, such as `ActivePageId()` and `SetPageDimmedNow()`, whose usages are reasonably obvious. In addition, `PageChanged()` is called immediately after focus is transferred from any control to a control on a different page. Its default implementation is empty and you should override it to perform any processing that may be necessary as the result of the page change.

4.10   Interface Class Usage in Dialogs

At the beginning of Section 4.8 we described, with the aid of a diagram, how `CEikDialog` was derived from `CCoeControl`. Figure 4.8 shows the diagram again, but in this section we want to concentrate

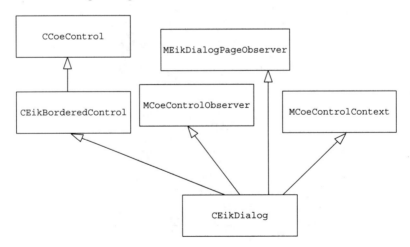

Figure 4.8   Generic derivation of `CEikDialog`

on the way in which `CEikDialog` derives from the three interface classes, `MCoeControlObserver`, `MEikDialogPageObserver` and `MCoeControlContext`.

4.10.1   `MCoeControlObserver`

At this point, there is nothing more to say about `MCoeControlOb-server`. Section 4.6 explains how controls use the `MCoeControlOb-server` interface's `HandleControlEventL()` function to handle certain types of event that a control might need to report to its observer, and that explanation applies to dialogs, as observers of their component controls.

4.10.2   `MEikDialogPageObserver`

As we have already seen, dialogs need to be aware of, and respond to, a wider range of events than those reported to a control observer, and support for those additional needs is provided by the `MEikDialog-PageObserver` class, whose class definition, in `eikdpobs.h`, is:

```
class MEikDialogPageObserver
    {
public:
    virtual void PrepareForFocusTransitionL()=0;
    virtual void PageChangedL(TInt aPageId)=0;
    virtual void LineChangedL(TInt aControlId)=0;
    virtual SEikControlInfo CreateCustomControlL(TInt aControlType)=0;
    // The following function is included in the interface
    // to support deprecated legacy code only.
    virtual void GetCustomAutoValue(TAny* aReturnValue,
        TInt aControlType,const CCoeControl* aControl)=0;
    };
```

You can see that there is little new to say here, since we have already described the usage of the first three of these functions and the last, `GetCustomAutoValue()`, is deprecated and should not be used in new code.

In UIQ, `GetCustomAutoValue()` is completely removed from the interface, but there is an additional function:

```
virtual void ProcessPageButtonCommandL(TInt aCommandId)=0;
```

We'll deal with this one first. It is not implemented in UIQ's `CEikDialog` class, so must be implemented for any multi-page dialog that has one or more buttons on any of its pages. The implementation should use the command ID parameter to determine which button was selected and take whatever action is appropriate.

The only remaining function is `CreateCustomControl()`, for which `CEikDialog` provides a default implementation that panics if it is called. The dialog framework calls this function during construction of the dialog, whenever the control to be constructed is of an unknown type. You must override this function if your dialog includes one or more custom controls.

4.10.3 `MCoeControlContext`

The purpose of the `MCoeControlContext` interface is to allow sharing of graphics settings between controls. The class is defined, in `coecc-ntx.h`, as:

```
class MCoeControlContext
    {
public:
    IMPORT_C virtual void ActivateContext(CWindowGc& aGc,
                                          RDrawableWindow& aWindow) const;
    IMPORT_C virtual void ResetContext(CWindowGc& aGc) const;
    IMPORT_C virtual void PrepareContext(CWindowGc& aGc) const;
    };
```

Unlike most interface classes, `MCoeControlContext` supplies default implementations of its member functions. `ActivateContext()` calls the passed graphics context's `Activate()` for the specified window, and then calls `PrepareContext()`. `ResetContext()` calls the passed graphics context's `Reset()` function, again followed by a call to `PrepareContext()`. The default implementation of `PrepareContext()` is empty.

You set a control's context by setting its `iContext` member, by using either of `CCoeControl`'s `SetControlContext()` or `CopyControlContextFrom()` functions, to contain a pointer to a class that implements the `MCoeControlContext` interface. Once that is done, the control framework ensures that `ActivateContext()` and `ResetContext()` are called before and after all drawing code.

`CEikDialog` overrides `PrepareContext()` to set the graphics context's pen color, brush style and brush color to suitable default values for controls within a dialog.

During its second-stage construction, a dialog sets its own context and the context of each of its controls, thus ensuring that they all share the graphics context defaults defined in the dialog.

4.11 Custom Controls in Dialogs

In addition to the stock controls that are available, you may need to include a control of your own. To do so, the steps that you need to follow are:

- First, write the control and test it outside a dialog to make sure it works the way you want.

- If necessary, in a .rh file, define a resource file struct associated with your control, specifying the member names and types for the resource initialization data.

- Choose a value for the control type that isn't used by any other control and add its definition to a .hrh file – a value of 0x1000 or more will certainly be safe.

- In the resource file for the dialog, include a DLG_LINE struct that specifies the new control type value and includes the control's resource struct, if it has one.

- If your control has a resource struct, implement the control's ConstructFromResourceL() function to read in initialization data from the struct in the application's resource file.

- Implement the dialog's CreateCustomControlL() function to test for the relevant control type, and construct and return an SEikControlInfo struct appropriate for your control.

The custom dialog's resource struct, in a .rh file, will normally provide default values for its members, as in the following example:

```
STRUCT MYCUSTOMCONTROL
    {
    WORD width = 100;
    WORD height = 50;
    }
```

The control type needs to be defined in a .hrh file since it needs to be #included in both the resource file and one or more C++ source files. It is typically defined as an enumeration:

```
enum
    {
    EMyCustomControl = 0x1000
    }

enum
    {
    EMyControlId
    }
```

We've also shown the enumeration that defines the control's ID, needed for any control within a dialog.

The dialog resource is for a Series 60 dialog, so we've omitted a specification of the dialog's title and used one of the standard Avkon button combinations. We've also chosen to replace the default value for the control's width.

```
RESOURCE DIALOG r_mycustomcontrol_dialog
    {
    buttons=R_AVKON_SOFTKEYS_OK_CANCEL;
    items=
        {
        DLG_LINE
            {
            type=EMyCustomControl;
            id=EMyControlId;
            control=MYCUSTOMCONTROL
                {
                width=200;
                };
            }
        };
    }
```

The custom control's `ConstructFromResourceL()` function is similar to a normal control's `ConstructL()` function, except that the data is read from the resource file. A control will need both a `Construct-FromResourceL()` and a `ConstructL()` if it is to be used both inside and outside a dialog. On entry to `ConstructFromResourceL()`, the passed resource reader is positioned to the start of the control's data and the function should consume all available data for the control.

```
void CMyCustomControl::ConstructFromResourceL(TResourceReader& aReader)
    {
// Read the width and height from the resource file.
    TInt width = aReader.ReadInt16();
    TInt height = aReader.ReadInt16();
    TSize controlSize (width, height);

    SetSize(controlSize);

    ActivateL();
    }
```

The default implementation of `ConstructFromResourceL()` is empty, so you don't need to implement it if your custom control has no associated resource struct.

You must, however, implement the dialog's `CreateCustomControlL()` function in each dialog that contains one or more custom controls, and it must be capable of creating every custom control that appears in the dialog. The function is called from the system's control factory whenever it is asked to construct a control of an unknown type. It is unusual in that it returns an `SEikControlInfo` *structure*, defined in `eikfctry.h` as:

```
struct SEikControlInfo
    {
    CCoeControl* iControl;
    TInt iTrailerTextId;
    TInt iFlags;
    };
```

In most circumstances it is sufficient to set the `iControl` element to point to the newly constructed control and set the remaining elements to zero, as in the following example.

```
SEikControlInfo CMyCustomControlDialog::CreateCustomControlL(TInt
    aControlType)
    {
    SEikControlInfo controlInfo;
    controlInfo.iControl = NULL;
    controlInfo.iTrailerTextId = 0;
    controlInfo.iFlags = 0;
    switch (aControlType)
        {
    case EMyCustomControl:
        controlInfo.iControl = new(ELeave) CMyCustomControl;
        break;
    default:
        break;
        }
    return controlInfo;
    }
```

5

Views and the View Architecture

The View architecture is the part of the UI Control Framework in Symbian OS that allows the operating system and other applications to communicate directly with an application's views. It is not necessary for applications to use the View architecture in order to provide UI controls. However, it is strongly recommended for the UIQ platform, and available for use for the Series 60 platform, through specific classes within an additional UI layer called Avkon.

In this chapter we will discuss how to implement views for controlling your application by making use of the View architecture. The method we will describe is the generic one used for Symbian OS v7.0s, although we'll also briefly mention some of the differences when implementing views for the Series 60 and UIQ platforms.

For further details on the specific implementation used for Avkon, refer to Chapter 3 and the white papers available for Series 60 on the Forum Nokia website, *http://www.forum.nokia.com*.

All the tasks necessary for creating views will be described in steps, using an example of a Symbian OS C++ application based on three different views: a List view, a Custom view and an Edit view. The full version of this application example can be found on the Symbian website at *http://www.symbian.com/books*.

5.1 Controlling Your Application with Views

The View architecture is part of the overall UI Control Framework for Symbian OS, and it allows applications to make and receive requests to show a particular view of their data. It also provides finer level of integration between the user interfaces of different applications, and enables users to navigate through the overall UI on the basis of the task they are performing, instead of having to load different applications.

This means that if you are developing a task-driven application that needs to use more than one screen within the UI, for example a Messaging/Contacts or Calendar application, one way of implementing this behavior in Symbian OS is to make use of the View architecture.

The View architecture allows you, for example, to create links directly from the email addresses in the Detail view of a Contacts application to a 'New email' view for that particular contact in your Messaging application. This enables the user to quickly move from one task to another when using the phone.

5.1.1   Defining Views

A view is essentially defined as a class that enables you to display application data. Views are owned by the application's main user interface class (app UI). This is derived from `CEikAppUi` (the framework provided by the Uikon Core API) and `CCoeAppUi` (the base class in the UI Control Framework):

- `CEikAppUi` – this class handles application-wide aspects of the application's user interface such as toolbar pop-up menus, opening and closing files and exiting the application cleanly. Every Uikon application should derive its own app UI class from `CEikAppUi` to handle application-wide user interface details.

- `CCoeAppUi` – this is a general-purpose application user interface class which handles application-wide aspects of the user interface, such as key-press events.

When the view is to be used as a control it also needs to be derived from `CCoeControl`, the control base class from which all controls are derived. (See Chapter 4 for more information on controls.)

The client/server process of sending and receiving requests to display a particular view is handled by the View server, which is the main part of the View architecture. This client-server interface (API) is not used directly by your application, but through wrapper functions within `CEikAppUi` (derived from `CCoeAppUi`), in the app UI framework.

So, for a class to be called a view, it must, therefore, also implement specific virtual functions available from the abstract view interface, `MCoeView`, as well as declare a View ID, so the View server can recognize that particular view:

- `MCoeView` – this class specifies an interface for views and should be used by all application views.

5.1.2   Advantages of Using Views

Using the View architecture adds the following advantages to your application:

- Application/view switching capability – the user can switch from one view to another, either within the same application or within another application.

- Support for saving data – by registering a view with the View server, the data for that view is always saved before the view is deactivated.

- Support for sending data – by packaging messages as descriptors identified by the UID, data can be sent from one view to another (within the same or different applications), via the View server.

5.2   View Architecture Components

The View architecture (Figure 5.1) depends heavily on both the Uikon and CONE components of Symbian OS (and Avkon for Series 60: see Chapter 3). These provide the connection with the View server.

When implementing views in your application, you need to know about the functions that are derived from the `MCoeView` interface and the control class `CCoeAppUi`, which are both part of CONE. You also

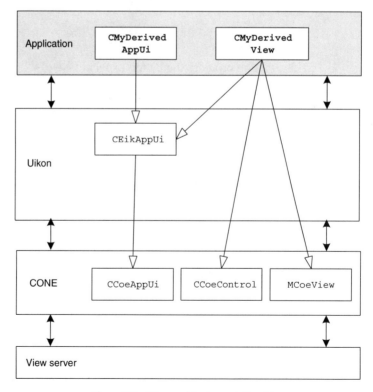

Figure 5.1   View architecture relationships

need to know how to use the view-specific functions contained within
CEikAppUi, which is part of Uikon.

The relationships within the View architecture are shown in Figure 5.1.

As you can see, a view within an application is created from the
CCoeControl and CEikAppUi classes. The View server knows about
the view via the MCoeView interface in CONE.

This means that the header files you need to include when constructing
your view class (in addition to standard application header files) are:

- coeview.h – allows you to override virtual functions from MCoe-
  View

- ccoecntrl.h – allows your view to be recognized as a control

- eikappui.h – allows you to override virtual functions within CEik-
  AppUi.

5.2.1   Functions Derived from MCoeView

The following virtual functions are defined in the abstract view interface
(MCoeView), and need to be implemented in the declaration for each
view class of your application:

```
virtual TvsViewId ViewId() const=0;
```

This function provides the ID for a specific view, so it will be recognized
by the View server.

```
virtual void ViewActivatedL(const TvwsViewId& aPrevViewId,
    Tuid aCustomMessageId, const TdesC8& aCustomMessage)=0;
```

This function activates a specific view (using the view ID) and provides
information about the previous view that was activated. It also allows you
to pass any message information from one view to another if required,
for example application-specific data such as a contact or calendar
appointment.

```
virtual void ViewDeactivated()=0;
```

This function deactivates the current view, before switching to another
view (within or outside the application).

```
virtual void ViewConstructL()=0;
```

This is the function to use when constructing your view and specifying its
content, by including standard control and drawing functions.

5.2.2   Functions Derived from `CEikAppUi`

The following functions, implemented in `CEikAppUi` (derived from `CCoeAppUi` in CONE), need to be called by your view class in order for the View server to activate, register and deregister the view in your application, and to know which view to use as the default (main) application view.

```
void ActivateViewL(const TvwsViewId& aViewId);
```

This function defines which view is to be activated next (using the view ID), when switching from the current view.

```
void ActivateViewL(const TvwsViewId& aViewId, Tuid aCustomMessageId,
    const TdesC8& aCustomMessage);
```

This function also defines which view is to be activated next, but it also allows you to pass any message information from one view to another if required, for example application-specific data such as a contact or calendar appointment. This data is sent in the form of a UID or descriptor message.

```
void RegisterViewL(McoeView& aView);
```

This function registers your view with the View server, so that it can be used within the View architecture framework.

```
void DeregisterViewL(const McoeView& aView);
```

This function allows your current view to be deregistered when it is deleted.

```
void SetDefaultViewL(const McoeView& aView);
```

This function defines which view is to be used as the default view, when first opening a multi-view application.

```
Void AddToStackL(const MCoeView& aView, CcoeControl* aControl,
    Tint aPriority=EcoeStackPriorityDefault,
    TInt aStackingFlags=EcoeStackFlagStandard);
```

This function adds a control to the view's control stack so that it can receive key events when the view is activated. It is used when the view is deactivated.

5.3   Implementing Views

When creating a view for your application, the overall process can be described as follows:

1. Create the view – this includes creating the view class and implementing the functions that enable view activation and deactivation (by linking to the correct header files).

2. Register the view – this means calling a specific function implemented in `CEikAppUi`, to make the view known by the View server.

3. Enable switching between views – this includes activating and deactivating views (through buttons, links or menus), as well as packaging data to be sent from one view to another, as either a UID or a descriptor message.

4. Deregister the view – this means calling the deregister function implemented in `CEikAppUi`, so the View server does not try to load the view when the application is closed.

A view can contain any type of data content as well as controls. In the example shown in this chapter, we have created a basic application based on three different views:

- one default view containing a list box (List view)
- one graphical view containing some drawing functions (Custom view)
- one detailed view containing editing functions (Edit view).

They are linked to each other through view switching.

5.4   Creating Views

As we are essentially using a view as a form of control, the first thing you need to do before creating a view is to make sure you have a custom control class in your application, derived from `CCoeControl` (for more information on controls, see Chapter 4). It is then easy to modify this class to contain view-specific behavior.

Once a view has been created you can add data content in the form of standard drawing functions such as list boxes, pictures, edit boxes, etc., which are applicable to all controls. For more information on drawing functions, see Chapter 4.

In our example for this chapter, we will first create a List view class, `CListView` (which contains a list box). This is going to be the default view for the application, as shown in Figure 5.2.

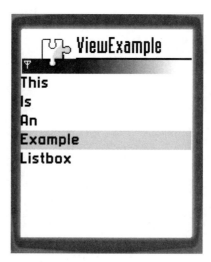

Figure 5.2 List view

The `CListView` class derived from `CCoeControl` forms the basis of user interaction with your application and should be defined in the custom control class of your application.

To implement the view-specific behavior, you need to add the relevant functions to the second-phase constructor `void ConstructL()` within your List view class (`CListView`):

```
class CListView : public CCoeControl
    {
public: // Construction and Destruction
    void ConstructL(); // Second phase constructor
    ~CListView();
protected: // Data
    CViewExTextListBox* iListBox;
    };
```

You also need to derive your view class from the abstract view interface `MCoeView`. To implement this interface, you need to include the file `ccoeview.h` at the beginning of your view header file, and then include the following four important functions, as described in Section 5.2.1:

- `virtual TVwsViewId ViewId() const;`

- `virtual void ViewConstructL();`

- `virtual void ViewActivatedL(const TVwsViewId& aPrevViewId, TUid aCustomMessageId, const TDesC8& aCustomMessage);`

- `virtual void ViewDeactivated();`

The modified version of the class should then look as follows:

```
class CListView : public CCoeControl, public MCoeView
    {
public: //Destruction
    ~CListView();
public: // From MCoeView
    virtual TVwsViewId ViewId() const;
public: // View Switching
    virtual void ItemSelected(TListItem aItem);
private:
    virtual void ViewConstructL();
    virtual void ViewActivatedL(const TVwsViewId& aPrevViewId,TUid
    aCustomMessageId,const TDesC8& aCustomMessage);
    virtual void ViewDeactivated();
private: // Data
    CViewExTextListBox* iListBox;
    };
```

Note that the second-phase constructor ConstructL() has been removed, and replaced by the new ViewConstructL() function. This function creates the view and will always be called automatically by the View server framework when the view is initialized.

The other three important virtual functions, which must be implemented, are as follows:

- ViewActivatedL() – performs view-specific actions (for example, displays contact details) when the view is activated.

- ViewDeactivated() – performs view-specific actions (for example, closes a database) when the view is deactivated.

- ViewId() – returns a unique ID of this view. The View ID is declared in the source file for the view, as follows:

```
// CONSTANTS
const TUid KUidViewEx = { 0x0257696A };
const TUid KUidListView = { 0x01000000 };
```

The next step is to define the content for the view, within the ViewConstructL() function. In this first view (CListView) of our example application we have decided to create a list box. The list box has been created using a class defined earlier as CNamesListArray. This shows the implementation of the list box:

```
// CListView

CListView::~CListView()
    {
    iEikonEnv->EikAppUi()->DeregisterView(*this);
    delete iListBox;
    }
```

```
void CListView::ViewConstructL()
    {
    iListBox=new (ELeave) CViewExTextListBox(this);
    iListBox->ConstructL(NULL);
    CNamesListArray* array=new (ELeave) CNamesListArray;
    array->ConstructL();
    iListBox->Model()->SetItemTextArray(array);
    iListBox->Model()->SetOwnershipType(ELbmOwnsItemArray);
    SetRect(iEikonEnv->EikAppUi()->ClientRect());
    iListBox->SetRect(iEikonEnv->EikAppUi()->ClientRect());
    iListBox->SetCurrentItemIndex(3);
    }

void CListView::ViewActivatedL(const TVwsViewId& /*aPrevViewId*/, TUid
/*aCustomMessageId*/, const TDesC8& /*aCustomMessage*/)
    {
    iEikonEnv->AddDialogLikeControlToStackL(iListBox);
    iListBox->ActivateL();
    iListBox->MakeVisible(ETrue);
    iListBox->DrawNow();
    }

void CListView::ViewDeactivated()
    {
    iListBox->MakeVisible(EFalse);
    iEikonEnv->RemoveFromStack(iListBox);
    MakeVisible(EFalse);
    }

TVwsViewId CListView::ViewId() const
    {
    TVwsViewId KListViewId(KUidViewEx, KUidListView );
    return KListViewId;
    }

void CListView::ItemSelected(TListItem aItem)
    {
    TVwsViewId KCustomViewId(KUidViewEx, KUidCustomView );
    TPckgBuf<TListItem> buffer(aItem);
    iEikonEnv->EikAppUi()->ActivateViewL(KCustomViewId, KUidCustomView,
        buffer);
    }
```

This completes the creation of our first view. The next step is to register it with the View server.

5.5 Registering Views

The aim of registering a view with the View server is to make sure that it can be used by your application (and other applications) at any time.

To enable this, you need to call the specific view registration functions (implemented in CEikAppUi) within the ConstructL() function of the app UI class for your application.

In our example shown below, the application UI class `CViewExAppUi` is derived from `CEikAppUi` directly and has a pointer to the `CListView` class as a data member:

```
void CViewExAppUi::ConstructL()
    {
    BaseConstructL();
    iListView = new (ELeave) CListView;
    RegisterViewL(*iListView);

    SetDefaultViewL(*iListView);
    }
```

As shown above, the `RegisterViewL()` function registers the view with the View server. This function is provided by `CCoeAppUi`, the parent class of `CEikAppUi`.

You can also define the List view to be the default view for this application, by using the `SetDefaultViewL()` function. This means that the View server framework will always automatically construct and activate this view each time the application is launched.

To make sure your view has a unique ID recognizable by the View server, you also need to implement the `ViewId()` function (derived from the `MCoeView` interface). This is really easy to do, as you need to return only one value, `TVwsViewId`.

`TVwsViewId` consists of the UID for the application containing the view, and the UID for the view to be activated, as shown in the following example:

```
TVwsViewId CListView::ViewId() const
    {
    TVwsViewId KListViewId(KUidViewEx, KUidListView );
    return KListViewId;
    }
```

Once you have implemented the `ViewId`, the registration of the List view with the View server is complete.

*You can obtain UIDs for your application by emailing **uid@symbiandevnet.com**, as described in Appendix 4. The ID for the view can be any 32-bit integer. The view ID makes the view unique within the application only, whereas the application UID makes the view unique within the whole system.*

Once the view has been registered with the View server, the next step is to enable view-specific behavior, such as switching between views by activating a new view while deactivating the current one.

5.6   Switching Between Views

The idea behind the view switching behavior is to allow the user to switch between views within the same or a different application at any time, typically by pressing a button, or making a selection from a menu, or following a link. This then activates a new view, while deactivating the current view.

To specify which view to activate, you should use the function `ActivateViewL()` from the application's app UI class. There are two different ways to use this function:

- for view switching purposes only. This is the most simple way, as you only need to pass the constant `TVwsViewId` for the view to be activated:

```
void CCoeAppUi::ActivateViewL(const TVwsViewId& aViewId)
```

- for view switching through message passing. This is a more powerful version of the same function, as you are able to pass data to the view to be activated:

```
void CCoeAppUi::ActivateViewL(const TVwsViewId& aViewId,
    TUid aCustomMessageId,const TDesC8& aCustomMessage)
```

The format of this data must be known in advance. In our example, we have created the following type definition to enable data to be passed:

```
typedef TBuf<20> TListItem;
```

This is a simple, modifiable string of 20 characters or less. The parameters for this version of `ActivateViewL()` are, therefore, as follows:

- `TVwsViewId& aViewId` – the ID of the view to be activated.

- `TUid aCustomMessageId` – an identifier or flag known to the view being activated. This typically allows you to specify an action to perform, the data being passed through the custom message. For example, a view receiving a contact could support the read-only display of that contact, or the ability to edit it directly. The custom message ID can be used to specify which action to perform.

- `TDesC8& aCustomMessage` – a descriptor containing data to be sent to the view about to be activated. Although usually defined as a descriptor of type `TDesC8`, a `TPckgBuf` is used in our example.

- TPckgBuf is derived from TDesC8 and acts as a package buffer for inter-process communication, wrapping up an object as a descriptor.

Because data is carried in a view, and activation is often transferred across process boundaries, it is necessary to transfer data in this manner.

For example, transferring a TListItem from one view to another would be achieved using the following code syntax:

```
TPckgBuf<TListItem> buffer=aListItem;
```

Another, more complex, view activation such as passing a text string can be achieved as follows:

```
...
const TUid KCustomMessageUid = { 0x00000001 };
const TUid KUidListView = { 0x01000000 };
const TVwsViewId KListViewId(KUidViewEx, KUidListView );
...

_LIT(KTextString, "Some Text");

...

TPckgBuf<TListItem> buffer=KTextString;
iEikonEnv->AppUi()->ActivateViewL(KListViewId, KCustomMessageUid, buffer);
```

5.6.1   Publishing View IDs

When switching from a view within one application to a view within another application, you are able to link only to those applications with a known UID and view ID. Likewise, if you want other applications to be able to link to views within your application, you will need to publish these IDs, for example by exporting them to additional header files. There is no standard way of publishing IDs, so you may be limited as to which type of applications you can link.

If you are using the view architecture only for the internal structure of your application, you do not need to publish the UID and view IDs.

5.6.2   Activating Views

To enable activation of a view, you need to implement the ViewActivatedL() function in the view class.

A typical example of using ViewActivatedL() is shown below:

```
void CListItem::ViewActivatedL(const TVwsViewId& aPrevViewId, TUid
aCustomMessageId, const TDesC8& aCustomMessage)
    {
```

```
if(aCustomMessageId==KCustomMessageUid)
    {
    TPckgBuf<TListItem> buffer;
    buffer.Copy(aCustomMessage);
    TListItem name=buffer();
    }
}
```

The `if` statement in the example above checks that the custom message ID is correct for this particular view. This prevents the view from being sent random data with which it is expected to be able to deal.

The first two lines inside the `if` statement construct a `TPckgBuf` for a `TListItem` using template specialization, and then fill it with the data contained by the descriptor `aCustomMessage`.

As the custom message ID has been tested and recognized, the view can assume that the data is in a specific format, ready to be drawn on screen.

The final line copies the data back in to a `TListItem` for use elsewhere in the application. This `TPckgBuf`/`CustomMessage` mechanism can be used to transfer any kind of data between views, not just text strings.

5.6.3 Deactivating Views

To enable deactivation of a view, you need to implement the `View-Deactivated()` function in the view class. Even though deactivation of a view is not always necessary, the `ViewDeactivated()` function must still be implemented (without data):

```
void CListView::ViewDeactivated()
    {
    iListBox->MakeVisible(EFalse);
    iEikonEnv->RemoveFromStack(iListBox);
    MakeVisible(EFalse);
    }
```

This function is also useful for performing application-specific cleanup (such as closing files or freeing memory).

5.7 Deregistering Views

When an application has been closed, the views within that application must be deregistered by the View server, so they are not assumed to be available for use.

To deregister your view, you simply use the `DeregisterView()` function (derived from `CCoeAppUi`) within the destructor of the app UI class, as follows:

```
CViewExAppUi::~CViewExAppUi()
    {
    DeregisterView(*iListView);
    delete iListView;
    }
```

Alternatively, you can allow the view to deregister itself upon destruction, as shown below:

```
CListView::~CListView()
    {
    iEikonEnv->EikAppUi()->DeregisterView(*this);
    delete iListBox;
    }
```

Once the application is opened again, and the view activated, the view is reregistered by the View server.

5.8   More on Views

So far in this chapter we have concentrated on the default view of our example application, the List view. However, the other views which are part of the example application are also worth mentioning briefly.

Even though the actual view-specific behavior for these views is constructed in the same way as for the List view (creating, registering, activating, deactivating and deregistering), the main aspect to note with these views is the actual links that enable view switching.

One view enables view switching within a drawing function (Custom view), and the other view enables view-switching behavior within an edit function (Edit view).

5.8.1   Custom View

The Custom view is a fairly simple class which receives data from the List view and draws it on the screen, as shown in Figure 5.3.

The particular point to make about this view is to show the implementation of the `ViewActivatedL()` function, as it includes data which is received as part of activating the view.

As described in Section 5.6, 'Activating views' and 'Deactivating views', the data to be passed is first copied from `aCustomMessage` into the Package buffer, and then extracted from this buffer and copied into `iName`, as seen below:

```
void CCustomView::ViewActivatedL(const TVwsViewId& /*aPrevViewId*/, TUid
/*aCustomMessageId*/, const TDesC8& aCustomMessage)
```

```
    {
    TPckgBuf<TListItem> buffer;
    buffer.Copy(aCustomMessage);
    iName=buffer();
    MakeVisible(ETrue);
    DrawNow();
    }
```

The Custom view is then drawn, and the data is displayed on screen
(`TListItem`).

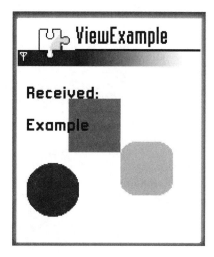

Figure 5.3   Custom view

To activate the next view, the 'Ok' button should be pressed. `CCus-
tomView` handles this as follows:

```
TKeyResponse CCustomView::OfferKeyEventL(const TKeyEvent& aKeyEvent,
    TEventCode  /*aType*/)
    {
    if(aKeyEvent.iCode==EKeyOK)
        {
        TPckgBuf<TListItem> buffer(iName);
        iEikonEnv->EikAppUi()->ActivateViewL(KEditViewId, KUidEditView,
            buffer);
        return EKeyWasConsumed;
        }
    return EKeyWasNotConsumed;
    }
```

In the example above, pressing the 'Ok' button takes you to the Edit view.

5.8.2   Edit View

The point to make about the Edit view (see Figure 5.4) is to show how
you can further specialize the behavior of an existing view, in this case

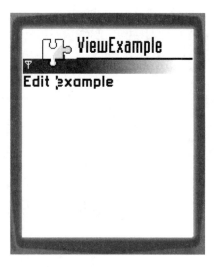

Figure 5.4   Edit view

the List view. The Edit view makes it possible for the user to edit data within a list and then switch back to the Custom view.

Within our example we have used all the standard behavior in the List view, to avoid writing the same code all over again. The view switching behavior is inherited from the List view, enabling the user to go back to the Custom view when pressing a link.

By creating a specialized class based on `CEikTextListBox`, we are overriding the default keyboard handling behavior, as can be seen below, in the `OfferKeyEventL()` function:

```
// CViewExTextListBox
class CViewExTextListBox : public CEikTextListBox
    {
public: //Construction
    CViewExTextListBox(CListView* aListView);
public:
    TKeyResponse OfferKeyEventL(const TKeyEvent& aKeyEvent,
        TEventCode aType);
private: // Data
    CListView* iListView;
    };

CViewExTextListBox::CViewExTextListBox(CListView* aListView)
    {
    iListView=aListView;
    }

TKeyResponse CViewExTextListBox::OfferKeyEventL(const TKeyEvent&
aKeyEvent,TEventCode aType)
    {
    if(aKeyEvent.iCode==EKeyOK)
        {
```

```
      TListItem item=Model()->ItemTextArray()->
         MdcaPoint(CurrentItemIndex());
      iListView->ItemSelected(item);
      return EKeyWasConsumed;
      }
   return CEikTextListBox::OfferKeyEventL(aKeyEvent, aType);
   }
```

Because the specific view functions within the List view are virtual, it is possible to override the behavior of the List view class within derived classes (such as in the Edit view class).

5.9   View-specific Behavior on UIQ and Series 60 Platforms

The method described for implementing views in this chapter is the Symbian OS generic method. There are, however, specific implementation aspects to consider, depending upon the target platform you are developing your application on. In this section we will discuss some of the view architectural behavior on UIQ and Series 60 platforms.

It is important that you know in advance which platform your application should target, as it greatly impacts the design and construction of your views, especially for Series 60.

For a more detailed explanation on how to implement the view architecture on these platforms, see the UIQ Control Framework section within the Developer Library on the UIQ SDK (which can be downloaded from ***http://www.symbian.com/developer***), or the Series 60 section on the Forum Nokia website, ***http://www.forum.nokia.com***.

5.9.1   Views on the UIQ Platform

The implementation of the view architecture for UIQ is basically the same as the generic method already described. The only additional difference is that UIQ provides support for two versions of the same view, in order to accommodate the fact that it is possible to have two user modes on the P800 and P900 – *flip open* and *flip closed*.

By creating two versions of the same view, for example CFlipClosedView and CFlipOpenView, you can apply different ways of representing data (drawing on screen) and controlling the application.

In the closed view, the visible screen will be smaller, for example, and the pen input is no longer activated, so you need to take into account the fact that you can show only a limited amount of data (for example, four list entries instead of eight).

If you are using multiple versions of different screen modes, then you must also implement a virtual function called ViewScreenModeCompatible, within the MCoeView interface.

This function needs to be implemented in both your view classes to ensure that none of them are activated in the wrong screen mode, as shown in the following example:

```
Tbool CFlipClosedView::ViewScreenModeCompatible(TInt aScreenMode)
    {
    // Assume 0 is the ID for the flip closed screen mode
    return (aScreenMode==0);
    }
```

The function compares the aScreenMode parameter to the screen mode your view is compatible with, and returns either ETrue or EFalse depending on the result. The default value for aScreenMode is 0.

Note: There is currently no standard list of integers and corresponding screen modes within UIQ, as this is specific to each phone. For example, for the P800 phone, there is a file called FcQikFlipDef.h which contains the enum for the different screen mode options:

```
enum TScreenMode
    {
    EScreenModeFlipOpen,
    EScreenModeFlipClosed
    };
```

5.9.2   Views on the Series 60 Platform

When implementing the view architecture on the Series 60 platform, the main deviation from the generic Symbian OS implementation is the use of the additional UI layer called *Avkon*.

The process of creating views using Avkon is as follows:

- Create the application views and app UI classes by deriving them from CAknView and CAknViewAppUi.

- Register the views with the app UI.

- Enable view-switching behavior by packaging data and sending it.

- Handle activation and deactivation of the views.

The major difference from the generic implementation is the addition of the Avkon-specific classes CAknView and CAknViewAppUi:

- CAknView is the base class for application views, and contains support for menus, softkeys, command handling, foreground/background handling, and object providers (similar to Uikon). It should be noted that CAknView is *not* derived from CCoeControl, as is the case for views when using the generic method. In Series 60, a view is more like a mini-app UI, and a derived view must create separate controls for user interaction.

- `CAknViewAppUi` is derived from `CAknAppUi`, which is the base class for view-based application UIs and provides functions for managing views.

The relationships within the view architecture are thus slightly different, as can be seen in Figure 5.5.

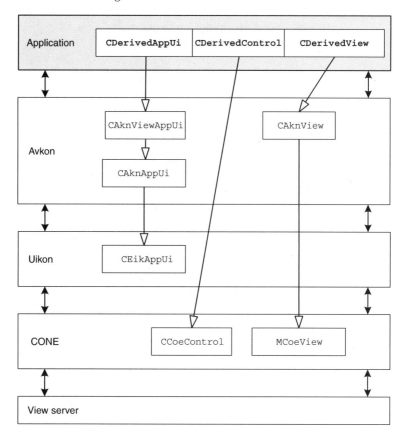

Figure 5.5   Avkon view architecture

To create a view using Avkon, you derive an app UI class from `CAknViewAppUi`, a view class from `CAknView` and any necessary control classes from `CCoeControl`.

In the `ConstructL()` function of the app UI, you create and register the view with the app UI as follows:

```
void CMyAppUi::ConstructL()
    {
    ...
    CAknView* view=CMyView::NewL();
    AddViewL(view); //The app UI takes ownership of the view
```

```
...
CleanupStack::Pop(view);
}
```

The app UI then owns the view.

The functions for activation and deactivation in Avkon are not the same as in the generic implementation. Instead of being derived from MCoeView, they are derived from CAknView, and they also have different names:

- CAknView::DoActivateL() – this corresponds to MCoeView::ViewActivatedL() and is used for handling activation of the current view. It may or may not pass data to the view. The DoActivateL() function would typically create a control, place it on the control stack and give it focus, as shown below:

```
void CMyView::DoActivateL(const TVwsViewId& /*PrevViewId*/, TUid
/*aCustomMessageId*/, const TDesC8& aCustomMessage)
    {
    //iControl will already exist if this is a reactivation
    if (!iControl)
        {
        iControlView=CMyControl::NewL(this);
        AppUi()->AddToStackL(*this, iControl);
        iControl->SetFocus(ETrue);
        }
    iControl->SetCustomMessageL(aCustomMessage);
    }
```

- CAknView::DoDeactivate() – this corresponds to MCoeView::ViewDeactivated() and is used for handling deactivation of the current view. The view is then deactivated when another view in the application is activated or the application is exited. The DoDeactivate() function would typically defocus and delete a control:

```
void CMyView::DoDeactivate()
    {
    if (iControl)
        {
        AppUi()->RemoveFromStack(iControl):
        iControl->SetFocus(EFalse);
        }
    delete iControl;
    iControl=NULL;
    }
```

5.9.2.1 More Differences

There are also other behavioral differences in Series 60 regarding view activation. These are:

- Support for multiple active views. The UI layer Avkon allows for one active view per application, instead of one active view in the whole system as for standard Uikon. An active view will be deactivated only when you switch to another view in the same application.

- Asynchronous view activation. In Uikon, when an application activates a view in another application, the `ActivateViewL()` function in the first application does not return data until the view in the second application is activated. If there is an error during activation, the view in the first application is shown. In Avkon, the `ActivateViewL()` function within an application returns data from another application sooner than in Uikon, and any errors to do with the view activation of the other application are handled by the UI framework. The framework will also restore the system to a sensible state.

More information on specific view architecture behavior for Series 60 and its implementation can be found in the white papers for Series 60 on the Forum Nokia website, ***http://www.forum.nokia.com***.

5.10 Summary

You should now know how to make use of the view architecture in Symbian OS, to enable users of your application to go from one view to another, within the same or other applications. This is very useful, for example, when developing task-driven applications such as Calendar, Messaging or Contacts applications.

In this chapter we have described how to create views as controls for your application by performing the following steps:

- creating your view based on a custom control class, and then implementing specific functions derived from the abstract view interface `MCoeView` within CONE, and `CEikAppUi` within Uikon

- registering the view with the View server, so it is visible to other views within the same or other applications

- enabling view-switching behavior, through activating a new view and deactivating the current view, as well as by sending data from one view to another

- deregistering the view with the View server.

We have also discussed briefly some of the aspects and differences which are worth considering before implementing views on the UIQ or Series 60 platform.

6

Files and the Filing System

This chapter provides an overview of a range of file-related topics, from the filing system itself to resource and bitmap files. With such a wide range of topics, it isn't possible to give a detailed description. The intention, instead, is to provide the essential information, together with some practical examples that will allow you to start making effective use of the system.

6.1 Filing System Services

This section describes the basic low-level services that underlie all other Symbian OS file-based services.

6.1.1 File Names and Their Manipulation

Symbian OS files are identified by a file specification, which may be up to 256 characters in length. As in DOS, a file specification consists of:

- a device, or drive, such as `c:`

- a path, such as `\Document\Unfiled\`, where the directory names are separated by backslashes ('\')

- a file name

- an optional file name extension, separated from the file name by a period ('.').

The filing system supports up to 26 drives, from `a:` to `z:`. On Symbian OS phones, the `z:` drive is always reserved for the system ROM and the `c:` drive is always an internal read-write drive – although on some phones it may have limited capacity. Drives from `d:` onwards may be internal, or may contain removable media. You should not assume that you can write to all such drives; many phones have one or more read-only drives, in addition to z:.

Subject to the overall limitation on the length of a file specification, a directory name, file name or extension may be of any length. The filing system preserves the case of such names, but all operations on the names are case-independent. Clearly, this means that you can't have two or more files in the same directory whose names differ only in the cases of some of their letters. File specifications may contain the wild cards '?' (a single character) and '*' (any sequence of characters) in any component other than the drive.

Although most Symbian OS applications do not do so, you are free to include a file name extension in any file's specification. Symbian OS applications do not normally rely on the extension to determine the file's type. Instead, they use one or more UIDs, stored within the file, to ensure that the file type matches the application.

File names are constructed and manipulated using the TParse class and its member functions. For example, to set an instance of TParse to contain the file specification C:\Documents\Oandx\Oandx.dat, you could use:

```
_LIT(KFileSpec, "C:\\Documents\\Oandx\\Oandx.dat");
TParse fileSpec;
fileSpec.Set(KFileSpec,NULL,NULL);
```

Following this code, you can call TParse's getter functions to determine the various components of the file specification. For example, filespec.Drive() contains the string "C:", and fileSpec.Path() contains "\Documents\Oandx\".

The Set() function takes three text parameters, the first being – as we have seen – a reference to a TDesC, containing the file specification to be parsed. The second and third parameters are *pointers* to two other TDesC descriptors, and either or both may be NULL. If present, the file specification pointed to by the second parameter (the *related* file specification) is used to supply any missing components in the first file specification. If used, the third parameter should point to a *default* file specification, from which any components not supplied by the first and second parameters will be taken. Any path, file name or extension may contain the wildcard characters '?' or '*', respectively representing any single character or any character sequence.

A TParse owns an instance of TFileName, which is a TBuf<256>. This is a large object and its use should be avoided if possible. If you can, create your own, smaller, buffer to contain the file specification and use a TParsePtr (referencing a modifiable buffer) or a TParsePtrC (referencing a constant buffer).

6.1.2 File Server Sessions

The Symbian OS file server provides the basic services that allow user programs to manipulate drives, directories and files, and to read and write data in files.

As with all servers, the file server uses session-based communication to convert a client-side operation into a message that is sent to the server. The requested function is performed in the server, and then any result is passed back to the client. So, in order to use the file server, you first need a connected file server session, represented by an instance of the RFs class.

The general pattern, ignoring error handling, for using the file server is:

```
RFs session;
session.Connect();
...
session.Close();
```

Between connecting and closing the RFs, you can use it to open any number of files or directories, or to perform any other file-related operations. If you wish, you may keep a file server session open for the lifetime of your application. It is preferable that you make sure that all opened file-based resources are correctly closed before you close the session. In any case, when the session is closed, the server will clean up any server-side resources associated with the session.

In fact, in a GUI application, you don't need to open a file server session, since the control environment already has an open RFs that you can access with iCoeEnv->FsSession(). Opening a file server session is an expensive operation so, if at all possible, you should use the existing session, rather than creating your own.

The RFs class provides many useful file system-related operations, including:

- making, removing, and renaming directories, using MkDir(), MkDirAll(), RmDir() and Rename()

- using Delete() and Rename() to delete or rename files

- reading and changing directory and file attributes by means of Att(), SetAtt(), Modified() and SetModified()

- notifying changes with NotifyChange() (and NotifyChange-Cancel())

- manipulating drives and volumes with Drive(), SetDriveName(), Volume() and SetVolumeLabel()

- peeking at file data without opening the file by using ReadFileSection()

- adding and removing file systems with `AddFileSystem()`, `MountFileSystem()`, `DismountFileSystem()` and `Remove-FileSystem()`.

These functions, and their use, are documented in any Symbian OS SDK.

Most `RFs`-related functions are *stateless* – that is, the results of a function call don't depend on any functions previously called. That's why you can usually use the control environment's `RFs` for your own purposes. However, `RFs` does have one item of state: its current directory. When you open an `RFs`, its current directory is normally set to `c:\`, but you can use `SetSessionPath()` to change the current directory used by an open `RFs`, or `SetDefaultPath()` to change the initial current directory for all future `RFs` objects.

> *The current directory includes the drive as well as directory names. So, unlike DOS, there is no concept of one current directory per drive.*

If you manipulate, or rely on, the current directory, make sure you use your own `RFs` rather than sharing one. In that case you will also have to handle errors associated with the session. How to handle them depends on how the session is stored. If it is declared as member data in one of your classes, with a class definition containing a line such as:

```
RFs iFs;
```

then it is sufficient to connect the file session (from a function that allows leaves to occur) with:

```
User::LeaveIfError(iFs.Connect());
```

and to close the session with:

```
iFs.Close()
```

in the class destructor.

Alternatively, if you declare the file server session on the stack, you will need to take a little more care. You will need to place the session on the cleanup stack and the best way is to use the cleanup stack's `CleanupClosePushL()`, as follows:

```
RFs myFs;
User::LeaveIfError(myFs.Connect());
```

```
CleanupStack::CleanupClosePushL(myFs);
...
// file session operations that may leave
...
CleanupStack::PopAndDestroy();
```

Remember that connecting a file server session is an expensive, time-consuming operation, so you should only do so if there is a good reason not to share the one that exists in the control environment.

6.1.3 Directories

A directory contains files and other directories, each of which is represented by a **directory entry**. The RDir class allows you to open a directory and read the entries it contains.

There are several different ways of reading the contents of a directory. The most straightforward, but not necessarily the most efficient, way is illustrated by the following example.

```
void ReadDirContentsL(RFs& aFs, const TDesC& aDirName)
    {
    RDir myDir;
    TInt err;
    User::LeaveIfError(myDir.Open(aFs, KDirName,
        KEntryAttNormal|KEntryAttDir));
    TEntry currentEntry;
    FOREVER
        {
        err=myDir.Read(currentEntry);
        if (err)
            {
            break; // No more entries, or some other error
            }
        // Process this entry
        }
    myDir.Close()
    if (err != KErrEof) // EOF signifies no more entries to read
        {
        User::LeaveIfError(err);
        }
    }
```

Each call to Read() reads a single entry, and is useful in cases where memory usage needs to be minimized. It is, however, inefficient to make multiple calls to server functions, so RDir also supplies an overload of Read() that, in a single call, reads all entries into a TEntryArray. As with many other functions that involve communication with a server, two further overloads supply asynchronous versions of the two types of Read(), to be used from an active object.

In addition to RDir's Read() functions, RFs provides a range of GetDir() functions that, in a single call, read a directory's content into a `CDir`. As you can see from the following code, which creates a list of entries similar to that of the previous example, `GetDir()` also allows you to specify a sort order for the resulting list.

```
void ReadDirContentsL(RFs& aFs, const TDesC& aDirName)
    {
    CDir* dirList;
    User::LeaveIfError(aFs.GetDir(aDirName,KEntryAttNormal,ESortByName,
        dirList));
    // Process the entries
    delete dirList;
    }
```

You can change attributes of directory entries, including the hidden, system, read-only, and archive bits. The only bit with a really unambiguous meaning is read-only. If a file is specified as read-only, then you won't be able to write to it, or erase it. The other attributes are supported for strict compatibility with VFAT, but they aren't important in Symbian OS, so it's probably best not to use them.

Symbian OS maintains directory entry timestamps in UTC, not local time, so that there is no risk of confusion being caused by a time zone change.

6.1.4   Files

Within a file server session, an individual file is represented by an `RFile` object, which provides the means of opening and manipulating the file. For example, to open a file and append data to it, you could use:

```
TInt WriteToFileL(RFs& aFs, const TDesC& aFileName, const TDesC8& aData)
    {
    RFile file;
    TInt err = file.Open(aFs,aFileName,EFileWrite);
    if (err==KErrNotFound) // file does not exist, so create it
        {
        err = file.Create(aFs,aFileName,EFileWrite);
        }
    User::LeaveIfError(err);
    CleanupStack::CleanupClosePushL(file);
    User::LeaveIfError(file.Seek(ESeekEnd,0));
    User::LeaveIfError(file.Write(aData));

    CleanupStack::PopAndDestroy(); // Closes the file
    }
```

This example attempts to open an existing file, but if the file does not exist, it is created. In addition to `Open()` and `Create()`, `RFile` provides two other means of opening a file:

- `Replace()`, which deletes an existing file and creates a new one
- `Temp()`, which opens a temporary file and allocates a name to it.

In this example, the file is opened as a binary file, with non-shared write access, but files can also be opened as text files, and a variety of other access modes are supported, including shared write and exclusive or shared read. You normally specify the access mode when you open the file, although you can use `ChangeMode()` to change it while the file is open. If you're using shared write access, you can use `Lock()` to claim temporary exclusive access to a region of the file, and then `UnLock()` it later.

When a file is first opened, the current read and/or write position is set to the start of the file. You can use `Seek()` to move to a different position, and the example uses it to move to the end of the file, ready to append more data. `RFile` provides a variety of overloaded `Write()` functions, all of which write 8-bit data from a `TDesC8`. Half of them are synchronous functions, such as the one used in the above example, and the others are asynchronous, for use from an active object.

There is a corresponding variety of overloaded `Read()` functions, again supplied in both synchronous and asynchronous versions, all of which read 8-bit data into a `TDes8`.

Unless you are reading or writing fairly small amounts of information, you should always use the asynchronous functions to ensure that the application remains responsive during potentially long read or write sequences.

As you can see from this brief discussion, `RFile` provides only the most basic of read and write functions, operating exclusively on 8-bit data. In consequence `RFile` is not well suited to reading or writing the rich variety of data types that may be found in a Symbian OS application. Far from being an accident, this is a deliberate part of the design of Symbian OS, which uses **streams** to provide the necessary functionality.

6.2 Streams

A Symbian OS stream is the external representation of one or more objects. The process of **externalization** involves writing an object's data to a stream, and the reverse process is termed **internalization**. The external representation may reside in one of a variety of media, including stores (which are described later in this chapter), files and in memory.

The external representation needs to be free of any peculiarities of internal storage – such as byte order, and padding associated with data alignment – that are required by the phone's CPU, the C++ compiler or the application's implementation. Clearly, it is meaningless to externalize a pointer; it must be replaced in the external representation by the data to which it points. The external representation of each item of data must have an unambiguously defined length. This means that special care is needed when externalizing data types such as `TInt`, whose internal

representation may vary in size between different processors and/or C++
compilers.

Storing multiple data items, which may be from more than one object,
in a single stream implies that they are placed in the stream in a specific
order. Internalization code, that restores the objects by reading from the
stream, must therefore follow exactly the same order that was used to
externalize them.

The concept of a stream is implemented in the two base classes
RReadStream and RWriteStream, with concrete classes derived from
them to support streams that reside in specific media. For example,
RFileWriteStream and RFileReadStream implement a stream
that resides in a file, and RDesWriteStream and RDesReadStream
implement a memory-resident stream whose memory is identified by a
descriptor. In this chapter the examples concentrate on file-based streams,
but the principles can be applied to streams using other media.

The following example externalizes a TInt16 to a file, which is
assumed not to exist before WriteToStreamFileL() is called.

```
void WriteToStreamFileL(RFs& aFs, TDesC& aFileName, TInt16* aInt)
    {
    RFileWriteStream writer;
    writer.PushL(); // writer on cleanup stack
    User::LeaveIfError(writer.Create(aFs, aFileName, EFileWrite));
    writer << *aInt;
    writer.CommitL();
    writer.Pop();
    writer.Release();
    }
```

Since the only reference to the stream is on the stack, and the following
code can leave, it is necessary to push the stream to the cleanup stack,
using the stream's (*not* the cleanup stack's) PushL() function. Once the
file has been created, the data is externalized by using operator<<().
After writing, it is necessary to call the write stream's CommitL() to
ensure that any buffered data is written to the stream. Only then can
you remove the stream from the cleanup stack, using the stream's Pop()
function. Finally, you need to call the stream's Release() to close the
stream and free the resources it has been using.

Instead of calling Pop() and Release(), you could make a single
call to the cleanup stack's PopAndDestroy(), which would achieve the
same result. I've used that option in the following example. Otherwise,
the code follows a similar pattern to the externalization example.

```
void ReadFromStreamFileL(RFs& aFs, TDesC& aFileName, TInt16* aInt)
    {
    RFileReadStream reader;
    reader.PushL(); // reader on cleanup stack
    User::LeaveIfError(reader.Open(aFs, aFileName, EFileRead));
```

```
    reader >> *aInt;
    CleanupStack::PopAndDestroy(); // reader
    }
```

These examples use operator<<() to externalize the data and operator>>() to internalize it. This is a common pattern that you can use for:

- all built-in types except those, like TInt, whose size is unspecified

- descriptors, although special techniques are needed to internalize the data if the length can vary widely

- any class that provides an implementation of ExternalizeL() and InternalizeL().

In Symbian OS, a TInt is only specified to be *at least* 32 bits and may be longer, so externalizing it with operator<<() would produce an external representation of indefinite size. If you try, the compiler will report an error. Instead, use your knowledge of the data it contains to determine the size of the external representation. For example, if you know that the value stored in a TInt can never exceed 16 bits, you can use RWriteStream's WriteInt16L() to externalize it and RReadStream's ReadInt16L() to internalize it:

```
TInt i = 1234;
writer.WriteInt16L(i);
...
TInt j = reader.ReadInt16L();
```

If you use operator<<() to externalize a descriptor, the external representation contains both the length and character width (8 or 16 bits) so that operator>>() has sufficient information to perform the appropriate internalization:

```
_LIT(KSampleText, "Hello world");
TBuf<16> text = KSampleText;
writer << text;
...
TBuf<16> newtext;
reader >> newtext;
```

This is fine, provided that the descriptor's content is reasonably small, and known not to exceed a certain fixed length. If the descriptor can potentially contain a large amount of data it would be wasteful always to have to allow for the maximum. To cater for this, HBufC has a NewL() overload that takes a read stream and a maximum length as parameters and constructs a buffer of precisely the correct size to contain the data:

```
const TInt KStreamsMaxReadBufLength=10000;

_LIT(KSampleText, "Hello world");
TBuf<16> text = KSampleText;
writer << text;
...
HBufC* newtext = HBufC::NewL(reader, KStreamsMaxReadBufLength);
```

You should set the maximum length to be at least as large as the greatest amount of data that you can reasonably expect to have to handle. If the reader attempts to read more than the specified maximum, `HBufC::NewL()` leaves with the error `KErrCorrupt`.

You can externalize and internalize any class that implements `ExternalizeL()` and `InternalizeL()`, which are prototyped as:

```
class Foo
    {
public:
    ...
    void ExternalizeL(RWriteStream& aStream) const;
    void InternalizeL(RReadStream& aStream);
    ...
    }
```

For such a class, externalization can use either:

```
Foo foo;
writer << foo
```

or:

```
Foo foo;
foo.Externalize(writer);
```

which are functionally equivalent.

Similarly, for internalization, you can use either:

```
Foo foo;
reader >> foo;
```

or:

```
Foo foo;
foo.Internalize(reader);
```

The noughts and crosses application contains some simple externalization and internalization code that uses these functions:

```
void COandXAppUi::ExternalizeL(RWriteStream& aStream) const
    {
```

```
    for (TInt i = 0; i<KNumberOfTiles; i++)
        {
        aStream.WriteInt8L(iTileState[i]);
        }
    aStream.WriteInt8L(iGameStatus);
    aStream.WriteInt8L(iCrossTurn);
    aStream.WriteInt8L(iAppView->IdOfFocusControl());
    }

void COandXAppUi::InternalizeL(RReadStream& aStream)
    {
    for (TInt i = 0; i<KNumberOfTiles; i++)
        {
        iTileState[i] = aStream.ReadInt8L();
        }
    iGameStatus = aStream.ReadInt8L();
    iCrossTurn = aStream.ReadInt8L();
    SetNavigationPaneL(iCrossTurn);
    iAppView->MoveFocusTo(aStream.ReadInt8L());
    }
```

It should be clear from this discussion that both << and >> can leave. You will need to use them with a trap if they occur in a non-leaving function.

The stream base classes provide a variety of `WriteXxxL()` and `ReadXxxL()` functions to handle specific data types, ranging from 8-bit integers (e.g. `WriteInt8L()`) to 64-bit real numbers (e.g. `ReadReal64L()`), and these functions get called when you use << and >> on built-in types. You should make explicit use of these functions only either when you have to, as in the above example to deal with `TInt`s, or when you want to be very specific about how your data is externalized.

The stream base classes also provide a range of `WriteL()` and `ReadL()` functions to handle raw data. You need to be careful when using them for the following reasons.

- The raw data is written to the stream exactly as it appears in memory, so you must make sure that it is already in an external format before calling `WriteL()`.

- You need to make sure that `ReadL()` reads exactly the same amount of data that was written by the corresponding `WriteL()`, perhaps by writing the length immediately before the data, or terminating the data with a uniquely recognizable delimiter.

- You must implement a way of dealing with the maximum expected length of the data, say, by using the strategy described earlier for `HBufC::NewL()`.

- The 16-bit `WriteL()` and `ReadL()` functions don't give you the standard Unicode compression and decompression that you get when you use `<<` and `>>`.

The full range of `WriteL()` functions is:

```
class RWriteStream
    {
public:
    ...
    IMPORT_C void WriteL(const TDesC8& aDes);
    IMPORT_C void WriteL(const TDesC8& aDes, TInt aLength);
    IMPORT_C void WriteL(const TUint8* aPtr, TInt aLength);
    ...
    IMPORT_C void WriteL(const TDesC16& aDes);
    IMPORT_C void WriteL(const TDesC16& aDes, TInt aLength);
    IMPORT_C void WriteL(const TUint16* aPtr, TInt aLength);
    ...
    }
```

The first of these writes out the whole of the data contained in the descriptor. The second two write aLength bytes, from the beginning of the descriptor, or the location indicated by the pointer, respectively. The 16-bit versions behave similarly, except that they write Unicode characters instead of bytes.

The corresponding set of `ReadL()` functions is:

```
class RReadStream
    {
public:
    ...
    IMPORT_C void ReadL(TDes8& aDes);
    IMPORT_C void ReadL(TDes8& aDes, TChar aDelim);
    IMPORT_C void ReadL(TDes8& aDes, TInt aLength);
    IMPORT_C void ReadL(TUint8* aPtr, TInt aLength);
    IMPORT_C void ReadL(TInt aLength);
    ...
    IMPORT_C void ReadL(TDes16& aDes);
    IMPORT_C void ReadL(TDes16& aDes, TChar aDelim);
    IMPORT_C void ReadL(TDes16& aDes, TInt aLength);
    IMPORT_C void ReadL(TUint16* aPtr, TInt aLength);
    ...
    }
```

The first of these functions reads `aDes.MaxLength()` bytes from the stream, exactly filling the passed descriptor. The second one reads the smaller of `aDes.MaxLen()` and the number of characters up to, and including, the first occurrence of the specified delimiter character. The following two functions read aLength bytes into the descriptor, or the memory location indicated by the pointer. The fifth function simply reads and discards aLength bytes.

As with `WriteL()`, the 16-bit versions have similar actions, but read Unicode characters rather than bytes.

6.3 Stores

A Symbian OS store is a collection of streams, and is generally used to implement the persistence of objects. The store class hierarchy is illustrated in Figure 6.1, where the concrete classes are indicated by displaying their names in bold text. The base class for all stores is `CStreamStore`, whose API defines all the functionality needed to create and modify streams within a store. The classes derived from `CStreamStore` selectively implement the API according to their needs.

As with streams, stores can use a variety of different media, including memory (`CBufStore`) a stream (`CEmbeddedStore`) and other stores (for example, `CSecureStore`, which allows an entire store to be encrypted and decrypted), but the most commonly used medium is a file.

The two file-based stores are `CDirectFileStore` and `CPermanentFileStore`. The main way in which they differ from each other is that `CPermanentFileStore` allows the modification of streams after they have been written to the store, whereas `CDirectFileStore` does not. This difference results in the two stores being used to store persistent data for two different types of application, depending on whether the store or the application itself is considered to contain the primary copy of the application's data.

For an application such as a database, the primary copy of the data is the database file itself. At any one time the application typically holds in memory only a small number of records from the file, in order to display and/or modify them. Any modified data is written back to the

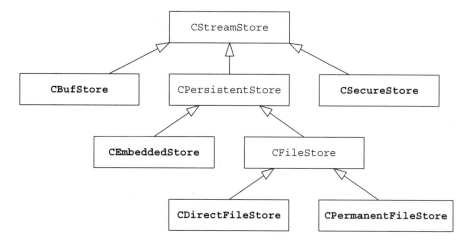

Figure 6.1 Store class hierarchy

file, replacing the original version. Such an application would use an instance of `CPermanentFileStore`, with each record being stored in a separate stream.

Other applications hold all their data in memory and, when necessary, load or save the data in its entirety. Such applications can use a `CDirectFileStore` since they never modify the store content, but simply replace the whole store with an updated version.

Another distinction that can be made between the different store types is whether or not they are **persistent**. A persistent store can be closed and, at a later time, opened again to read its content. The data in such a store therefore persists after a program has closed it and even after the program itself has terminated. A `CBufStore` is not persistent, since its in-memory data will be lost when the store is closed, but a file-based store ought to be persistent.

The persistence of a store is implemented in `CPersistentStore` abstract class. It does so by defining a **root stream**, which is always accessible on opening the store. The root stream, in turn, contains a **stream dictionary** that contains pointers to the remaining streams, as illustrated in Figure 6.2, thereby providing access to the rest of the data in the store.

The physical structure of a direct file store consists of:

- a 16-byte header, containing three UIDs, whose meanings are explained below, followed by a four-byte checksum

- a four-byte stream position, indicating the start of the root stream, which normally contains the stream dictionary

- a sequential list of the data for each stream, usually ending with the root stream.

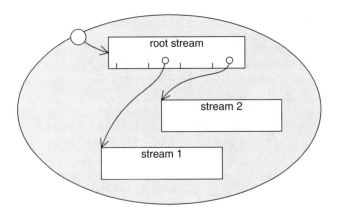

Figure 6.2 Logical view of a persistent store

The stream dictionary starts with an externalized `TCardinality` of the number of associations in the dictionary, followed by the associations themselves. Each association occupies eight bytes and consists of a stream's UID and a four-byte stream position, indicating the start of the relevant stream.

TCardinality is a useful class that provides a compact externalization of numbers (such as the count of elements in an array) which, though potentially large, normally have small values.

The structure of a permanent file store follows the same logical pattern, but the physical structure is more complex since it has to allow for records to be replaced or deleted.

6.3.1 Creating a Persistent Store

The following code illustrates how to create a persistent store. The example creates a direct file store, but creating a permanent file store follows a similar pattern.

```
void CreateDirectFileStoreL(RFs& aFs, TDesC& aFileName, TUid aAppUid)
    {
    CFileStore* store =
        CDirectFileStore::ReplaceLC(aFs,aFileName,EFileWrite);
    store->SetTypeL(TUidType(KDirectFileStoreLayoutUid,KUidAppDllDoc,
        aAppUid));

    CStreamDictionary* dictionary = CStreamDictionary::NewLC();

    RStoreWriteStream stream;
    TStreamId id = stream.CreateLC(*store);
    TInt16 i = 0x1234;
    stream << i;
    stream.CommitL();
    CleanupStack::PopAndDestroy(); // stream

    dictionary->AssignL(aAppUid,id);

    RStoreWriteStream rootStream;
    TStreamId rootId = rootStream.CreateLC(*store);
    rootStream << *dictionary;
    rootStream.CommitL();
    CleanupStack::PopAndDestroy(); // rootStream

    CleanupStack::PopAndDestroy(); // dictionary

    store->SetRootL(rootId);
    store->CommitL();

    CleanupStack::PopAndDestroy(); // store
    }
```

To create the store itself we use:

```
CFileStore* store = CDirectFileStore::ReplaceLC(aFs,aFileName,EFileWrite);
```

The call to `ReplaceLC()` will create the file if it does not exist, otherwise it will replace any existing file. In a real application, it might be more convenient to store the pointer to the file store object in member data, rather than on the stack.

You need to set the store's type, which we do by calling:

```
store->SetTypeL(TUidType(KDirectFileStoreLayoutUid,KUidAppDllDoc,
    aAppUid));
```

The three UIDs in the `TUidType` respectively indicate that the file contains a direct file store, the store is a document associated with a Symbian OS Unicode application and that it is associated with the particular application whose UID is `aAppUid`. For the file to be recognized as containing a direct file store, it is strictly necessary only to specify the first UID, leaving the other two as `KNullUid`, but including the other two allows an application to be certain that it is opening the correct file.

To create a permanent file store you would use:

```
CFileStore* store =
    CPermanentFileStore::CreateLC(aFs,aFileName,EFileWrite);
store->SetTypeL(TUidType(KPermanentFileStoreLayoutUid,KUidAppDllDoc,
    aAppUid));
```

We have used the store's `CreateLC()` function here, because you wouldn't normally want to replace a permanent file store.

Once you have created a file store, you can find its layout UID by calling `store->Layout()` which, depending on the file store's class, returns either `KDirectFileStoreLayoutUid` or `KPermanentFileStoreLayoutUid`.

We then create a stream dictionary, for later use:

```
CStreamDictionary* dictionary = CStreamDictionary::NewLC();
```

Creating, writing and closing a stream follows a similar pattern to the one we saw earlier, in the description of streams:

```
RStoreWriteStream stream;
TStreamId id = stream.CreateLC(*store);
TInt16 i = 0x1234;
stream << i;
stream.CommitL();
CleanupStack::PopAndDestroy(); // stream
```

An important difference is that, to write a stream to a store, you must use an instance of RStoreWriteStream, whose CreateL() and CreateLC() functions return a TStreamId. Once writing the stream is complete, you need to add an entry to the stream dictionary, making an association between the stream ID and an externally known UID:

```
dictionary->AssignL(aAppUid,id);
```

If, as in this case, your application writes all the application's document data into a single stream, it is safe to use the application's UID.

Once all the data streams have been written and added to the stream dictionary, you need to create a stream to contain the stream dictionary, and mark it as being the root stream:

```
RStoreWriteStream rootStream;
TStreamId rootId = rootStream.CreateLC(*store);
rootStream << *dictionary;
rootStream.CommitL();
CleanupStack::PopAndDestroy(); // rootStream
...
store->SetRootL(rootId);
```

All that remains is to commit all the changes made to the store and then to free its resources which, in this case, is done by the call to the cleanup stack's PopAndDestroy().

```
store->CommitL();
CleanupStack::PopAndDestroy(); // store
```

If the store is not on the cleanup stack you should simply use:

```
destroy store;
```

The store's destructor takes care of closing the file and freeing any other resources.

If you create a permanent file store you can, at a later time, reopen it and add new streams, or replace or delete existing streams. To ensure that the modifications are made efficiently, replaced or deleted streams are not physically removed from the store, so the store will increase in size with each such change. To counteract this, the stream store API includes functions to compact the store, by physically removing replaced or deleted streams.

You also need to be especially careful that you do not lose a reference to any stream within the store. This is analogous to a memory leak within an application, and results in the presence of a stream that can never be accessed or removed. Arguably, losing access to a stream is more serious than a memory leak, since a persistent file store outlives the application that created it. The stream store API contains a tool, whose central class is `CStoreMap`, to assist with stream cleanup.

You can find more information on these topics in the SDK.

6.3.2   Reading a Persistent Store

The following code opens and reads the direct file store that was created in the previous example.

```
void ReadDirectFileStoreL(RFs& aFs, TDesC& aFileName, TUid aAppUid)
    {
    CFileStore* store = CDirectFileStore::OpenLC(aFs,aFileName,EFileRead);

    CStreamDictionary* dictionary = CStreamDictionary::NewLC();

    RStoreReadStream rootStream;
    rootStream.OpenLC(*store, store->Root());
    rootStream >> *dictionary;
    CleanupStack::PopAndDestroy(); // rootStream

    TStreamId id = dictionary->At(aAppUid);

    CleanupStack::PopAndDestroy(); // dictionary

    RStoreReadStream stream;
    stream.OpenLC(*store, id);
    TInt16 j;
    stream >> j;
    CleanupStack::PopAndDestroy(); // stream

    CleanupStack::PopAndDestroy(); // store
    }
```

After opening the file store for reading, and creating a stream dictionary, you need to open the root stream, with:

```
rootStream.OpenLC(*store, store->Root());
```

after which you can internalize its content to the stream dictionary. You then extract one or more stream IDs, using the dictionary's `At()` function before using an `RStoreReadStream` to open each stream and internalize its content as appropriate for the application concerned.

6.3.3   Embedded Stores

Not all stores are as simple as the ones described. As illustrated in Figure 6.3, a store may, in fact, contain an arbitrarily complex network

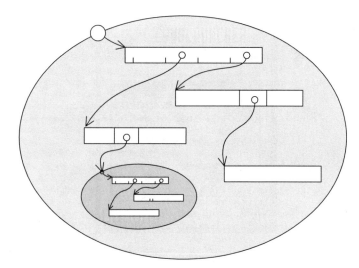

Figure 6.3 A more complex store

of streams. Any stream may contain another stream – by including its ID – and a stream may itself contain an *embedded* store.

It may be useful to store a collection of streams in an embedded store: from the outside, the embedded store appears as a single stream, and can, for example, be copied or deleted as a whole, without the need to consider its internal complexities. An embedded store cannot be modified, and thus behaves like a direct file store – which means that you can't embed a permanent file store.

The following two code examples illustrate writing and reading a permanent file store that contains an embedded file store.

```
void COandXAppUi::WriteEmbeddedFileStoreL(RFs& aFs, TDesC& aFileName)
    {
    CFileStore* mainStore = CPermanentFileStore::ReplaceLC(aFs,
        aFileName, EFileWrite);
    mainStore->SetTypeL(KPermanentFileStoreLayoutUid);

    RStoreWriteStream hostStream;
    TStreamId embeddedStoreId = hostStream.CreateLC(*mainStore);

    // construct the embedded store
    CPersistentStore* embeddedStore = CEmbeddedStore::NewLC(hostStream);
    RStoreWriteStream subStream;
    TStreamId id = subStream.CreateLC(*embeddedStore);
    TInt16 i = 0x1234;
    subStream << i;
    subStream.CommitL();
    CleanupStack::PopAndDestroy(); // subStream
    embeddedStore->SetRootL(id);
    embeddedStore->CommitL();
    CleanupStack::PopAndDestroy(); // embeddedStore
```

```
    hostStream.CommitL();
    CleanupStack::PopAndDestroy(); // hostStream

    RStoreWriteStream mainStream;
    TStreamId mainId = mainStream.CreateLC(*mainStore);
    TInt16 j = 0x3456;
    mainStream << j;
    mainStream.CommitL();
    CleanupStack::PopAndDestroy(); //mainStream

    RStoreWriteStream rootStream;
    TStreamId rootId = rootStream.CreateLC(*mainStore);
    TInt16 k = 0x5678;
    rootStream << k;
    rootStream << mainId;
    rootStream << embeddedStoreId;
    rootStream.CommitL();
    CleanupStack::PopAndDestroy(); //rootStream

    mainStore->SetRootL(rootId);
    mainStore->CommitL();
    CleanupStack::PopAndDestroy(); // mainStore
    }
```

To create the embedded store, you first create a stream in the main store, as normal:

```
RStoreWriteStream hostStream;
TStreamId embeddedStoreId = hostStream.CreateLC(*mainStore);
```

Then you create the embedded store, hosted in that stream:

```
// construct the embedded store
CPersistentStore* embeddedStore = CEmbeddedStore::NewLC(hostStream);
```

You create and write streams within the embedded store, and set the root stream, in exactly the same way as for any other store:

```
RStoreWriteStream subStream;
TStreamId id = subStream.CreateLC(*embeddedStore);
TInt16 i = 0x1234;
subStream << i;
subStream.CommitL();
CleanupStack::PopAndDestroy(); // subStream
embeddedStore->SetRootL(id);
embeddedStore->CommitL();
```

In this example, the IDs of the other streams are stored directly in the root stream's data, rather than using a stream dictionary:

```
RStoreWriteStream rootStream;
TStreamId rootId = rootStream.CreateLC(*mainStore);
```

```
TInt16 k = 0x5678;
rootStream << k;
rootStream << mainId;
rootStream << embeddedStoreId;
rootStream.CommitL();
CleanupStack::PopAndDestroy(); //rootStream
```

Here is the code to read the file that was written in the previous example:

```
void COandXAppUi::ReadEmbeddedFileStoreL(RFs& aFs, TDesC& aFileName)
    {
    CFileStore* store =
        CPermanentFileStore::OpenLC(aFs,aFileName,EFileRead);

    RStoreReadStream rootStream;
    rootStream.OpenLC(*store, store->Root());
    TInt16 k;
    TStreamId mainId;
    TStreamId embeddedStoreId;
    rootStream >> k;
    rootStream >> mainId;
    rootStream >> embeddedStoreId;
    CleanupStack::PopAndDestroy(); // rootStream

    RStoreReadStream mainStream;
    mainStream.OpenLC(*store, mainId);
    TInt16 j;
    mainStream >> j;
    CleanupStack::PopAndDestroy(); // mainStream

    RStoreReadStream hostStream;
    hostStream.OpenLC(*store, embeddedStoreId);
    CPersistentStore* embeddedStore = CEmbeddedStore::FromLC(hostStream);
    RStoreReadStream subStream;
    subStream.OpenLC(*embeddedStore, embeddedStore->Root());
    TInt16 i;
    subStream >> i;
    CleanupStack::PopAndDestroy(3); // subStream, embeddedStore,
                                    // hostStream

    CleanupStack::PopAndDestroy(); // store
    }
```

After opening the file, open the root stream and read its data, including the IDs of the other two streams:

```
RStoreReadStream rootStream;
rootStream.OpenLC(*store, store->Root());
TInt16 k;
TStreamId mainId;
TStreamId embeddedStoreId;
rootStream >> k;
rootStream >> mainId;
rootStream >> embeddedStoreId;
CleanupStack::PopAndDestroy(); // rootStream
```

To read the embedded store, first open the host stream, using the ID from the root stream:

```
RStoreReadStream hostStream;
hostStream.OpenLC(*store, embeddedStoreId);
```

Then open the embedded store it contains, using `CEmbeddedStore`'s `FromLC()` function:

```
CPersistentStore* embeddedStore = CEmbeddedStore::FromLC(hostStream);
```

Then we can open the embedded store's root stream and read it like any other stream:

```
RStoreReadStream subStream;
subStream.OpenLC(*embeddedStore, embeddedStore->Root());
TInt16 i;
subStream >> i;
```

6.3.4 Stores and the Application Architecture

In UIQ, the application architecture provides support for an application to save its data in a direct file store. The majority of Series 60 applications don't need to access an application document file and such access is disabled. If necessary, you can enable it in a Series 60 application by supplying an implementation of the application document's `OpenFileL()` function, as illustrated by the noughts and crosses application, whose document class definition, in `oandxdocument.h`, contains the following:

```
class COandXDocument : public CAknDocument
    {
    ...
public: // from CAknDocument
    ...
    // Restore CEikDocument's implementation of OpenFileL()
    CFileStore* OpenFileL(TBool aDoOpen,const TDesC& aFilename,RFs& aFs);
    ...
    };
```

The `OpenFileL()` function simply calls `CEikDocument`'s `OpenFileL()`:

```
CFileStore* COandXDocument::OpenFileL(TBool aDoOpen,
                                      const TDesC& aFilename,RFs& aFs)
    {
    return CEikDocument::OpenFileL(aDoOpen,aFilename,aFs);
    }
```

When support for the application's document file is enabled, all you need to supply are implementations of the document's `StoreL()` and `RestoreL()` functions, which are prototyped as:

```
public:
    void StoreL(CStreamStore& aStore,CStreamDictionary& aStreamDic) const;
    void RestoreL(const CStreamStore& aStore,const CStreamDictionary&
        aStreamDic);
```

and whose default implementations are empty. The noughts and crosses application implements these functions as:

```
void COandXDocument::StoreL(CStreamStore& aStore,
                            CStreamDictionary& aStreamDic) const
    {
    TStreamId id = iAppUi->StoreL(aStore);
    aStreamDic.AssignL(KUidOandXApp, id);
    }

void COandXDocument::RestoreL(const CStreamStore& aStore,
                            const CStreamDictionary& aStreamDic)
    {
    TStreamId id = aStreamDic.At(KUidOandXApp);
    iAppUi->RestoreL(aStore, id);
    }
```

The responsibility for storing and restoring the application's data is transferred to the AppUi, which uses code similar to that described in Sections 6.3.1 and 6.3.2:

```
TStreamId COandXAppUi::StoreL(CStreamStore& aStore) const
    {
    RStoreWriteStream stream;
    TStreamId id = stream.CreateLC(aStore);
    stream << *this; // alternatively, use ExternalizeL(stream)
    stream.CommitL();
    CleanupStack::PopAndDestroy();
    return id;
    }

void COandXAppUi::RestoreL(const CStreamStore& aStore, TStreamId
    aStreamId)
    {
    RStoreReadStream stream;
    stream.OpenLC(aStore,aStreamId);
    stream >> *this; // alternatively use InternalizeL(stream)
    CleanupStack::PopAndDestroy();
    }
```

These, in turn, use the AppUi's `InternalizeL()` and `External-izeL()` functions that were described in Section 6.2:

```
void COandXAppUi::ExternalizeL(RWriteStream& aStream) const
    {
    for (TInt i = 0; i<KNumberOfTiles; i++)
        {
        aStream.WriteInt8L(iTileState[i]);
        }
    aStream.WriteInt8L(iGameStatus);
    aStream.WriteInt8L(iCrossTurn);
    aStream.WriteInt8L(iAppView->IdOfFocusControl());
    }

void COandXAppUi::InternalizeL(RReadStream& aStream)
    {
    for (TInt i = 0; i<KNumberOfTiles; i++)
        {
        iTileState[i] = aStream.ReadInt8L();
        }
    iGameStatus = aStream.ReadInt8L();
    iCrossTurn = aStream.ReadInt8L();
    SetNavigationPaneL(iCrossTurn);
    iAppView->MoveFocusTo(aStream.ReadInt8L());
    }
```

The application architecture takes care of almost everything else, including creating the store when the application is run for the first time and reading the file when the application is run on subsequent occasions.

In addition to creating the root stream and the stream containing the stream dictionary, it also adds an application identifier stream. This stream contains the application's UID and the application file name (for example, `"OANDX.APP"`) and can be used to verify that the file is a valid one for the application to open and read. The application identifier stream is identified in the stream dictionary by a UID of `KUidAppIdentifierStream`, and is created by code of the form:

```
void WriteAppIdentifierL(CFileStore* aStore,
                         CStreamDictionary* aDict,
                         TUid aAppUid,
                         TDesC& aAppName)
    {
    TApaAppIdentifier appIdent(aAppUid, aAppName);
    RStoreWriteStream stream;
    TStreamId id = stream.CreateLC(*aStore);
    stream << appIdent;
    stream.CommitL();
    CleanupStack::PopAndDestroy(); // stream
    aDict->AssignL(KUidAppIdentifierStream,id);
    }
```

In Series 60 applications, you need to ensure that the file is updated when the application is closed. You do this by calling the AppUi's `SaveL()` function in the exit cases of `HandleCommandL()`:

```
void COandXAppUi::HandleCommandL(TInt aCommand)
    {
    switch(aCommand)
        {
        case EEikCmdExit:
        case EAknSoftkeyExit:
            {
            SaveL();

            Exit();
            }
            break;
    ...
    }
```

UIQ applications don't normally exit, unless they are forced to do so by the memory manager. If the memory manager needs to recover memory, it calls an application document's `MSaveObserver::SaveL()` function, passing it a value of `MSaveObserver::EReleaseRAM`. On receipt of this value, an application is expected to save its state and data *and then close down*, and this is the default action for all applications that derive their document class from `CQikDocument`. Any application that ignores this request – say, by deriving its document from `CEikDocument`, or by providing its own implementation of `MSaveObserver::SaveL()` – is liable to be forced to close, without any further opportunity to save its state and data. System applications are never sent an `EReleaseRAM` request and they, together with any application that has marked itself as Busy within the last 60 minutes, will not be forced to close.

6.4   Using .ini Files

In addition to an application's document file, the application architecture provides support for an application to open, read and modify a second file. By convention, the file has the same name as the application, but has a `.ini` extension; hence such files are known as `.ini` files.

The intention is that a `.ini` file should be used to store global settings and preferences that are independent of the document data that the application is processing. An application can then, in principle, open a different document without affecting the existing preference settings.

In practice, applications running on mobile phones tend to use only a single document and therefore have no real need to use `.ini` files. In recognition of this fact, the Series 60 UI disables the application architecture's support for `.ini` files, but it is a simple matter to restore it if necessary, by replacing `CAknApplication`'s `OpenIniFileLC()` with a version that calls `CEikApplication`'s `OpenIniFileLC()`, such as:

```
CDictionaryStore* CMyAppApplication::OpenIniFileLC(RFs& aFs) const
    {
    return CEikApplication::OpenIniFileLC(aFs);
    }
```

As well as opening the application's `.ini` file for exclusive read/write access – and pushing it to the cleanup stack – this function will create the `.ini` file if it does not exist, and will replace a corrupt file. The function is called by the application architecture, for example to record the last opened file, and can be called by your application code.

The `.ini` file is represented by an instance of `CDictionaryFile-Store`, which derives from the `CDictionaryStore` base class. These classes are slightly misleadingly named since they have no connection with stream dictionaries. Also, since they do not derive from `CStream-Store`, they do not represent stream stores. Despite this, there are similarities of usage: the dictionary store contains streams associated with UIDs, and you read and write the streams by means of `RDic-tionaryReadStream` and `RDictionaryWriteStream` classes, in the same way as you would with a stream store.

However, a significant difference is that a dictionary store never contains more than a simple list of streams, unlike the complex network that is possible in a stream store. Furthermore, you use UIDs to access the streams directly, as illustrated in Figure 6.4, rather than via a stream dictionary.

The following simple examples, designed to be used as member functions of an application's AppUi, illustrate writing to and reading from the application's `.ini` file.

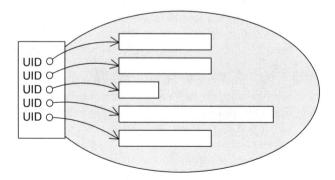

Figure 6.4   A dictionary store

```
void WriteToIniFileL(RFs& aFs)
    {
    CDictionaryStore* iniFile = Application()->OpenIniFileLC(aFs);
    RDictionaryWriteStream stream;
    stream.AssignLC(*iniFile, TUid::Uid(0x101ffac5)); // direct access by
                                                      // UID
    TUint16 i = 0x3456;
    stream << i;
    stream.CommitL();
    iniFile->CommitL();
    CleanupStack::PopAndDestroy(2); // stream and iniFile
    }

void ReadIniFileL(RFs& aFs)
    {
    CDictionaryStore* iniFile = Application()->OpenIniFileLC(aFs);
    RDictionaryReadStream readStream;
    readStream.OpenLC(*iniFile, TUid::Uid(0x101ffac5)); // direct access
                                                        // by UID
    TInt16 i;
    readStream >> i;
    CleanupStack::PopAndDestroy(2); // readStream and iniFile
    }
```

6.5   Resource Files and Bitmaps

Symbian OS keeps resource files and bitmaps separate, using different
tools to generate them. The resource compiler and the bitmap converter
both generate binary data files that are delivered in parallel with an
application's executable. Windows, on the other hand, uses a resource
compiler that supports icons and graphics as well as text-based resources,
and which builds the resources right into the application executable.

Unlike Windows developers, Symbian OS developers target a wide
range of hardware platforms. Resource file and bitmap formats are independent of the target platform, but executables may require a different
format for each platform. Keeping the resources and bitmaps separate
from each other and from the application's executable introduces a layer
of abstraction that reduces the effort required by third-party developers
to move applications between different hardware platforms. In addition,
it provides good support for localization. Separating text from graphics
and the executable facilitates the process of translation and allows a
multilingual application to be supplied as a single executable, together
with a number of language-specific resource files.

6.5.1   Resource Files

The resource compiler generates a Symbian OS resource file from a plain
text source file, or resource script, which – by convention – has a `.rss`

extension. The first stage of processing uses a standard C preprocessor, so a resource script has the same lexical conventions as a C program, including source file comments and C preprocessor directives.

The output of the compilation process is a binary resource file with the same name as the source file, but a `.rsc` extension, and a generated header file with a `.rsg` extension. The `.rsg` file contains a list of defined symbolic constants, one for each resource in the resource file. You #include this file in your application source files to provide access to the resources.

6.5.1.1 *Source File Syntax*

The built-in resource data types are:

Data type	Description
BYTE	A single byte that may be interpreted as a signed or unsigned integer value.
WORD	Two bytes that may be interpreted as a signed or unsigned integer value.
LONG	Four bytes that may be interpreted as a signed or unsigned integer value.
DOUBLE	Eight-byte real, for double precision floating point numbers.
TEXT	A string terminated by a null. This is deprecated; use LTEXT instead.
LTEXT	A Unicode string with a leading length byte but no terminating null.
BUF	A Unicode string with no leading byte or terminating null.
BUF8	A string of 8-bit characters with no leading byte or terminating null.
BUF<n>	A Unicode string of up to n characters with no leading byte or terminating null.
LINK	The ID of another resource (16 bits).
LLINK	The ID of another resource (32 bits).
SRLINK	A 32-bit self-referencing link that contains the resource ID of the resource in which it is defined. Its value is assigned automatically by the resource compiler.

These types are used to specify the data members of a resource, as described later.

A resource file may contain statements of the following types:

Statement type	Description
NAME	Defines the leading 20 bits of any resource ID. Must be specified prior to any RESOURCE statement.
STRUCT	Defines a named structure for use in building aggregate resources.
RESOURCE	Defines a resource, which may optionally be given a name.
ENUM/enum	Defines an enumeration, and supports a syntax similar to C's.
CHARACTER_SET	Defines the character set for strings in the generated resource file. If not specified, cp1252 is the default.

6.5.1.2 *STRUCT Statements*

A STRUCT statement is of the form:

```
STRUCT struct-name [ BYTE | WORD ] { struct-member-list }
```

The struct_name specifies a name for the struct. The name must start with a letter and be in uppercase characters. It may contain letters, digits and underscores, but not spaces. The optional BYTE or WORD keywords are intended for use with structs that have a variable length. They have no effect unless the struct is used as a member of another struct, when they cause the data of the struct to be preceded by a length BYTE or length WORD, respectively.

The struct_member_list is a list of member initializers, terminated by semicolons and enclosed in braces { }. A member may be one of the built-in types, a previously defined struct or an array. It is common to supply members with default values, which are frequently either a numeric zero or an empty string. The following example, taken from eikon.rh, illustrates many of the features of a STRUCT statement. Note that an array member does not specify either the number or type of its elements.

```
STRUCT DIALOG
    {
    LONG flags=0;
    LTEXT title="";
    LLINK pages=0;
    LLINK buttons=0;
    STRUCT items[];   // an array
    LLINK form=0;
    }
```

As in this case, STRUCT statements are conventionally placed in separate files with a .rh (resource header) extension and #included in the resource script. You may find it instructive to review the contents of the various .rh files that you will find in the \epoc32\include directory of an installed SDK.

6.5.1.3 *RESOURCE Statements*

The RESOURCE statement is probably the most important, and certainly the most frequently used, statement. Such statements are of the form:

```
RESOURCE struct_name [ id ] { member_initializer_list }
```

The struct_name refers to a previously encountered STRUCT statement. In most application resource scripts, this means a struct defined in a #included .rh file. The id is an optional symbolic resource ID. If specified, it must be in lowercase, starting with a letter. Otherwise it may contain letters, digits and underscores, but not spaces. The member_initializer_list consists of a list of member initializers, separated by semicolons and enclosed in braces { }.

The following example, taken from the noughts and crosses application's resource script, shows a resource constructed using the DIALOG struct that was shown earlier.

```
RESOURCE DIALOG r_oandx_first_move_dialog
    {
    title = "First player";
    flags = EAknDialogSelectionList;
    buttons = R_AVKON_SOFTKEYS_OK_CANCEL;
    items =
        {
        DLG_LINE
            {
            type = EAknCtSingleListBox;
            id = EOandXPlayerListControl;
            control = LISTBOX
                {
                flags = EAknListBoxSelectionList;
                array_id = r_oandx_player_list;
                };
            }
        };
    }
```

In this case the items array contains only one item which is, itself, a DLG_LINE struct. Note that it is not necessary for the resource to declare the initializers in the same order as they were specified in the corresponding struct. The resource in this example does not provide initalizers for the DIALOG struct's pages and form items, so their values will take the default values that are specified in the struct.

The punctuation rules for a RESOURCE statement, although not obvious at first sight, are quite simple:

1. Assignment statements, of any type, are terminated by a semicolon.

2. Items in a list, except for the last, are terminated by commas.

3. In any other situation, no punctuation is required.

For example:

```
RESOURCE ARRAY r_oandx_player_list
    {
    items =
        {
        LBUF
            {
            txt= "\tNoughts first"; // Rule 1
            },                      // Rule 2
        LBUF
            {
            txt= "\tCrosses first"; // Rule 1
            }                       // Rule 2
        };                          // Rule 1
    }                               // Rule 3
```

6.5.1.4   *ENUM Statements*

To ensure that your resource script and C++ program use the same values for symbolic constants, the resource compiler supports enum (and #define) definitions of constants, with a syntax similar to that used by C++. By convention, these definitions are contained in .hrh include files. The .hrh extension is intended to convey that the file is suitable for inclusion either as a .h file in C++ source, or as a .rh file in resource scripts.

6.5.1.5   *The NAME Statement*

A resource script must contain a single NAME statement, which must appear before the first RESOURCE statement. The NAME keyword must be followed by a name, in upper case, containing a maximum of four characters, for example:

```
NAME OANX
```

An application identifies each resource by a symbolic ID that is published in the generated .rsg header file. The leading 20 bits of the ID are generated from the name supplied in the NAME statement and thereby identify the resource file that contains the resource. The remaining 12 bits of the ID identify an individual resource within the file – which means that a resource file is limited to containing not more than 4095 resources.

However, a further consequence is that an application can access resources from more than one resource file, provided none of the resources that the application accesses have identical names. You can generally ensure that your application meets this condition by not using names used by system resource files. For example, you should avoid names that begin with A (in Series 60) and Q (in UIQ). If you also avoid the names EIK, CONE, BAFL, and other Symbian OS component names, you should be safe.

6.5.1.6 *Localization of Resource Strings*

Several of the example applications used in this book leave their text strings in the resource script file. That is because they are aimed at developers, and there is no intention to translate them into other languages. Keeping the text within its resource helps a programmer to understand the purpose and structure of the resource.

In a real application, different considerations apply, especially if you are intending to translate your application into one or more different languages. A translator will not necessarily be a programmer and will probably find it difficult to preserve the general structure of a resource script through the translation process. For this reason alone, it is good practice to store all resource text strings in a separate file.

All you have to do is to replace each text string in the resource script with a symbolic identifier and, in a separate file, associate the identifier with the original string.

For example, instead of your resource script containing:

```
RESOURCE MENU_PANE r_oandx_menu
    {
    items =
        {
        MENU_ITEM
            {
            command = EOandXNewGame;
            txt = "New game";
            }
        };
    }
```

you would modify the resource to something like:

```
RESOURCE MENU_PANE r_oandx_menu
    {
    items =
        {
        MENU_ITEM
```

```
         {
         command = EOandXNewGame;
         txt = text_new_game_menu;
         }
     };
 }
```

and, in a separate text localization file, you would use the `rls_string` keyword to associate the symbolic ID with the string:

```
rls_string text_new_game_menu  "New game"
```

Obviously, you need to #include the text localization file in the resource script.

You can include C- or C++-style comments in the text localization file to inform the translator (and yourself!) of the context in which each text item appears, and to give information about any constraints, such as the maximum length permitted for a string. In fact, there is a double advantage; the translator sees only the text and the comments, and the programmer sees a resource script free from comments that could disguise the structure of a resource.

There is no specific convention about the file name extension to be used for text localization files. At Symbian, developers tend to use a `.rls` extension (standing for 'resource localizable strings') but Series 60 suggests the use of a `.loc` extension.

6.5.1.7 Compiling a Resource File

The resource compiler is normally invoked as part of the application build process, either from within an IDE or from the command line. As mentioned earlier, in addition to creating the binary resource file, the resource compiler also generates a header file, with a `.rsg` extension, containing symbolic IDs for every resource contained in the file. You need to #include this file in the application's source, which is why the resource compiler is run *before* you run the C++ compiler when building a Symbian OS program. The `.rsg` file is always generated to `\epoc32\include`, but if the generated file is identical to one already in `\epoc32\include`, the existing one isn't updated. That means you don't have to recompile the application if you make a change to the resource file – such as altering a text string – that doesn't alter the resource IDs.

Uikon has a wide range of specific resources (especially string resources) that are accessible via the resource IDs listed in `eikon.rsg`.

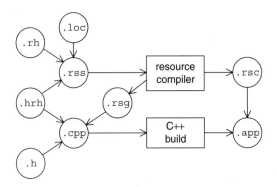

Figure 6.5 Overview of the build process

Figure 6.5 shows the overall build process, indicating the relationship between the various file types involved.

Application-specific input files to the resource compilation process include:

Appname.rss	The application's resource script
Appname.rh	A resource header containing application-specific resource structs
Appname.hrh	An application-specific header containing symbolic constants and other data that is needed in both the resource script and one or more C++ source files
Appname.loc	The application's localizable strings

You may also need to include system .rh files, such as eikon.rh, and other UI-specific .rh files.

In addition to appname.hrh and the generated appname.rsg file, C++ files might need to include system files such as eikon.rsg and eikon.hrh, together with one or more UI-specific .rsg and .hrh files.

6.5.1.8 *Resource File Structure*

The resource compiler builds the resource file sequentially, appending successive resources to the end of the file, in the order of definition, and finally writes an index at the end of the file. Each index entry is a word that contains the offset, from the beginning of the file, to the start of the appropriate resource. The index contains, at its end, an additional entry that contains the offset to the byte immediately following the last resource, which is also the start of the index. An individual resource is simply binary data, whose length is found from its own index entry and that of the following resource.

Where possible, resource files compress Unicode text so that each character occupies a single byte. A resource's content is divided into alternating runs of compressed and uncompressed data, starting with a compressed run (which is usually of zero length). Each run is preceded by its length which, for runs not exceeding 255 bytes, is stored in a single byte.

Except for the effects of compression, each resource just contains the individual elements, listed sequentially. A WORD, for example, occupies two bytes and an LTEXT is represented by a byte specifying the length, followed by the text. Where there are embedded structs, the effect is to flatten the structure. The interpretation of the data is the responsibility of the class or function that uses it.

6.5.1.9 Reading Resources

The application framework opens the application's resource file during the startup of the application. The most straightforward way to read one of its resources is to call CCoeEnv's AllocReadResourceL() or AllocReadResourceLC() functions, which take a resource ID as a parameter. Both functions allocate an HBufC big enough for the uncompressed resource and read it in. CCoeEnv also supplies variants of these two functions to read a resource as explicit 8-bit or 16-bit descriptors, or descriptor arrays.

All resource reading functions will uncompress any compressed data.

Alternatively, you could use CCoeEnv's ReadResource() function, which simply reads a resource into an existing descriptor, and panics if the read fails.

A further method is to use TResourceReader, which is a stream-oriented reader for the data of an individual resource. TResourceReader functions are provided that correspond to each of the resource compiler's built-in data types. You can use CCoeEnv's CreateResourceReaderLC() function to create and initialize a TResourceReader object. There is an example at the end of this chapter that illustrates how to use a TResourceReader object to load a bitmap.

The most fundamental way to read a resource is to use an instance of the RResourceFile class. This class provides the means to open a named resource file and read its resources.

6.5.2 Bitmaps

Adding graphical icons to your program can make a big difference to the end user's impression of your application, as well as enhancing usability. The noughts and crosses application uses its own specific icons in three ways:

- to represent the application in the application selector (or shell)
- to provide an application-specific image in the application's context pane
- in the navi pane, to indicate whose turn it is to play.

The last two of these are illustrated in Figure 6.6.

Figure 6.6   Icon use in noughts and crosses

In all cases, the first step of the process is to create Windows bitmaps, using any image editing program that can generate files in Windows bitmap (.bmp) format. Figure 6.7 shows the bitmap, oxs_icon.bmp, to be displayed in the Series 60's application selector. It needs to be 42 pixels wide and 29 pixels high. The real bitmap doesn't have a black border – this has been added to the figure so that you can see the bitmap's extent.

Figure 6.7   Application selector icon bitmap

For each icon that you wish to display, you need a corresponding two-color *mask* bitmap, of the same dimensions. The mask bitmap corresponding to the application selector icon is shown in Figure 6.8.

Figure 6.8   Application selector mask bitmap

Again, there's a black border around the bitmap to indicate its extent. Only the regions of the icon bitmap corresponding to the black regions of the mask are displayed. Everything else is ignored; the white areas of the mask are effectively transparent, regardless of the color of the corresponding parts of the icon bitmap.

Having created the bitmaps, you now have to build them into your application. Depending on their intended purpose, icons can be dealt with in one of two possible ways:

- Icons representing the application as a whole – that is those that will be displayed either in the application selector, or in the application's context pane – are included in an application information (.aif) file. They are handled by system code and need no further cooperation from the application.

- Images that are explicitly referenced by the application are combined in a multi-bitmap (.mbm) file. The .mbm file is referenced in the application's resource file and you will, in general, need specific application code to handle the images.

6.5.2.1 Icons

As its name suggests, a .aif file supplies information about the application and its capabilities. It can specify the icon(s) and caption used to represent the application in the shell, and also provides information about the application's capabilities, such as whether its documents can be embedded into another application. Most applications use the default capabilities that are specified in aiftool.rh. To be recognized, a .aif file must have the same name as the application's program file, but with a .aif extension, and must reside in the same directory.

The .aif file is constructed from one or more icon bitmap files, together with their associated mask bitmap files, and a resource file that contains an AIF_DATA resource. The noughts and crosses application's .aif resource file, oandxaif.rss, contains the following:

```
#include <aiftool.rh>

RESOURCE AIF_DATA
    {
    app_uid = 0x101ffabe;
    caption_list=
        {
        CAPTION
            {
            code = ELangEnglish;
            caption = "OandX";
            }
        };
    num_icons = 2;
    embeddability = KAppNotEmbeddable;
    newfile = KAppDoesNotSupportNewFile;
    }
```

You create a .aif file by running the aif generation tool, aiftool. While it is possible to run this tool from the command line, it is usually more convenient to invoke it as part of the build process. You do this by including an AIF statement in the application's .mmp file. The noughts and crosses application uses the following statement:

```
AIF OandX.aif ..\aif OandXAif.rss \
    c12 oxicon44.bmp oxicon44m.bmp oxs_icon.bmp oxs_iconm.bmp
```

The first three parameters respectively specify the name of the generated .aif file, the relative path to the source files and the name of the resource script. The next parameter specifies a color depth of 12 bits (see below) and the remaining ones list the icon and mask bitmap file pairs that are to be included. The number of pairs should match the num_icons value specified in the resource file. A Series 60 application should be supplied with two icons, one 44 × 44 pixels and the other 42 × 29 pixels.

If, as in this case, the statement extends over more than one line, use a backslash to indicate that the following line is a continuation.

You need to specify the color depth to be used when converting the icon bitmaps. Symbian OS provides support from 1-bit color up to 24-bit color, and individual phones will vary in terms of the color depth supported by their hardware. Most current Symbian OS phones support at least 12-bit color ($2^{12} = 4096$ colors) on screen, and support for 16-bit color ($2^{16} = 65\,536$ colors) is becoming more common. However, there is a trade-off between battery-life and RAM usage. Higher color depth images (and icons) consume more power when displayed, occupy more disk space when stored *and* use more RAM when loaded. For this reason, many application icons use an 8-bit color depth (giving $2^8 = 256$ colors), striking a suitable balance between aesthetics and file size/power consumption.

6.5.2.2   *Images that are Loaded by the Application*

To illustrate how such images are handled, we can use the bitmaps that are displayed in the noughts and crosses application's navi pane, to indicate whose turn it is. The Series 60's navi pane is 260 pixels wide and 16 pixels high, so we need to construct a pair of bitmaps with these dimensions, with the nought and the cross in a suitable position. The two bitmaps are illustrated in Figure 6.9 and they are outlined for clarity. (If you were to see these in color, the nought would be red, the cross would be green.)

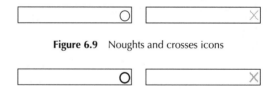

Figure 6.9   Noughts and crosses icons

Figure 6.10   Noughts and crosses masks

As shown in Figure 6.10, each has a corresponding two-color mask bitmap of the same dimensions, with the region to be displayed – this time, just the symbols themselves – drawn in black.

In this case, after creating the bitmaps, you need to convert them into the specific multi-bitmap (.mbm) file format used by Symbian OS. To do this you use the bitmap converter tool, bmconv, which combines a number of bitmaps in a single .mbm file. It also generates an associated

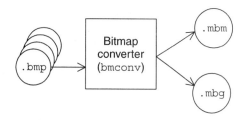

Figure 6.11 Bitmap conversion process

.mbg header file, for inclusion in other source files, containing an enumeration that provides symbolic identifiers for the bitmaps in the .mbm file. The process is illustrated in Figure 6.11.

You can use the bmconv tool as a stand-alone application, to convert (in either direction) between .bmp and .mbm files but, as with aiftool, you can modify the project's .mmp file to configure the build to include the generation of a .mbm file. The noughts and crosses application uses the following text:

```
START BITMAP oandx.mbm
HEADER
TARGETPATH \system\apps\oandx
SOURCEPATH .
SOURCE c8 navio.bmp
SOURCE  1 naviomask.bmp
SOURCE c8 navix.bmp
SOURCE  1 navixmask.bmp
END
```

The statements have the following meanings:

START BITMAP	Marks the start of the bitmap conversion data and specifies the multi-bitmap filename.
HEADER	Specifies that a symbolic ID file, in this case oandx.mbg, is to be created (in the \epoc32\include folder).
TARGETPATH	Specifies the destination location for the multi-bitmap file.
SOURCEPATH	Specifies the location of the original bitmap files.
SOURCE	Specifies the color depth, and the names of one or more Windows bitmap files to be included.
END	Marks the end of the bitmap conversion data.

Symbian OS developers conventionally specify only one bitmap file per SOURCE statement and, although the ordering does not really matter, specify each mask file immediately after its corresponding bitmap file.

The format of the generated `oandx.mbg` file is as follows:

```
// OANDX.mbg
// Generated by BitmapCompiler
// Copyright (c) 1998-2001 Symbian Ltd.  All rights reserved.
//

enum TMbmOandx
    {
    EMbmOandxNavio,
    EMbmOandxNaviomask,
    EMbmOandxNavix,
    EMbmOandxNavixmask
    };
```

The naming convention here is clear enough: there's an enumeration whose type name includes the `.mbm` file name, and an list of enumerated constants whose names are constructed from the `.mbm` file name *and* the source bitmap file name.

To specify an image in the resource file, you need the name and location of the `.mbm` file that contains it and references to both the image and the corresponding mask within the file. We have chosen to use a `NAVI_IMAGE` resource struct, defined in `avkon.rh` – and shown below – but any resource struct that contains the same three items would do.

```
STRUCT NAVI_IMAGE
    {
    LTEXT bmpfile="";
    WORD bmpid=0xffff;
    WORD bmpmask=0xffff;
    }
```

The relevant resources in the noughts and crosses application's resource file are:

```
#include <oandx.mbg>
...
RESOURCE NAVI_IMAGE r_oandx_navio_image_resource
    {
    bmpfile = "\\system\\apps\\oandx\\oandx.mbm";
    bmpid   = EMbmOandxNavio;
    bmpmask = EMbmOandxNaviomask;
    }

RESOURCE NAVI_IMAGE r_oandx_navix_image_resource
    {
    bmpfile = "\\system\\apps\\oandx\\oandx.mbm";
    bmpid   = EMbmOandxNavix;
    bmpmask = EMbmOandxNavixmask;
    }
```

The final step is to load the resource into the application. The noughts and crosses application uses a `TResourceReader` to read the relevant resource:

```
void COandXAppUi::SetNavigationPaneL(TBool aCrossTurn)
    {
    delete iNaviDecorator;
    iNaviDecorator = NULL;
    TResourceReader reader;
    iCoeEnv->CreateResourceReaderLC(reader,
                                    aCrossTurn ?
                                    R_OANDX_NAVIX_IMAGE_RESOURCE :
                                    R_OANDX_NAVIO_IMAGE_RESOURCE
                                    );
    iNaviDecorator = iNaviPane->CreateNavigationImageL(reader);
    CleanupStack::PopAndDestroy();  // resource reader
    iNaviPane->PushL(*iNaviDecorator);
    }
```

Since the call to `CreateNavigationImageL()` takes a `TResource-Reader`, noughts and crosses doesn't need to read the file name and the IDs of the two bitmaps. If your application needs to do so, you can use code of the form:

```
HBufC* bitmapFile = reader.ReadHBufCL(); // bmp filename
TInt bitmapId = reader.ReadInt16();      // bmp id
TInt maskId = reader.ReadInt16();        // bmp mask id
```

Then, to load the image itself into a `CFbsBitmap`, you could use:

```
CFbsBitmap* image = new(ELeave) CFbsBitmap();
CleanupStack::PushL(image);
User::LeaveIfError(image->Load(bitmapFile,bitmapId,ETrue));
CleanupStack::Pop(); // image
```

with similar code to load the mask.

7

Multimedia Services

This chapter provides you with an overview of the multimedia services available to software engineers developing on Symbian OS v7.0s and beyond.

We'll begin with an overview of the multimedia architecture, including a brief outline of how it has evolved. This is followed by a description of the separate parts of the multimedia subsystem. The remainder of the chapter is split into sections describing the various services that are available.

7.1 The Multimedia Component Architecture

Before discussing the current state of multimedia, we'll look at how multimedia has evolved in Symbian OS.

Historically, the multimedia subsystem has comprised separate APIs for dealing with the following multimedia areas:

- audio – playback, recording and manipulation
- still images – decoding, encoding and manipulation.

This has not changed in Symbian OS v7.0s, even though the underlying subsystem has changed considerably. The multimedia APIs in Symbian OS v7.0s are largely a superset of the APIs present in earlier releases and, in some cases, new and legacy methods exist concurrently.

In order to explain why many radical improvements were made to the multimedia subsystem in Symbian OS v7.0s, it is necessary to give an overview of the subsystems in previous revisions of the operating system and the problems which these faced. This section gives a brief overview of the Media Server in v6.1 and v7.0 and why this has changed.

Symbian OS C++ for Mobile Phones, Volume 2.   Edited by Richard Harrison
© 2004 Symbian Software Ltd   ISBN: 0-470-87108-3

The Media Server

In Symbian OS v6.1 and v7.0, all multimedia processing was performed through the Media Server. This is a standard Symbian OS server and is a single process that provides all multimedia functionality. The server supported the playback and recording of audio, along with the encoding/decoding and manipulation of still images. The operating system shipped with an array of audio and image format support, which could be extended by writing plugins.

In order to use the server a user could either explicitly instantiate a connection to the server, or allow the client APIs to provide a connection automatically. Each client user would supply an observer class to the server, which would allow the server to communicate and pass messages to the user application or library.

The server keeps a list of client objects and concurrently cycles through the multimedia requests. This means that a number of different client users can use the server at the same time and, for example, enables a user application to playback audio whilst simultaneously decoding images for display. Although this sounds ideal, the practical issues in producing this kind of behavior were complicated and fraught with difficulties.

If, for example, a user actually did want to use two parts of the server's functionality at the same time, the latency created by having one process control both could make the system virtually unusable. For instance, a processor-intensive task – such as decoding an image – would prevent the usage of any real-time multimedia task, like audio streaming, and the audio stream would very quickly underflow at the device driver. The situation was made worse when poorly written third-party plugins were used (which were often converted from non-Symbian code and contained lengthy procedures that consumed large amounts of processor time). The plugin framework itself was extremely complicated to write for, which did not improve the situation.

These sorts of problems, coupled with the fact that connecting to the server could take, in the worst case, several seconds, meant improvements were necessary.

The Start of a New Era

Symbian began to write an entirely new multimedia subsystem that would successfully allow different areas of the subsystem to be used simultaneously, and also provide a lightweight framework that mobile phone manufacturers and third parties could extend easily. The new system was not based solely on one server and instead split up the various multimedia sections to use separate servers and processes.

The new subsystem used multiple concurrent multimedia threads, with none of the side effects that were seen in the Media Server. It made use of many of the same client APIs as the Media Server, but took advantage of a new plugin-resolving methodology, known as ECom. This allowed plugins to be written and incorporated by mobile phone manufacturers and/or third parties – during phone manufacture or after a phone shipped.

The new subsystem was so successful that it was quickly integrated into Symbian OS v7.0s, which was just beginning its development lifecycle. As development of Symbian OS v7.0s progressed, the subsystem evolved into what is now known as the Multimedia Framework (MMF).

7.2   The Multimedia Framework (MMF)

The MMF or Multimedia Framework provides the framework by which multimedia is handled in Symbian OS v7.0s. The framework itself is extremely lightweight and provides a multithreaded system, based on ECom, which retains a subset of the original APIs from v6.1 and v7.0, but also provides numerous enhancements.

The basic structure of the MMF consists of a client API layer, a controller framework, controller plugins and lower-level subsystems. Figure 7.1 gives an overview of the architecture.

In Figure 7.1, the controller framework can be thought of as the client/server layer, which provides the interface to the MMF Controller Plugins. A description of each part of the subsystem follows below.

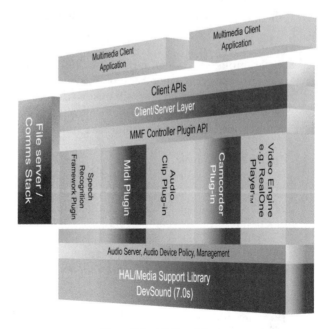

Figure 7.1   MMF architecture

7.2.1   Client APIs

The client APIs (or 'Application APIs', as they are sometimes known) are at the highest level of abstraction in the subsystem. These APIs provide a developer with the basic capabilities of the MMF. In Symbian OS v7.0s there are client APIs for the generation of sine-wave tones, audio and video clip manipulation, and audio streaming. In all but two types of the client APIs, the controller framework is used to obtain and control the relevant controller plugins for performing a multimedia task. The exceptions to this are the audio streaming APIs and the tone generation API, which both avoid the usage of the controller framework and, instead, interface directly with the DevSound layer – a low-level API which is discussed briefly later in this chapter.

7.2.2   The Controller Framework

The controller framework provides the framework for the support of multimedia plugins within the MMF. The controller framework provides all client/server communication between the client API layer of the MMF and the actual MMF controller plugins and is split into the following logical areas.

Controller Plugin Resolver

The controller plugin resolver allows the system to select the best plugin for a particular application. This is exposed as a set of APIs that clients (of which the MMF Client APIs are one) can use to obtain plugins.

Controller Proxy

The controller proxy handles all threading and inter-thread communication across the controller framework. This is a simple layer over standard Symbian OS client/server mechanisms and provides the means by which the client and server sides of the Controller API communicate. This is not usable directly but is mentioned here for completeness.

Controller API layer

The controller API layer consists of matching classes on the client and server sides of the controller proxy. The client-side class, RMMF-Controller, is used by an application (or the MMF) to access the functionality of a controller plugin. The server-side class, CMMFController, is the class that controller plugin writers must derive from in order to implement their plugins – this class also performs all unpacking of messages sent to the controller plugins.

Custom Commands

The controller plugin API is, by design, limited to very basic data flow operations. In order to extend this API, a custom command framework was put in place, which allows clients to access controller-specific extensions. In the general case, this framework is used by the MMF to define APIs for handling common commands (known as the 'Standard Custom Commands') such as setting the volume or retrieving the balance. Manufacturer-specific APIs on a controller plugin can also be exposed using this technique.

7.2.3 Controller Plugins

Controller plugins provide specific multimedia functionality to the MMF. The basic task of a controller is to channel data from one or more sources, transform the data into a different format and then pass the data to one or more sinks. The sources are typically files, microphones or cameras and the sinks are typically files, speakers or screens. A controller plugin will usually support playing or recording of one or more multimedia formats, such as mp3 or avi. It will be able to understand the raw data contained in the relevant sources and sinks – and will read data from a source, apply any required data transformations and then write the data to the sink.

The architecture of a controller plugin can be somewhat complicated and is beyond the scope of this book. However, the minimal implementation of a controller plugin follows the steps outlined below.

- Implement the MMF Controller Plugin API. As described in the Controller Framework overview above, all MMF controller plugins are derived from the `CMMFController` base class. This base class provides the controller API, and also provides functionality such as the instantiation of controller plugins via the ECom plugin framework. The minimal set of functions which need to be overridden are declared as pure virtual functions in this base class and include functions such as `PlayL()` and `StopL()`.

- Any functions required from the Standard Custom Commands set have to be implemented. This is achieved by deriving the controller from an appropriate 'Custom Command Implementor' class and then using the `CMMFController` function, `AddCustomCommandParserL()`, to register the controller as being able to handle that command.

- The MMF provides a set of base classes to aid with the writing of controller plugins. These include data source and sink classes, and buffer classes. Data sources and sinks are themselves ECom plugins, and encapsulate the use of files, descriptors and audio input and output. Buffers are used to contain data as it is transferred from a source to a sink. There are a number of buffer types, all of which are derived from `CMMFBuffer`. The buffer classes differ mainly with regard to how the data is stored, for example whether it is stored in a descriptor or in a kernel-side buffer.

- In order to work with the controller framework plugin resolver, every controller plugin (and indeed every ECom plugin) requires an ECom plugin resource file. The controller framework plugin resolver uses this to identify the multimedia capabilities of the plugin, such as whether it is capable of playing or recording, and what formats are supported.

You can find more information about ECom plugins in the SDK.

An MMF Controller plugin may have a source or sink that is hardware specific. This is often a device driver, or a custom API, for the hardware involved and is implemented by the licensee and/or the hardware manufacturer. The only exception to this in Symbian OS v7.0s is the DevSound API, which has a standard implementation, used by the reference audio controller.

7.2.4   Lower-Level Subsystems

There are a number of subsystems that reside at the bottom of the MMF architecture. A licensee will implement these when a new phone is created and they are generally hardware-specific. What the subsystems contain is largely dependent upon the licensee's requirements and these layers will dictate the majority of the multimedia capabilities of the phone. These lower layers are not generally relevant to a third-party user but, as an example, we briefly describe one of the main subsystems, DevSound.

DevSound

A licensee or hardware manufacturer implements the CMMFDevSound class when Symbian OS is ported to a new platform. Its API provides the lowest level of abstraction for audio playback and recording in the MMF, supplying a hardware-independent layer above the audio device drivers. The DevSound API provides support for playing and recording audio data, playing audio tones, and the modification of audio hardware properties such as volume and balance.

In addition, DevSound contains a way of arbitrating between different clients trying to use the sound hardware simultaneously. This arbitration layer deals with the priorities of sounds (set in the Client APIs) whose individual priority levels are defined by the licensee. A typical usage would be to determine how DevSound should deal with a high-priority client, such as a telephone ring tone, when it tries to interrupt the playback of a low-priority client, such as an MP3 player.

It is also possible to include the use of 'priority preferences', which can be requested in a similar way to the priority and are used to decide what to do with a lower-priority sound if a higher-priority sound is pending. For example, it can choose to stop the sound of lower priority, or mix it with the higher-priority sound.

7.2.5   The Image Conversion Library (ICL)

In previous versions of Symbian OS all image conversion and manipulation had been performed via client APIs interacting with the Media Server. This was no longer possible since the Media Server had been removed, so a new library had to be created to handle this functionality. At around the same time that work was progressing on the MMF, another team was set up to provide a new framework that would provide similar image conversion functionality to that of the Media Server, but in a more lightweight manner. This became known as the image conversion library (ICL).

The ICL provides a lightweight library, based on active objects, to support encoding and decoding image files, and the rotation and scaling of bitmaps. The library is a client-side framework that is, if required, capable of decoding or encoding image files in separate threads. In a similar manner to the MMF, third parties can develop plugins to extend the supported image formats. The ICL utilizes the ECom framework to identify the correct plugins and thus all ICL plugins are also ECom plugins.

The basic architecture comprises a client API layer, the ICL framework, and the ICL plugins themselves. Illustrated in Figure 7.2, it also shows how data is transferred through the ICL.

The bitmap transformation classes are not shown in Figure 7.2 since they are self-contained and do not make use of any ICL plugins.

Client APIs

The client APIs form the top-level abstraction of the ICL. In Symbian OS v7.0s there are client APIs for the decoding and encoding of images, as well as the rotation and scaling of bitmaps. Each of these is described in more detail later in this chapter.

ICL framework

The ICL framework provides a communication layer between the client APIs and the actual ICL plugins, and all plugin control is performed via this layer. The underlying architecture is subdivided into two further layers.

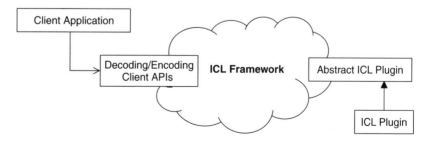

Figure 7.2   ICL architecture

- The **relay** layer provides a connection between the underlying client API classes and the core framework. The relay is mainly used to provide thread encapsulation for the ICL when threaded decoding or encoding is employed. All inter-thread communication is performed by this layer, allowing the Client API classes and the core framework to be ignorant of all threading.

- The **core framework** is essentially a centralized place for storing data and for achieving the functionality associated with the ICL itself. In normal usage, it is the core framework that performs all plugin instantiation and control. This framework communicates with the ICL plugins via the abstract plugin API that all ICL plugins have to implement.

ICL plugins

The ICL plugins provide the actual decoding and encoding functionality to the ICL. The ICL framework provides four abstract classes from which all ICL plugins are derived. These are `CImageDecoderPlugin` and `CImageEncoderPlugin` for decoding and encoding respectively, and corresponding codec classes, `CImageReadCodec` and `CImageWrite-Codec`. The intended split in responsibilities is as follows.

- The decoder and encoder plugin classes are designed to deal with the interface to the ICL framework, the retrieval and writing of image headers, and additional 'non-frame' data, such as text fields.

- The read and write codec classes are designed to deal with the main decoding and encoding stages for individual frames.

Provided the virtual functions of these classes are implemented, there is no reason why a plugin writer could not have more complicated processing chains inside a plugin. For example, a codec class could forward messages to one or more sub-codec classes, to provide specialized processing.

7.2.6   Onboard Camera API (ECam)

The ECam API constitutes the final piece of the multimedia architecture. This is a header file (with appropriate library definition files) shipped by Symbian to define a standard API for the usage of camera hardware. The ECam API allows an application to access and control any camera hardware that is attached to a device. It provides functions to query the status of the camera, adjust camera settings and capture images and video.

The actual implementation of the underlying camera software is performed by the licensee or hardware manufacturer and is entirely optional. If implemented, however, it opens up the possibility for cross-platform camera clients to interface with the camera hardware on any given device.

7.3   Using the MMF

Having introduced the architecture of the MMF, we can now examine how to use its various subsystems. We'll now look at each of the main APIs in turn, providing examples where appropriate, and discussing key concepts as they arise.

The Observer Pattern

Before proceeding, it is worth mentioning a design pattern common across the entire MMF. This is known as the 'observer pattern' and uses observer classes to provide feedback to the API user. These classes are implemented as 'mixin' classes and define a set of pure virtual observer functions. This pattern is not restricted to the MMF, but is widely used in Symbian OS. In Chapter 4, for example, we saw the use of observer classes to report significant events to a control.

In order to use an MMF client API, you need to create an object derived from the relevant observer class, supply an implementation of its virtual functions, and pass a reference to this object to the MMF client utility. As is shown in the following examples, it is common to derive the MMF client itself from the observer class.

In the MMF, observer functions typically provide notification that a utility has been created, or that one of a range of specific actions has begun or ended, and supply error codes and state information. It is worth noting that the observer pattern relies on the usage of active objects and that an active-scheduler must be running – as it is in a standard GUI application – in order for the client APIs to work correctly.

7.4   Using Audio

The MMF Audio APIs can be used to perform simple tone generation, audio clip manipulation (playing, recording, and converting) and audio streaming (playing and recording).

When using audio in software for the first time it can be hard to understand what is actually happening and how the audio is stored. Some basic concepts are shown in Figure 7.3.

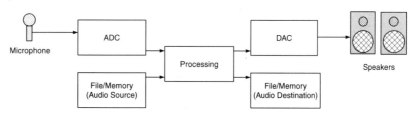

Figure 7.3   Analog to digital and digital to analog conversion

Analog to Digital Conversion

The first step in transferring audio to a digital form is generally analog to digital conversion. This is performed by microchips known as 'Analog to Digital Converters' or ADCs. For example, in the case of a microphone recording, the microphone converts the sound waves to an electrical signal, and the ADC stage then converts this signal into binary data. The binary data can be processed and stored in memory or saved to a file.

Digital to Analog Conversion

The reverse process is known as digital to analog conversion. In this step, a binary sequence of audio data is converted by microchips known as 'Digital to Analog Converters' or DACs. These chips output an analog electrical signal, which can then be turned into sound waves by hardware such as speakers.

Digital Audio Formats

All digital audio is stored in a particular format, which describes a number of details about the audio. This information includes the sample rate at which it was recorded (the number of samples taken per second by the ADC), the number of channels, and whether the audio has been encoded into a state that would require a separate decoding stage. Uncompressed audio is known as Pulse Code Modulation (PCM) audio, and each audio sample is directly measurable as the amplitude of a given audio signal at a given time. For example, a digital PCM sample of a sine wave at 50 Hz might appear as shown in Figure 7.4.

For the purpose of this diagram, a 16-bit PCM encoding format has been used. This means that the digital audio sample has to be stored in 16 bits, with values ranging from −32 768 to +32 767. These values correspond to the negative and positive peaks of the sine wave. Each point on the graph corresponds to a sample that would need to be

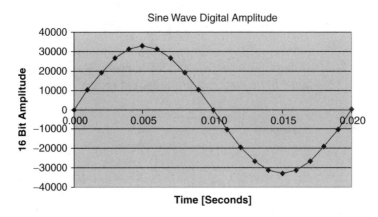

Figure 7.4   A digitally sampled sine wave

stored. In PCM audio, these samples would be stored without any further processing, and hence each sample would occupy 16 bits, or two bytes. How the audio data is stored in other formats is dependent upon the additional processing that is applied to the signal. For example, the audio data for a compressed signal may include the coefficients of the transformation that has been applied.

Audio Clips

A clip can be considered as a 'package' of audio or video. It has a clearly defined beginning and end, and is usually packaged with information describing its format, such as the sample rate and the encoding method. Clips are generally contained either in files or in descriptors, although some of the client APIs also allow for the possibility of treating a URL as a clip.

Audio Streams

An audio stream consists of a continuous flow of audio data, with no clearly defined beginning or end. The flow of audio data can be intermittent, and in chunks of variable size. An example of a stream is Real Audio being streamed from the Internet.

7.4.1 Playing Tones

The tone player utility class, CMdaAudioToneUtility, is defined in mdaaudiotoneplayer.h. It provides the ability to play individual monophonic sinewave tones, user-defined dual-tones (not available on the Nokia 6600 phone), dual-tone multi-frequency (DTMF) tones, or a tone sequence. In a typical application, DTMF tones are used during a phone call to simulate phone key presses, and tone sequence files are used to produce ring tones.

A client that wishes to utilize the tone player utility must first derive an observer class from the MMdaAudioToneObserver, whose class definition is shown below:

```
class MMdaAudioToneObserver
  {
public:
  virtual void MatoPrepareComplete(TInt aError) = 0;
  virtual void MatoPlayComplete(TInt aError) = 0;
  };
```

Create the tone player utility by calling its NewL() function, supplying a reference to the observer class as a parameter. Playing a tone then requires the following two stages.

1. Prepare the tone by calling the appropriate 'Prepare' function. The client is informed that the preparation is complete via the observer's

`MatoPrepareComplete()` function. If the preparation completed successfully, this function is called with an error value of `KErrNone`, otherwise it will report the relevant error code, such as `KErrNot-Supported` if tone playback is not supported on this device.

2. If preparation was successful, the client can start playing the tone, or perform other processing on the tone, such as setting the volume. If an error occurred then these functions must not be called since the underlying subsystems will not have been correctly initialized.

The client is informed when the tone has completed playing by means of a call to the observer's `MatoPlayComplete()` function.

This set of events is common to each of the three tone-playing APIs. The only difference is in the data that is passed as a parameter to the relevant 'prepare' function.

The `audio4` example application illustrates how to use the tone-playing utility, and plays a tone at 2600 Hz for three seconds.

Following the observer pattern described earlier, the application's engine class implements the audio tone observer's functions:

```
class CAudio4Engine : public CBase, public MMdaAudioToneObserver
  {
  ...
public: // from MMdaAudioToneObserver
  void MatoPrepareComplete(TInt aError);
  void MatoPlayComplete(TInt aError);
  ...
  };
```

The application engine's `ConstructL()` is straightforward:

```
void CAudio4Engine::ConstructL()
  {
  iUtility=CMdaAudioToneUtility::NewL(*this);
  iUtility->
  PrepareToPlayTone(KFrequency,TTimeIntervalMicroSeconds(KDuration));
  }
```

It first creates an instance of `CMdaAudioToneUtility` by calling its `NewL()` function, passing a reference to itself as the observer. The tone is then prepared by calling the utility's `PrepareToPlayTone()` function, passing the tone's frequency and duration in hertz and microseconds respectively. In this example, the values `KFrequency` and `KDuration` are `2600` and `3000000`. You should note that, at any given time, only one prepared tone can exist. If you call `PrepareToPlayTone()` a second time, the new tone will replace the previous one.

When the tone utility has prepared the tone, it calls `MatoPrepare-Complete()`:

```
void CAudio4Engine::MatoPrepareComplete(TInt aError)
  {
  if (aError==KErrNone)
    {
    iUtility->SetVolume(iUtility->MaxVolume());
    iState=EReady;
    }
  }
```

If the preparation was successful, the error code passed to this function will be KErrNone. Any other value indicates an error – for example, if the sound device is in already in use by a process with a higher priority, the value will be KErrInUse.

It is important to note that the time taken for the tone utility to prepare the tone and call MatoPrepareComplete() is undefined. In some devices this could happen instantaneously, whereas in others the delay could be quite noticeable. Therefore you should not use the single tone playing API to play a tune that consists of a sequence of tones. In this case you should instead use the tone sequence API.

Now that the tone is prepared the client can play it by calling Play():

```
void CAudio4Engine::PlayL()
  {
  iUtility->Play();
  iState=EPlaying;
  }
```

When the tone has completed playing, the MatoPlayComplete() function will be called:

```
void CAudio4Engine::MatoPlayComplete(TInt /*aError*/)
  {
  iState=EReady;
  }
```

A successful playback will result in a KErrNone error code being passed to this function. In this example the error code is ignored, but a real application would take note of an error, perhaps by attempting to start playing the tone again.

You can repeat the playing of the tone by making multiple calls to the Play() function without having to call PrepareToPlayTone() again. However, if you want to play a different type of tone, then you must make another call to PrepareToPlayTone().

It is interesting to note that, unlike many of the other utilities that we shall see later, the tone utility has no Stop() function. Instead, you should use the CancelPlay() function to stop a tone which has started playing:

```
void CAudio4Engine::StopL()
  {
  iUtility->CancelPlay();
  ...
  }
```

Canceling the tone before it has finished playing prevents the observer's `MatoPlayComplete()` function being called.

When you have finished using the tone utility, all necessary audio device cleanup is performed by simply destroying the tone generator object.

This example illustrates how to play single sinewave tones, but the same principles apply for playing dual tones, DTMF tones and tone sequences. There is a corresponding 'prepare' function for each type of tone and, once prepared, the tones can be played using `Play()` and stopped using `CancelPlay()`. The same observer functions that are described above are called in each case.

As a guide, the 'prepare' functions for the other tone types take the following information:

- dual tone – the frequencies of the two tones, with a single duration

- DTMF – a reference to a descriptor that contains a list of DTMF tones

- sequence from file – a reference to the filename of the tone sequence file

- sequence from descriptor – a reference to the descriptor containing a tone sequence

- fixed sequence – an integer that references a hard-coded predefined tone sequence.

Symbian OS does not specify the formats used to define tone sequences, so the formats will, in general, differ from phone to phone. A DTMF tone sequence is generally specified by a text string that contains characters from the set `0123456789#*`, but this format could also be device-specific. Refer to a phone-specific SDK for details of the precise formats that are used in a particular phone.

7.4.2 Playing Clips

The audio player utility `CMdaAudioPlayerUtility` is defined in `mdaaudiosampleplayer.h` and provides the ability to play sampled audio data, which can be supplied in a file (e.g. a WAV file) or a descriptor (`TDes8` or `TDesC8`) or can be located at a specified URL address.

The audio player utility supplies standard functionality to play and stop playing the audio data, to set the volume and balance, and so on.

It also provides a number of more advanced operations, such as volume ramping, the setting of repeats, metadata manipulation and the passing of custom commands to the current MMF controller.

Metadata is non-audio data contained within the audio clip. For example, a WAV file can contain information about when the file was created, and by whom.

When deciding which utility to use to play a clip, you can choose either the audio player or the audio recorder – described in the following section – which contains the functionality for both the playing *and* recording of clips. In general, you would use the audio recorder utility, which provides the most flexible option, but using the audio player utility is the simplest way to play a clip.

A client that wishes to utilize the audio player utility must first derive an observer class from the `MMdaAudioPlayerCallback` class, whose definition is:

```
class MMdaAudioPlayerCallback
  {
public:
  virtual void MapcInitComplete(TInt aError,
        const TTimeIntervalMicroSeconds& aDuration) = 0;
  virtual void MapcPlayComplete(TInt aError) = 0;
  };
```

The standard way to create the audio player utility is to use its `NewL()` function which, again, takes a reference to the observer class. As with the tone-playing utility, playing the clip then follows two discrete stages.

1. The clip must be opened using the relevant 'open' function from those described below. Completion of this stage is indicated by a call to the observer's `MapcInitComplete()` function. If the clip has been opened successfully, this function is called with an error code of `KErrNone`, but if the clip format is not recognized, the error code will be `KErrNotSupported`.

2. If the clip was opened successfully, the client is free to either start playing the clip or to manipulate the settings, such as the volume. When the clip has completed playing, the observer's `MapcPlay-Complete()` function will be called.

This set of events is common to playing audio from each of the three possible sources, the only difference being the data that is passed to the relevant 'open' function – a file name for `OpenFileL()`, a descriptor for `OpenDesL()` and a URL for `OpenUrlL()`.

Using the OpenUrlL() function, you can also specify an Internet access point ID. This instructs the controller to use the specified access point instead of its default. For more information on this type of parameter, refer to the CMMFUrlSink class in mmfurl.h.

If you know that the audio utility is only going to be used to process one specific type of clip, then it is possible to create an instance by using one of the NewDesPlayerL(), NewDesPlayerReadOnlyL(), or NewFilePlayerL() functions instead of calling NewL(). You additionally supply these functions with a descriptor containing either the audio data itself, or the filename of the audio clip. They encapsulate both the instantiation and opening phase of the player utility so that, upon completion, they will have attempted to open the supplied clip and will have called the observer's MapcInitComplete() function with the relevant error code.

The audio1 example illustrates how to open a WAV file and then play it. The application engine's SetFileL() function creates an instance of the audio player utility class, using the utility's NewFilePlayerL() function, as it is created with the explicit intention of playing a single file:

```
void CAudio1Engine::SetFileL(const TDesC& aFileName)
  {
  if (iUtility)
    {
    delete iUtility;
    }
  iUtility=CMdaAudioPlayerUtility::NewFilePlayerL(aFileName, *this);
  }
```

The engine itself is derived from the observer class. NewFilePlayerL() not only creates the utility but also attempts to open the clip, resulting in a call to the observer's MapcInitComplete() function, passing an error code and the audio clip's duration.

```
void CAudio1Engine::MapcInitComplete(TInt aError,
            const TTimeIntervalMicroSeconds& /*aDuration*/)
  {
  if (aError==KErrNone)
    {
    iState=EReady;
    iAppUi.UpdateViewL();
    }
  }
```

In this example the duration is ignored but in a real application it could be used, for example, to display the clip's remaining playing time.

It is important that you should not call functions such as Play(), SetVolume() or SetRepeats() until MapcInitComplete() has been called with an error code of KErrNone, otherwise a panic will occur. This kind of error is often the most common mistake made when using these functions. A useful way of ensuring that this does not happen is to call `Play()` directly from the `MacpcInitComplete()` function.

If the clip has been opened successfully you can play it using the `Play()` function.

```
void CAudio1Engine::PlayL()
    {
    iUtility->Play();
    iState=EPlaying;
    }
```

When the clip has completed playing, the `MapcPlayComplete()` function will be called.

```
void CAudio1Engine::MapcPlayComplete(TInt /*aError*/)
    {
    iState=EReady;
    }
```

A successful playback will result in the function being called with an error code of `KErrNone`. As in the previous example, the error code is ignored. A more realistic application might, for example, respond to an error such as `KErrInUse` by making another attempt to play the clip.

After completion, you can replay the clip by calling `Play()` again, without having to call an 'open' function. Before playing a new clip you must close the current clip by calling the utility's `Close()` function and use the relevant 'open' function to open the new one. If, as in this example, you created the utility to play a specific type of clip then you will have to create a new instance in order to play a different type of data.

Before playing the clip you can set a 'play window' by means of the `SetPlayWindow()` function, supplying the start and finish of the window in microseconds. Once you have done this, playing is restricted to the portion of the clip within the window. By default, the play window is the whole length of the clip.

While the clip is playing, you can stop it by using the utility's `Stop()` and `Pause()` functions. The `Stop()` function will reset the clip position to the start of the current play window, not to the start of the whole clip.

7.4.3 Recording Clips

The audio recorder utility `CMdaAudioRecorderUtility`, defined in `mdaaudiosampleeditor.h`, is derived from `CMdaAudioClipUtility`. It contains a superset of the capabilities provided by the `CMdaAudioPlayerUtility`, adding the ability to record a clip as well as to play it. As with the player utility, the audio data can be processed in a file (e.g. a WAV file), a descriptor (`TDes8` or `TDesC8`) or at a remote URL.

As mentioned earlier, this is the class that is generally chosen when playing or recording clips.

In addition to supplying the basic functionality of playing, recording, stopping, setting the volume, setting the balance, and so on, it also allows you to crop data from the current clip, from the current read/write position to either the beginning or the end of the data. As with the player utility, it provides a number of more advanced operations, such as volume ramping, setting repeats, metadata manipulation and the passing of custom commands to the current MMF controller.

With one exception (explained later) the playing functionality of the audio recorder utility is identical to that of the player utility, so here we'll concentrate on the recording functionality.

As with the other utilities, a client that wishes to utilize the audio recorder utility must first derive an observer class from the `MMdaObjectStateChangeObserver`, whose class definition is:

```
class MMdaObjectStateChangeObserver
  {
public:
  virtual void MoscoStateChangeEvent(CBase* aObject,
                    TInt aPreviousState,
                    TInt aCurrentState,
                    TInt aErrorCode)=0;
  };
```

The usage of this observer class is slightly more involved than with those of the player and tone utilities. `MoscoStateChangeEvent()` is called whenever the state of the recorder utility object changes, passing a pointer to the utility object itself, and the previous and current states, as well as an error code. Since an object pointer is supplied, one observer can observe more than one utility object.

The possible states of operation for the recorder utility are enumerated, in the `CMdaAudioClipUtility` parent class, as:

```
enum TState
  {
  ENotReady = 0,
  EOpen,
  EPlaying,
  ERecording
  };
```

The audio recorder utility is created using its `NewL()` function which, as usual, takes a reference to the observer as a parameter. The recording of audio then follows a similar set of steps to that of playing using the player utility.

1. The clip must be opened using the relevant 'open' function. On completion, the observer's `MoscoStateChangeEvent()` function

will be called with a current state of EOpen and an error code which, if the clip has been opened successfully, will be KErrNone.

2. If the opening was successful, the client is free either to start recording to the clip or to manipulate the settings. As with the player utility, you should not call these functions if an error state has occurred. When recording to the clip has started, the observer's MoscoStateChangeEvent() function will be called with a current state of ERecording. When recording of the clip is completed, the observer's MoscoStateChangeEvent() function is called again. The error code will be KErrNone if recording was terminated by a call to Stop(), KErrEof if the clip has recorded to a user-defined length limit, or another error code if an error condition has occurred.

As with the player utility, this set of events is common to each of the three audio recording clip types and, again, the only difference is in the data that is passed to the appropriate 'open' function. Compared with the clip player utility, there is a wider variety of 'open' functions. Some of these take a TMdaClipFormat parameter (this data structure is discussed briefly in the later section on audio conversion) and some allow the caller to select a specific controller plugin, rather than allowing the MMF architecture to choose one that it considers suitable. Finally, the recorder utility provides some 'open' functions that take legacy Media Server arguments. In general, if a destination file that does not yet exist is supplied as a parameter to an 'open' function, the file will be created, using the specified file extension as a clue to the format to use.

The audio2 example application illustrates the use of both the recorder and convert utilities, and makes direct use of the controller framework, as described in Section 7.6. This section looks at its use of the recorder utility to open a WAV file and record to it.

In the audio2 example's engine, the ConstructL() function creates an instance of the audio recorder utility class and, in a pattern that should by now be becoming familiar, indicates that the engine itself is the utility's observer:

```
void CAudio2Engine::ConstructL()
  {
  delete iUtility;
  iUtility=NULL;
  iUtility = CMdaAudioRecorderUtility::NewL(*this);
  iUtility->OpenFileL(KFileName);
  }
```

After creating the utility, ConstructL() opens a file for recording, which results in the observer's MoscoStateChangeEvent() function being called with a current state of EOpen and, if all went well, an error code of KErrNone.

```
void CAudio2Engine::MoscoStateChangeEvent(CBase* aObject,
                    TInt aPreviousState,
                    TInt aCurrentState,
                    TInt aErrorCode)
  {
  if (aErrorCode!=KErrNone)
    {//message
    iState=ENotReady;
    return;
    }

  if (aObject == iUtility)
    {
    switch (aCurrentState)
      {
    case CMdaAudioClipUtility::EOpen:
      iState=EReady;
      break;
    case CMdaAudioClipUtility::ERecording:
      iState=ERecording;
      break;
    case CMdaAudioClipUtility::EPlaying:
      iState=EPlaying;
      break;
    default:;
      }
    }
  else //must be converter
    {
    ...
    }
  }
```

In this example, the states reported by MoscoStateChangeEvent()
are used to maintain an internal state machine. Any error sets the internal
state to ENotReady, which is used to signify that the utility classes can
not be used.

Before starting to record, it is possible to set up one or more recording
parameters, using functions such as:

- SetDestinationDataTypeL()

- SetDestinationFormat()

- SetDestinationBitRate()

- SetDestinationSampleRateL()

- SetDestinationNumberOfChannels()

- SetMaxWriteLength().

The destination data type is supplied as a TFourCC, which specifies the
correct encoding method to use from those defined in MmfFourCC.h.

The destination format is specified as one of the UIDs defined in mmfFor-matImplementationUIDs.hrh, and determines the file's data header and structure type. Interestingly, the maximum write length is set in kilobytes and not in microseconds as might be expected. The parameters for the remaining functions are generally integers and are self-explanatory.

You can query the utility to identify suitable parameters for virtually all of the functions listed above, using the utility's various GetSupportedXxxL() functions. The only exception to this is for the setting of the maximum write length. However, there is a RecordTimeAvailable() function, which returns the maximum time in microseconds that is available for recording to the current clip.

If the clip has been opened successfully, you can record to it using the Record() function, which again results in a call to the observer's MoscoStateChangeEvent() function, with a state of ERecording.

```
void CAudio2Engine::StartRecordingL()
  {
  iUtility->SetAudioDeviceMode(CMdaAudioRecorderUtility::EDefault);
  iUtility->RecordL();
  }
```

Before calling Record(), the example application calls SetAudioDeviceMode(), which is used to specify the device(s) from which the sound will be recorded and/or to which it will be played. On a typical phone you can record from either the phone's microphone or any current telephony call, and you can play the sound via the phone's earpiece or its main loudspeaker. The example code specifies the default source, which will generally record from a telephony call, if there is one in progress, or from the phone's microphone if there is not. The available modes are defined in MdaAudioSampleEditor.h and are discussed in the Symbian OS SDK documentation.

Assuming that recording was successfully initiated, you can stop recording by calling the utility's Stop() function:

```
void CAudio2Engine::StopL()
  {
  if (iState==ERecording)
    {
    iUtility->Stop();
    iRecorded=ETrue;
    }
  ...
  iState=EReady;
  }
```

The observer's MoscoStateChangeEvent() function will be called again, with a current state of ERecording and an error code signifying success or failure.

As with the playing utility it is possible to pre-empt the recording with a higher-priority audio client.

When you stop recording, the write position is held at the current recording location, rather than being reset to the beginning of the file. Hence, if you call `RecordL()` again, the new data will be appended to the current clip and will not replace it. If necessary, you can set the read/write position by calling the utility's `SetPosition()` function. If you don't call the `Stop()` function, recording will continue until the end of the file or descriptor – or a user-defined limit – is reached, or an error occurs.

There is a slight difference between using the audio player utility and using the audio recorder utility to play a clip. This is due to the recorder utility's observer class being state-based, in contrast to the simple two-function class used by the player utility. When initiating playback using the recorder utility, the client will receive two observer function calls – at the start and end of playback – as opposed to the single call at the end of playback that is provided by the player utility.

If you are intending to write applications for a range of Symbian OS phones, you need to be aware that (at time of writing) there is a source compatibility break in the audio recorder utility between Symbian OS v7.0s and earlier versions. In previous revisions, using the `Stop()` function to stop clip playback through the recorder utility would keep the play position at the current read position, but in Symbian OS v7.0s the play position is reset to the start of the current play window (or the beginning of the file if no play window has been set). This difference may be resolved in a later release of Symbian OS v7.0s. Make sure you consult the documentation supplied with each version of Symbian OS to determine the current behavior.

7.4.4 Format Conversion

The audio convert utility class, `CMdaAudioConvertUtility`, is derived from `CMdaAudioClipUtility` and defined in `mdaaudiosample-editor.h`. It provides the ability to convert an audio clip from one format to another.

As with the record utility, the convert utility can read and write either files or descriptors, and you can either leave the utility to select an appropriate controller or make an explicit selection yourself.

Although the convert utility does not allow for the playing or recording of data, its syntactical operation closely follows that of the record utility. Like the recorder utility, it allows you to crop an audio clip between the current read/write position and either the beginning or end of the data. In addition, you can select the destination settings, discover the maximum record time available for the converted clip and set a maximum size for the converted data.

Just as with the recorder utility, you need to derive an observer class from the `MMdaObjectStateChangeObserver` interface class, and create an instance of the audio convert utility by means of its `NewL()` function. The conversion of an audio clip then follows the steps below.

1. Open a clip using the relevant 'open' function, which results in a call to the observer's `MoscoStateChangeEvent()` function with a current state of `EOpen`.

2. If the clip was opened successfully, you can then use the utility's `GetSupportedXxxL()` functions to determine the available target data settings and call the corresponding `SetXxxL()` functions to set up the required format for the destination data. These are directly equivalent to the similar recorder utility functions discussed in the previous section. As with the other utilities, if an error state has occurred, you should not call any of these functions.

3. Start the conversion by using `ConvertL()`, which results in a call to the observer's `MoscoStateChangeEvent()` function with a current state of `EPlaying`. Possible errors include `KErrNotSupported`, which may occur if you do not check that the utility supports the requested output format.

4. On completion of the conversion, the observer's `MoscoStateChangeEvent()` function is called again. The error code will be `KErrNone` if conversion was terminated by a call to `Stop()`, `KErrEof` if the conversion reached a user-defined limit to the length of the converted clip, or some other value if an error condition has occurred.

As mentioned earlier, a number of the 'open' functions allow the selection of a specific controller for both the input and output stages. If an 'open' function is supplied with a file destination that doesn't already exist, the file will be created by the MMF, using the specified file extension as a clue as to which format to use.

A number of the 'open' functions take `TMdaClipLocation` parameters. This is an abstract base class and, at time of writing, there are two derived classes that you can use:

*   `TMdaDesClipLocation` – contains parameters specifying a descriptor location

*   `TMdaFileClipLocation` – contains parameters specifying a file location.

Some of the 'open' functions also take `TMdaClipFormat` and `TMdaPackage` parameters, which respectively specify the format and codec to use for the conversion (at both the source and destination stages).

As mentioned in the previous section, the `audio2` example application uses the convert utility to convert from the recorded WAV file into any format supported by the MMF. The target formats are enumerated using the controller framework functions discussed in Section 7.6.

Following the standard pattern, the application engine is also the utility's observer. Its `OpenConverterL()` function creates an instance of the audio convert utility class by calling `NewL()`:

```
_LIT(KFileName,"rec.wav");

void CAudio2Engine::OpenConverterL(TPtrC aExt)
  {
  ...
  iConverter=CMdaAudioConvertUtility::NewL(*this);
  _LIT(KConvertName,"conv");

  TBuf<32> fileName;
  fileName=KConvertName;
  fileName.Append(aExt);

  RFs session;
  session.Connect();
  session.Delete(fileName);
  session.Close();

  iConverter->OpenL(KFileName, fileName);
  }
```

The function then constructs a name for the conversion file from the hard-coded text `"conv"` and the caller-supplied file extension. A filing system session is opened in order to delete any previous file of this name – a real application would inform the user that there was a pre-existing file. Finally, the convert utility is opened, specifying the source filename, `KFileName`, and the newly constructed destination filename.

If the converter is opened successfully, the observer's `MoscoState-ChangeEvent()` is called, and the relevant section of this function's code is:

```
void CAudio2Engine::MoscoStateChangeEvent(CBase* aObject,
                    TInt aPreviousState,
                    TInt aCurrentState,
                    TInt aErrorCode)
  {
  ...
  else //must be converter
    {
    switch (aCurrentState)
      {
      case CMdaAudioClipUtility::EOpen:
        {
        if (iState==EGettingConversionFormats)
          {
```

```
        GetFormats2L();
        iState=EReady;
        break;
        }
    else
        {
        TState state=iState;
        iState=EReady;
        if (state!=EConverting)
            {
            Convert2L();
            }
        break;
        }
    }
case CMdaAudioClipUtility::ERecording:
    iState=EConverting;
    break;
case CMdaAudioClipUtility::EPlaying:
    iState=EConverting;
    break;
default:;
    }
    }
}
```

The function is called with a current state of EOpen and an error code of KErrNone, and calls GetFormats2L() to fetch a list of available destination types.

```
void CAudio2Engine::GetFormats2L()
    {
    //ask the destination which types it supports
    TRAPD(err,iConverter->
    GetSupportedDestinationDataTypesL(*iDataTypes));
    ...
    TRAP(err,iConverter->
    GetSupportedConversionSampleRatesL(*iSampleRates));
    ...
    TRAP(err,iConverter->
    GetSupportedConversionNumberOfChannelsL(*iChannels));
    ...
    iConversionTypes=ETrue;
    }
```

This function fills three arrays with the supported destination data types, sample rates and channels respectively. In the application, the user can then pick the required destination parameters. Once the parameters have been selected, the clip is closed and reopened, causing the observer's MoscoStateChangeEvent() to be called again but this time it runs the Convert2L() function:

```
void CAudio2Engine::Convert2L()
    {
```

```
if (iCodecIndex!=-1)//selections were made
   {
   TFourCC cc=(*iDataTypes)(iCodecIndex);
   iConverter->SetDestinationDataTypeL(cc);
   TInt s=(*iSampleRates)(iSampleRatesIndex);
   iConverter->SetDestinationSampleRateL(s);
   TInt c=(*iChannels)(iChannelsIndex);
   iConverter->SetDestinationNumberOfChannelsL(c);
   }
iConverter->ConvertL();
}
```

This function sets the conversion parameters and starts the conversion
which, in this example, will continue until the file has been completely
processed. As with the recorder utility, it would also be possible to set
a maximum write length of the destination file, using `SetMaxWrite-
Length()`, or to use `Stop()` to halt the conversion. The observer's
`MoscoStateChangeEvent()` function is called, with a current state of
`ERecording`, at the start and at the end of the conversion.

7.4.5 Audio Streaming

The audio streaming utility classes allow you to stream audio from an
input or to an output at a low latency, avoiding the overhead of the MMF
controller framework.

An audio stream is a continuous flow of audio data, which has no
clearly defined beginning or end. A streaming client is able to start
relaying small chunks or packets of data into an audio stream and the
audio will be processed as and when it is received, instead of being
processed only when all the data has been received.

Suppose that there is an audio file – located on a remote server on the
Internet – that a user wants to listen to. Suppose also that the file size is
3 MB, that it has been recorded using a low bit-rate of 12 kbps and that
the user's device has a download speed of 28.8 kbps.

It would be possible to fully download this file and then use the
clip playing APIs for playback. However, at the user's download speed of
28.8 kbps (3.6 kBps) it would take approximately 15 minutes to download.
A much better approach would be to download the data in small chunks
and play this back as the data is received, in real time. If the user
subsequently decided that they did not want to download the full file
before it had finished playing, it would be possible to simply abort the
download. This, in simple terms, is the kind of functionality that the
streaming clients provide.

Symbian OS supplies two audio streaming APIs, one for streaming
to an output sound device, and another for streaming from an input
sound device. These APIs are virtually identical, except that they process
data in opposite directions. In consequence, this section gives a detailed
description of the output streaming API (which is the one that is more

commonly used) and then just explains the differences in processing when using the input streaming API.

7.4.5.1 Double-Buffering

One final point to discuss before examining the streaming APIs themselves is the concept of 'double-buffering'. This is a methodology that is very useful when dealing with the real-time processing of data, and is widely used across the audio processing code in Symbian OS (as well as in other areas, such as graphics). The basic strategy with regard to audio is as follows.

1. Prepare a buffer of audio data for an audio stream.

2. Pass this buffer to the audio stream and immediately start preparing another buffer.

3. When the audio stream requires more data, pass the second buffer to the stream, whilst refilling the first.

Thus, as soon as more data is required, it is made available to the stream – without having to do any extra processing. This approach is generally good practice when streaming data, and can be extended to multiple buffers if required – for example, audio streaming using eight buffers simultaneously is not uncommon. As we shall see, the Symbian OS audio streaming APIs make this buffering scheme quite easy to use.

7.4.5.2 Output Streaming

The audio output stream `CMdaAudioOutputStream` class is defined in `mdaaudiooutputstream.h` and provides the ability to stream audio to an output device. The audio data must be in the PCM16 format, although the sample rate and number of channels can be specified. As with the clip-based utilities, you can adjust the volume and balance of the audio output.

The observer class for the audio output stream utility is `MMdaAudio-OutputStreamCallback`, whose definition is:

```
class MMdaAudioOutputStreamCallback
  {
public:
  virtual void MaoscOpenComplete(TInt aError) = 0;
  virtual void MaoscBufferCopied(TInt aError,
             const TDesC8& aBuffer) = 0;
  virtual void MaoscPlayComplete(TInt aError) = 0;
  };
```

As usual, you create an instance of the audio output stream utility by calling its `NewL()` function, supplying a reference to its observer. The streaming of audio through this utility then follows the steps below.

1. First open the audio stream by means of the utility's `Open()` function. The success or failure is reported by the error code passed to the observer's `MaoscOpenComplete()` function.

2. If the stream was opened successfully, you can pass the first audio buffer (stored in a `TDesC8` descriptor) to the streaming utility, using its `WriteL()` function. You must not modify the contents of this buffer until you are informed – as described later – that the buffer's content has been successfully copied.

3. The recommended way of using the utility is then to pass one or more additional buffers of audio data to the streaming utility, again using the `WriteL()` function, up to a user-defined limit (for example, eight buffers). The streaming client utility stores these buffers in a FIFO queue and passes them to the sound device as soon as they are required, with very low latency. This effectively creates a pre-buffer, which can provide some protection against data underflows.

4. When a buffer has been used by the sound device, a reference to the used audio buffer is passed, together with an error code, to the client application in a call to the observer's `MaoscBuffer-Copied()` function. If the buffer has been copied successfully, the error code will be `KErrNone`. If you call the utility's `Stop()` function before all the buffers have been copied to the audio stream, then `MaoscBufferCopied()` will be called, with an error code of `KErrAbort`, for each unused buffer.

5. When you stop supplying data to the streaming utility, and once the queued buffers have been used up, the sound device will underflow. The client application is notified by means of a call to the observer's `MaoscPlayComplete()` function with an error code of `KErrUnderflow`. If, at any time, the client streaming utility runs out of data to supply to the sound device, then this will generate a call with the same error. This is usually caused by the client application not filling and supplying buffers to the `WriteL()` function quickly enough.

At any time during this process, you can call the utility's `GetBytes()` function to determine the total number of bytes processed by the audio stream. It is important to note that this does not necessarily mean that this many bytes have actually been played – the sound driver itself will usually maintain a buffering scheme, so the value may also include the as yet unplayed content of any internal buffers.

The audio output stream's `Open()` function takes a pointer to a `TMda-Package` as a parameter, to set the sample rate and number of channels.

You can, if you wish, pass a NULL value, and set the sample rate and number of channels at a later time, by means of the utility's SetAudio-Properties() function. If you choose to set this data via the Open() function, you should supply it with a pointer to a TMdaAudioDataSettings class. This class is derived from the TMdaPackage abstract base class, and is defined in mda\common\audio.h.

The nature of an application has a strong effect on the decision of how many buffers to use. A program that requires tight synchronization with the sound being played will need to use a small number of buffers, to ensure the delay between writing the audio data and playing it remains short. In contrast, an application that simply plays an audio file from a remote site will be able to use a large number of buffers. In this case, the priority is to avoid data underflow; the delay between writing data to a buffer and playing the buffer's content is not a significant issue.

The audio3 example illustrates two different ways of using the audio output stream utility. The first is intended to demonstrate the concept of streaming and uses a single buffer, which is refilled each time MaoscBufferCopied() is called. In a real application you should not rely on this call to supply a new buffer, but should supply buffered data as soon as it is available.

The second part of the example illustrates a more realistic use of audio streaming. It simulates a real application in a situation where packets of audio data are arriving from the network at random intervals and are buffered as and when they arrive.

The single-buffer case is based on the example's CSynchronousExampleStream class, whose ConstructL() is responsible for creating and opening an instance of the audio output stream utility:

```
void CSynchronousExampleStream::ConstructL(CAudio3Engine* aEngine)
  {
  iEngine=aEngine;
  iStream=CMdaAudioOutputStream::NewL(*this);
  iStream->Open(NULL);
  }
```

In opening the stream, we use the option mentioned earlier, of passing a NULL parameter to the Open() function and setting the audio properties at a later time. We actually set them in the observer's MaoscOpenComplete() function, which is called on completion of the opening of the audio stream:

```
void CSynchronousExampleStream::MaoscOpenComplete(TInt aError)
  {
  if (aError==KErrNone)
    {
    iStream->SetAudioPropertiesL(
      TMdaAudioDataSettings::ESampleRate8000Hz,
```

```
    TMdaAudioDataSettings::EChannelsMono);
  iStream->SetVolume(iStream->MaxVolume());
  iEngine->SetState(CAudio3Engine::EReadySynchronous);
  }
}
```

In addition to setting the audio properties, using enumerated values from the TMdaAudioDataSettings class mentioned above, MaoscOpen-Complete() also sets the volume.

In this example, the MaoscOpenComplete() function does nothing if the opening of the audio stream is unsuccessful. In such a case, and depending on the nature of the error, a real application might try to open the stream again.

When the application starts playing, it first opens the source file and then calls the NextBuffer() function:

```
void CSynchronousExampleStream::NextBuffer()
  {
  if (iFileError==KErrNone)
    {
    iFileError=FillBuffer();
    TRAPD(err,iStream->WriteL(iBuffer));
    if (err!=KErrNone)
      {
      Stop();
      }
    }
  }
```

NextBuffer() calls FillBuffer(), to fill the audio buffer, and then passes the buffer to the output stream, using its WriteL() function. If this function returns an error, the stream is immediately stopped. Otherwise, depending upon how many layers of buffering are present in the sound driver itself, you could start hearing the sound immediately.

When the output stream has finished with the audio buffer it calls the observer's MaoscBufferCopied() function. Since there is only one buffer being used, the buffer reference passed to this function is ignored.

```
void CSynchronousExampleStream::MaoscBufferCopied(TInt aError,
                      const TDesC8& /*aBuffer*/)
  {
  if (aError==KErrNone || aError==KErrAbort) //user selected "Stop"
    {
    NextBuffer();
    }
  else
    {
    User::Invariant();
    }
  }
```

The error code will be KErrNone if all the data in the buffer has been played successfully, or KErrAbort if the utility's Stop() function was called. In either case, the buffer is refilled by means of a call to NextBuffer().

When playing of the stream stops, the utility calls the observer's MaoscPlayComplete() function:

```
void CSynchronousExampleStream::MaoscPlayComplete(TInt aError)
  {
  if (aError==KErrNone || aError==KErrCancel) //user selected "Stop"
    {
    iEngine->SetState(CAudio3Engine::EReadySynchronous);
    PlayEnded(aError);
    }
  else
    {
    User::Invariant();
    }
  }
```

The error code will be KErrCancel if Stop() has been called, KErrUnderflow if the audio streaming utility has run out of data, or another value if an error has occurred.

Since a stream has no clearly defined end point, the audio streaming utility is unable to differentiate between the data source being exhausted and the client application not supplying buffers quickly enough. A KErrUnderflow error will therefore signify a true error only if the application still has data to supply to the streaming utility.

The application's CAsynchronousExampleStream class provides the second example of audio streaming. As mentioned earlier, this example simulates a fully asynchronous streaming client, in which buffers are passed to the streaming utility as and when they are available. The class is an active object, using a timer to simulate the arrival of packets. The application maintains an array of buffers, which are filled and emptied asynchronously by the timer and the sound device respectively.

As in the synchronous case, the ConstructL() function creates and opens an instance of an audio output stream utility. When the user initiates asynchronous playing, the timer is started, which results in a call to the class's RunL() function with an internal state of ETimer.

```
void CAsynchronousExampleStream::RunL()
  {
  switch (iState)
    {
    case ETimer:
      //A buffer is available to read from the source
      if (iBufList(iEnd).Length()==0)
        {
        iFile.Read(iBufList(iEnd),iStatus);
        iState=EReading;
```

```
        SetActive();
        break;
        }
    else
        {
        //... but ignore it because there are no buffers free
        TInt pos=KBufferSize;
        iFile.Seek(ESeekCurrent, pos);
        RandomDelay();
        break;
        }
case EReading:
    //Got the buffer from the source, so send it
    if (iBufList(iEnd).Length())
        {
        TRAPD(err,iStream->WriteL(iBufList(iEnd)));
        if (++iEnd==KNumBuf)
            {
            iEnd=0;
            }
        iState=ETimer;
        RandomDelay();
        break;
        }
    else
        {
        iFinished=ETrue;
        }
    }
}
```

If there is a free buffer, `RunL()` sets up an asynchronous read of a packet of data from the audio source file into the buffer, otherwise the audio packet is skipped. In a real application that relied on audio synchronization, this would be the equivalent of dropping audio frames in order to minimize the latency of the audio playback.

When the asynchronous file read request has completed, `RunL()` is called again, this time with an internal state of `EReading`. An empty buffer indicates that the file contains no more data, in which case `RunL()` simply sets `iFinished` to `ETrue` to mark the end of streaming. Otherwise, it uses `WriteL()` to transfer the buffer to the audio streaming utility and then calls `RandomDelay()` to run the timer again.

When the output stream has finished with the audio buffer it calls the observer's `MaoscBufferCopied()` function:

```
void CAsynchronousExampleStream::MaoscBufferCopied(TInt aError,
                    const TDesC8& aBuffer)
    {
    if (aError==KErrNone || aError==KErrAbort)//user selected "Stop"
        NextBuffer(aBuffer);
    else
        User::Invariant();
    }
```

This behaves exactly as it does in the synchronous case, except that the buffer reference is passed to `NextBuffer()`.

```
void CAsynchronousExampleStream::NextBuffer(const TDesC8& aBuffer)
    {
    //assume the buffer is already there!
    ASSERT(iBufList(iReturned)==aBuffer);
    iBufList(iReturned).SetLength(0);
    if (++iReturned == KNumBuf)
        {
        iReturned=0;
        }
    }
```

In the asynchronous case, `NextBuffer()` does not need to provide another buffer of data, since that has already been done by the `RunL()` function. All it has to do is to empty the buffer and mark it as being returned, so that it is available for reuse in a later call to `RunL()`.

Obviously, a real-world application that is downloading an audio file over a network link and streaming it to the audio streaming utility would not need a timer to generate interrupts and simulate packet transfers. Instead, the packets would arrive from the network at varying times and would be transferred to the client streaming utility immediately, using `WriteL()`.

Ideally, such an application would prepare a number of audio buffers before transferring any data to the streaming utility. It could then make a number of `WriteL()` calls in quick succession until the application's audio buffers are exhausted of data. The application could then prepare further buffers while the streaming utility relays its existing data to the sound driver. This approach minimizes the chance of an underflow condition and allows the user to remain largely unaware of the latency involved in the audio packet transfers from the network.

7.4.5.3 Input Streaming

The audio input stream utility `CMdaAudioInputStream` is defined in `mdaaudioinputstream.h` and provides the ability to stream audio from a sound input device. As with output streaming, the audio data must be in the PCM16 format, and you can specify the sample rate and number of channels, and adjust the volume and balance of the audio stream. In fact, the input streaming utility works in exactly the same way as the output streaming utility (but in reverse). In consequence, there is no need to describe it in detail and we will just concentrate on the differences between the two utilities.

The observer class for the audio input stream utility is `MMdaAudioIn-putStreamCallback`, whose definition is:

```
class MMdaAudioInputStreamCallback
  {
public:
  virtual void MaiscOpenComplete(TInt aError) = 0;
  virtual void MaiscBufferCopied(TInt aError,
               const TDesC8& aBuffer) = 0;
  virtual void MaiscRecordComplete(TInt aError) = 0;
  };
```

As you can see, the only difference between this observer and that of the audio output stream is that the `MaiscPlayComplete()` function is replaced by `MaiscRecordComplete()`.

To use it, create an instance of the audio input stream utility using its `NewL()` function, exactly as for the output stream. The streaming of audio through this utility then takes place in the following stages.

1. Open the audio input stream by calling its `Open()` function. As with the output stream, you can use `Open()` to set the sample rate and number of channels, or you can do this later by means of a call to `SetAudioProperties()`. On completion of this process, the utility calls the observer's `MaiscOpenComplete()` function and, if the utility was opened successfully, the error code will be `KErrNone`.

2. Now use the utility's `ReadL()` function to pass it one or more audio buffers. As with the output streaming utility, you should not modify the content of any buffer until you are informed that the buffer has been filled. The streaming utility stores the buffers in a FIFO queue and fills them as soon as audio data is available from the sound device.

3. Each time a buffer is filled, the utility calls the observer's `Maisc-BufferCopied()` function, passing a reference to the relevant buffer and an error code.

4. When recording from the sound device has completed, the utility calls the observer's `MaiscRecordComplete()` function. If recording was terminated by a call to the utility's `Stop()` function, `Maisc-BufferCopied()` will be called with an error of `KErrAbort` for every outstanding buffer in the utility's FIFO. If, however, the input streaming utility has run out of buffers to fill with audio data, then `MaiscRecordComplete()` will be called with an error code of `KErrOverflow`. Such an overflow error condition usually arises because the application is not using `ReadL()` to supply empty buffers quickly enough.

As with the output streaming utility, you can call the utility's `GetBytes()` function at any time to determine the current number of bytes rendered (recorded) by the hardware.

Audio data is streamed from the audio hardware in real time. In consequence, there are no synchronization issues with passing a large number of empty buffers to the input streaming utility, so as to avoid an overflow condition. However, as with any Symbian OS application, you should use your judgment with regard to how many buffers are allocated, to avoid excessive memory use. Remember that an 'empty' buffer is actually a descriptor with a specific length (its user-defined maximum length) and its current data length set to zero, rather than a NULL buffer.

7.5 Using Video

The MMF is a generic controller framework that is not limited to the processing of audio, and Symbian OS v7.0s also contains client APIs for the manipulation (playing and recording) of video clips. Since a video controller needs to be highly device-specific, and depends heavily on the precise features of a phone's camera and screen hardware, Symbian OS v7.0s does not include one, but expects an appropriate controller to be provided by the manufacturer of each phone that provides video services.

This section begins by discussing some of the terms and concepts of video usage, then gives an overview of a sample architecture involving video, and finally introduces the client APIs themselves.

Digital Video Formats

Digital video is basically a set of video frames that are encoded (possibly together with audio) in a source medium, such as an AVI file. One of the simplest examples of a video clip of this form would be a set of uncompressed bitmap frames which have been interleaved (see below) with a stream of uncompressed PCM16 audio. This would be referred to as an uncompressed video format. However, streaming an uncompressed format such as this would require a very large bit-rate, so most video formats offer some form of compression for both the video frames and the audio stream. Often, this results in a video format that requires a large amount of processor power to decompress in real time. In consequence, you have to pick video formats carefully, so as to make the software usable on low-powered hardware. However, some phones now have custom digital signal processing (DSP) hardware built into them, allowing highly compressed video to be used without a large burden on the main processor. You should always check the phone's capabilities, and make use of such 'hardware-accelerated' video processing formats if they are available.

Audio/Video Interleaving

Generally, a video stream will contain both audio and video data in an interleaved format. Interleaving (or multiplexing) is the process by which two or more streams of data can be combined into a single stream. How this is accomplished is entirely dependent upon the destination format of the data. The simplest form of audio/video interleaving would be to alternate frames of video with frames of the equivalent duration of audio. So, for example, if a video clip is recorded at 30 frames per second, the format could contain a sequence of items, each consisting of a video frame, followed by 1/30 second of audio data.

Video Controller Integration

In Symbian OS, the video controller is responsible for all the processing that is required to channel video from a source to a sink. It therefore has to be integrated with the range of sources and sinks that a particular phone provides.

In order to achieve the maximum efficiency, a video controller will usually be device-specific, communicating directly with hardware device drivers and utilizing any hardware accelerated capabilities available. In consequence, as with the controller itself, Symbian OS does not provide any standard sources or sinks for video, and phone manufacturers are expected to provide their own. However, if the video controller is using other, standard source or sink plugins, such as an audio output sink, or a file-based source, then the controller also has to be able to process the buffer types that these plugins understand.

As can be seen from the video controller architecture illustrated in Figure 7.5, it is also possible for the video controller to interface with the ECam API, instead of interfacing directly to camera hardware APIs or device drivers. If ECam (an optional on-board camera API, described in Section 7.8) has been implemented on the phone, then the ability to use this API provides an abstraction to the camera hardware and thus increases the portability of the controller. However, if a given phone does not have an ECam implementation, then the controller will have to interface with the camera hardware directly in order to support recording from a camera.

Video Synchronization

Most video controllers support the simultaneous playback of both audio and video They therefore have to ensure that the video stream is kept synchronized with the audio stream, otherwise audible items, such as speech, will not match what is seen in the video frames. There is a variety of ways in which this synchronization can be maintained, but a common way is to synchronize the video output with the sound device, dropping video frames where necessary, so as to keep the audio stream intact. In order to achieve this, the video controller can use a common clock source for both the audio and the video stages.

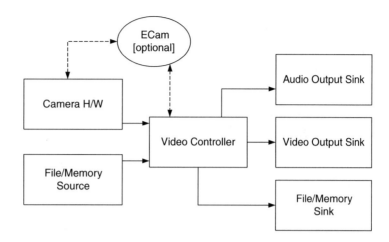

Figure 7.5 A video controller architecture

7.5.1 Video Playback

The video player utility `CVideoPlayerUtility` is defined in `Video-Player.h`, and provides the ability to play video clips.

The functionality of the video player utility is broadly similar to that of the audio player utility. For example, it allows the source data to be supplied in a file, in a `TDesC8` descriptor or at a specified URL address, it includes basic playback functionality and it allows you to modify a variety of playback settings.

It also supplies video-specific functionality, such as frame grabbing and setting the clipping region, frame size, and cropping region. Video frames are grabbed as `CFbsBitmaps`. It also provides more advanced functionality, such as rebuffering and rebuffering notification (useful when rendering a video stream). The video player utility is, in principle, format agnostic, but is restricted to using only those formats that are supported by the video controllers on the device.

To use the video player, derive an observer from `MVideoPlayer-UtilityObserver`, whose class definition is:

```
class MVideoPlayerUtilityObserver
  {
public:
  virtual void MvpuoOpenComplete(TInt aError) = 0;
  virtual void MvpuoPrepareComplete(TInt aError) = 0;
  virtual void MvpuoFrameReady(CFbsBitmap& aFrame,TInt aError) = 0;
  virtual void MvpuoPlayComplete(TInt aError) = 0;
  virtual void MvpuoEvent(const TMMFEvent& aEvent) = 0;
  };
```

As with most other multimedia utilities, you create an instance of the video player utility by calling its `NewL()` function, which is prototyped as:

```
IMPORT_C static CVideoPlayerUtility* NewL(
        MVideoPlayerUtilityObserver& aObserver,
        TInt aPriority,
        TMdaPriorityPreference aPref,
        RWsSession& aWs,
        CWsScreenDevice& aScreenDevice,
        RWindowBase& aWindow,
        const TRect& aScreenRect,
        const TRect& aClipRect);
```

This is more complex than the corresponding function prototypes for the audio utilities, and needs some explanation. The first parameter is the usual reference to the observer class, which should be a familiar concept by now. The next two parameters supply the desired video player priority – a topic that is explained later, in Section 7.6.1. You can change the priority at a later time by means of a call to the utility's SetPriorityL() function.

The next three parameters specify the window and screen parameters that the player should use. The first is a reference to a window server session, which the player utility can use to access the services of the window server. You can create such a session using code of the following form:

```
RWsSession ws;
User::LeaveIfError(ws.Connect());
```

In addition to the window server session, the video player utility requires a reference to a screen device, so that it can use direct screen access (DSA). A suitable screen device object can be created using the following code:

```
CWsScreenDevice screen = new (ELeave) CWsScreenDevice(ws);
User::LeaveIfError(iScreen->Construct());
```

The last of these three parameters is a reference to the window to which the video player utility should draw. The window would normally be created as part of one of the application's views, which are described in Chapters 4 and 5.

The final two parameters define the nature of the image that will appear on the phone's screen. The screen rectangle specifies – in screen coordinates – the area, within the previously specified window, to be used to display the video image. The video player calculates the best possible fit of a video frame in this rectangle, while maintaining the original aspect ratio (unless different scaling options have been set). The clip rectangle, again in screen coordinates, specifies the region to actually draw. Only the region of the video frame that is included in the intersection of the clip rectangle and the screen rectangle will actually appear.

The relationship between the source video frame, the screen rectangle and the clip rectangle is illustrated in Figure 7.6. This diagram also shows

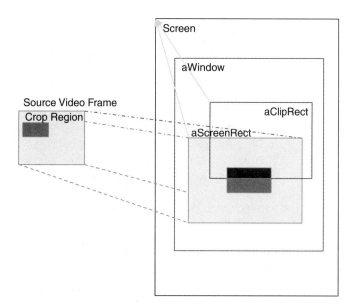

Figure 7.6 Displaying a video frame

a crop region within the source frame. If you set a crop region, by means of a call to the utility's `SetCropRegionL()` function, only that region of the video frame will be displayed, centered in the screen rectangle. If you use a crop region in conjunction with a scale factor you can, for example, expand the crop region to fill the window rectangle, effectively zooming in on a certain area.

You can change the values of any or all of the five window-related parameters at a later time by means of a call to the utility's `SetDisplayWindowL()` function.

Once you have created an instance of the video player utility, basic playback of a clip requires the following steps.

1. First open the clip, using one of `OpenFileL()`, `OpenDesL()` or `OpenUrlL()`. Each of these functions may optionally be passed a controller UID, to request a specific controller. Otherwise, the MMF will attempt to locate a suitable controller, using data from the specified clip. On completion of the opening process, the observer's `MvpuoOpenComplete()` function is called.

2. If the opening was successful, indicated by `MvpuoOpenComplete()` being passed an error code of `KErrNone`, you may execute any controller-dependent configuration custom commands (which are defined by the controller supplier). You must also prepare the video controller for playback, using the utility's `Prepare()` function. On completion of playback preparation, the utility calls the observer's `MvpuoPrepareComplete()` function.

3. Assuming that playback preparation was successful, indicated by `MvpuoPrepareComplete()` being passed an error code of `KErr-None`, the video player utility is now in a configured state and you are free to query and set the video playback properties, such as the scaling factor and crop region.

4. Now start video playback by calling the utility's `Play()` function. Once playback has completed – either by means of a call to `Stop()` or because the end of the clip has been reached – the utility calls the observer's `MvpuoPlayComplete()` function. The passed error code will be `KErrNone` if playback has been successful.

The parameterless overload of `Play()` plays the whole clip, from beginning to end. If you wish to play only a part of the clip, you can use the overload that is prototyped as:

```
void Play(const TTimeIntervalMicroSeconds& aStartTime,
    const TTimeIntervalMicroSeconds& aEndTime);
```

Frame grabbing is another useful piece of functionality that the video player utility provides. You can use the `GetFrame()` function to capture the current frame of a playing video into a bitmap. When you call this function, the video controller prepares a `CFbsBitmap` containing the current frame of video and passes it to the observer in a call to the `MvpuoFrameReady()` function. If a frame grab was not possible, this will be reflected by the error code that is also supplied to this function. Ownership will likely stay with the client, but check your controller documentation to make sure.

The video player utility supports the usage of custom commands and these can be passed to the controller synchronously or asynchronously, using the various overloads of `CustomCommandSync()` and `CustomCommandAsync()` that are defined in `VideoPlayer.h`. Custom callbacks are also supported by the observer's `MvpuoEvent()` function. The usage of this function is not defined by the video player utility and thus a video controller supplier may use it to return manufacturer-specific information. The event is returned as a `TMMFEvent`, which contains a UID specifying an event type and an error code.

The video player utility also supports video clip reloading or rebuffering by allowing you to request notification each time a video clip that is being streamed by the controller is being reloaded or rebuffered. The prototype of the relevant function is:

```
void RegisterForVideoLoadingNotification(MVideoLoadingObserver&
    aCallback);
```

The function takes a reference to an observer class derived from `MVideoLoadingObserver`, whose class definition is:

```
class MVideoLoadingObserver
  {
public:
  virtual void MvloLoadingStarted() = 0;
  virtual void MvloLoadingComplete() = 0;
  };
```

If requested, the video player utility calls the observer's `MvloLoadingStarted()` function each time reloading or rebuffering of a clip starts, and calls `MvloLoadingComplete()` when the process has completed. While reloading or rebuffering is taking place, you can obtain a progress report by calling the utility's `GetVideoLoadingProgressL()` function. It fills an integer with a percentage value, indicating what proportion of the process has already been performed.

7.5.2   Video Recording

The video recorder utility `CVideoRecorderUtility` is defined in `VideoRecorder.h` and provides the ability to record video clips.

The functionality of the video recorder utility is broadly similar to that of the audio recorder utility. For example, it allows the data to be recorded to a file, a `TDesC8` descriptor or a specified URL address, it includes basic recording functionality and it allows you to modify a variety of audio and video recorder settings. As with the video player utility, the video recorder utility supports the passing of custom commands to the controller. The video recording type can be set to any of the supported video formats.

To use the video player, derive an observer from `MVideoRecorderUtilityObserver`, whose class definition is:

```
class MVideoRecorderUtilityObserver
  {
public:
  virtual void MvruoOpenComplete(TInt aError) = 0;
  virtual void MvruoPrepareComplete(TInt aError) = 0;
  virtual void MvruoRecordComplete(TInt aError) = 0;
  virtual void MvruoEvent(const TMMFEvent& aEvent) = 0;
  };
```

Create an instance of the video recorder utility by means of a call to its `NewL()` function which, like the other multimedia utilities, takes a reference to the observer class as a parameter. Recording video then takes place in the following stages.

1. First open the clip, using the relevant 'open' function. These are more complicated than with most of the other client utilities and are described below. On completion of the opening process, the utility calls the observer's `MvruoOpenComplete()` function.

2. If the opening was successful, indicated by an error code of `KErr-None` being passed to `MvruoOpenComplete()`, you are then free to execute any controller-dependent custom commands that have been defined by the controller supplier, or to set up the video properties, such as the frame rate, the maximum clip size and whether audio is to be enabled.

3. Before starting recording, as with the video player utility, you have to prepare the video controller by calling the utility's `Prepare()` function. Once the preparation has completed, the utility calls the observer's `MvruoPrepareComplete()` function.

4. Provided that an error code of `KErrNone` is passed to `MvruoPre-pareComplete()`, the video recorder utility is now in a configured state and you can start video recording by calling the utility's `Record()` function. When recording has completed, either by a call to the utility's `Stop()` function or by reaching a predefined maximum clip size, the utility calls the observer's `MvruoRecord-Complete()` function. The passed error code will be `KErrNone` if recording has been successful.

As with the video playing utility, the video recorder utility has three 'open' functions, `OpenFileL()`, `OpenDesL()` and `OpenUrlL()`, which take, as their first parameter, a reference to the appropriate data source (and, in the case of opening a URL, a second parameter that specifies an Internet access point ID). However, all three functions take five further parameters, as illustrated by the prototype of the `OpenFileL()` function:

```
void OpenFileL(const TDesC& aFileName,
        TInt aCameraHandle,
        TUid aControllerUid,
        TUid aVideoFormat,
        const TDesC8& aVideoType = KNullDesC8,
        TFourCC aAudioType = KMMFFourCCCodeNULL);
```

The first additional parameter is the camera handle, which specifies which camera to record from. In general, the camera handle will be device-specific, but if the manufacturer has implemented ECam, and the video controller is known to use ECam, then you can get the handle by calling `CCamera::Handle()`.

The controller UID parameter has the same meaning as in the video player utility. If you specify the UID of a known video controller, then the video recorder utility will attempt to use that controller. Otherwise, the MMF will attempt to use the required destination data types and formats to select a suitable controller.

The third additional parameter is the required video format, specified by a UID. The final two parameters are the video type and the audio recording type. The video type is specified by a MIME type, contained in a descriptor, and the audio type is specified in a `TFourCC` (see Section 7.4.4). As indicated by the prototype, if the video and audio type parameters are not specified, the video controller will use suitable default values.

You can also set the audio and video types by calling the utility's `SetAudioTypeL()` and `SetVideoTypeL()` functions when the video recorder utility is either open or prepared. To enable audio recording, call the utility's `SetAudioEnabledL()` function when the recorder utility is either prepared or recording.

To obtain lists of the supported audio and video types, you can call the utility's `GetSupportedAudioTypesL()` and `GetSupportedVideoTypesL()` functions. The audio types are written to an `RArray` of `TFourCC` classes and the video types are written – as MIME types – to an 8-bit descriptor array (`CDesC8Array`). The caller owns both arrays and is therefore responsible for destroying them when they are no longer needed.

Synchronous or asynchronous communication of custom commands to the video controller is supported in the same way as for the video player utility. Custom callbacks are also supported by the observer's `MvruoEvent()` function. Again, as in the video player, this function's usage is not defined by the video player utility and thus a video controller supplier may use it to return manufacturer-specific information. The event is returned as a `TMMFEvent`, which contains a UID specifying an event type and an error code.

7.6 Controller Framework API

The client APIs described in the previous sections provide the main functionality that a multimedia application developer will require. This section introduces the controller and controller framework APIs, which map directly onto the controller plugins themselves. Much of the usage of these APIs is beyond the scope of this book, so we discuss two simple cases: the enumeration of supported formats; and the direct usage of a controller through the `RMMFController` API itself.

Enumerating Formats
Many multimedia applications will need to find out which controller plugins are available to perform a particular operation. You might want to specify the selection criteria in a variety of ways and, once the set

of suitable controller plugins has been selected, you may need to query them for further details of the services that they support.

The `CMMFControllerPluginSelectionParameters` class provides a straightforward means of enumerating the available controller formats for an operation, based on a set of user-supplied parameters. Its two functions, `SetRequiredPlayFormatSupportL()` and `SetRequiredRecordFormatSupportL()`, set up the class to search for certain play and record formats respectively. Each of these functions takes a reference to a `CMMFFormatSelectionParameters` class, which allows you to specify a set of parameters to identify suitable formats. Once you have supplied two suitable format selection classes, then a call to the `ListImplementationsL()` function fills a `RMMFControllerImplInfoArray` with a list of suitable controllers.

A real application would identify the required formats by specifying the MIME type of the clip, an extract of header data, or a file extension. The `CMMFFormatSelectionParameters` class provides a set of `SetMatchToXxxx()` functions to specify each of these three types of selection criterion. If you choose not to call any of these functions, effectively leaving the selection class blank, then the call to `ListImplementationsL()` will return all available controllers for the given task.

The `audio2` example application contains code for enumerating controllers, in its engine's `RecordTypesL()` function. The function starts by creating an `RMMFControllerImplInfoArray` to hold a list of controllers, and pushing it to the cleanup stack:

```
void CAudio2Engine::RecordTypesL(CDesCArray& aTypeArray)
  {
  RMMFControllerImplInfoArray controllers;
  CleanupResetAndDestroyPushL(controllers);

  CMMFControllerPluginSelectionParameters* cSelect =
    CMMFControllerPluginSelectionParameters::NewLC();
  CMMFFormatSelectionParameters* fSelectRead =
    CMMFFormatSelectionParameters::NewLC();
  CMMFFormatSelectionParameters* fSelectWrite =
    CMMFFormatSelectionParameters::NewLC();
```

Following this, it creates instances of the controller plugin selection class and two format selection classes, for reading and writing respectively. At this point it would be possible to populate the format selection classes by calling their `SetMatchXxx()` functions, to select controllers by MIME type, file extension or header data. The example code leaves them unpopulated, so as to return all available controllers.

If you wish, you can also specify settings that modify plugin selection – say, to indicate a preferred supplier.

Now set the required media IDs, which specify the type of plugin to return:

```
RArray<TUid> mediaIds;
mediaIds.Append(KUidMediaTypeAudio);
CleanupClosePushL(mediaIds);

cSelect->SetMediaIdsL(mediaIds,
  CMMFPluginSelectionParameters::EAllowOnlySuppliedMediaIds);
```

By placing only KUidMediaTypeAudio in the array passed to Set-
MediaIdsL(), the example specifies that the search should be restricted
to plugins that support audio. Supplying a value of EAllowOnlySup-
pliedMediaIds in the second parameter means that the search is further
restricted to plugins that support *only* audio and therefore excludes, for
example, ones that support both audio and video.

Having set up the required selection criteria, the next step is to obtain
the list of matching controllers:

```
cSelect->SetRequiredPlayFormatSupportL(*fSelectRead);
cSelect->SetRequiredRecordFormatSupportL(*fSelectWrite);
// need to specify this as otherwise the supported record
// formats won't be retrieved
cSelect->ListImplementationsL(controllers);
```

The first two calls pass the required play and record format criteria to the
selector class and the call to ListImplementationsL() retrieves the
list of supported controllers into the controller implementation information array that was created earlier.

Finally, the function loops through the controller list and retrieves a
list of supported file extensions from each controller:

```
TBuf<10> buf;
// Loop through each returned plugin and
// get their record formats
for (TInt i=0; i<controllers.Count(); i++)
  {
  const RMMFFormatImplInfoArray& recordInfo =
    controllers(i)->RecordFormats();

  // Get array of supported file extensions
  for (TInt j=0; j<recordInfo.Count(); j++)
    {
    const CDesC8Array& extensions =
      recordInfo(j)->SupportedFileExtensions();

    // and finally add each extension to the array
    for (TInt k=0; k<extensions.Count(); k++)
      {
      buf.Copy(extensions(k));
      aTypeArray.AppendL(buf);
      }
    }
  }
  CleanupStack::PopAndDestroy(5);
}
```

The call to `RecordFormats()` returns an instance of `RMMFFormatImplInfoArray` (which is an array of pointers to instances of `CMMFFormatImplementationInformation`). The elements of this array can be queried to find the supported MIME types, header data or file extensions. The example uses calls to `SupportedFileExtensions()` to extract a list of all supported file extensions from each element of the array. Each list of file extensions is first copied into a temporary buffer and then appended to the list argument passed to the function.

This example uses a limited-size temporary buffer, which is fine for sample code. A real application should allow the buffer to grow in size as necessary. Also, in a real application, the UIDs of all suitable controllers could be collected and stored for later use.

Using Controllers Directly

Direct use of the MMF controllers is somewhat complicated and is mostly beyond the scope of this book, so this section provides a brief overview of the steps involved. The controller APIs are fully documented in the Symbian OS SDK documentation and you are advised to consult this for further help.

A typical usage of the MMF controllers might follow the following sequence.

1. Obtain the UID of a suitable controller, for example by enumerating the controllers that support a certain format of audio.

2. Open the controller using its `Open()` function.

3. Add a source to the controller to support the required input type. For example, a file source will be able to read data from a file.

4. Add a sink – such as an audio output sink – to the controller to support the required output type.

5. Prime the controller. This step initializes the controller and makes sure that it ready to play or record. It may involve starting DSP tasks, or the allocation of internal buffers for the controller to use.

6. Adjust controller settings, such as the volume.

7. Start playback or recording.

8. Stop and close down the controller.

The following code examples illustrate this sequence. They are provided purely to help you to understand MMF controller functionality and do not contain references to all of the classes that would be required. In most applications it would be preferable to use the client APIs rather than calling these functions directly.

```
// Find UID of required controller
_LIT(KTestFileName, "test.wav");
TUid controllerUid = ChooseControllerL(KTestFileName);

// Open the chosen controller plugin
RMMFController controller;
User::LeaveIfError(controller.Open(controllerUid));
CleanupClosePushL(controller);
```

The ChooseControllerL() function is assumed to be picking a controller that supports the playback of 'test.wav'. The returned UID is then used to open the controller via the controller's Open() function.

```
// Now add the source and sink
TMMFFileConfig sourceCfg;
sourceCfg().iPath = _L("test.wav");
User::LeaveIfError(controller.AddDataSource(KUidMmfFileSource,
    sourceCfg));
User::LeaveIfError(controller.AddDataSink(KUidMmfAudioOutput,
    KNullDesC8));
```

The call to the controller's AddDataSource() function adds a suitable data source that supports the reading of 'test.wav'. This is followed by the addition of a data sink that is capable of outputting audio to, say, a loudspeaker. The various data source and sink types are defined in MmfDataSourceSink.hrh.

```
// The controller plugin is now ready to prime
User::LeaveIfError(controller.Prime());
User::LeaveIfError(controller.Play());
// The plugin should now be playing
```

Once the controller has been primed using the Prime() function, you are free to start playback or recording. In this example, the controller is being used to play a test file, so it calls the controller's Play() function. Many of the controller API functions may return an error, so it is advisable to check for errors by using the User::LeaveIfError() function.

```
// Stop the controller and clean up
User::LeaveIfError(controller.Pause());
User::LeaveIfError(controller.Stop());

CleanupStack::PopAndDestroy();
```

Finally, once all required operations have been performed, you can stop the controller by calls to the Pause() and Stop() functions as appropriate. If you have no further use for the controller, then it should be deleted. If the controller is pushed to the cleanup stack, as in this example, then the cleanup stack will perform all necessary actions.

7.6.1   Priorities

The descriptions of the various MMF client APIs contain passing references to setting priorities. Most of the client utilities allow you to specify both a priority and a priority preference, either when instantiating the utility itself, or by a later call to a `SetPriority()` function.

The priority is specified as a `TInt` in a range decided by the phone manufacturer, but the interpretation of its value is determined by the UI or, in some cases, by the controller itself. The priority is usually used to determine which services should have access to the hardware if two or more services require simultaneous access.

The priority preference specifies what actions should be taken if such a conflict occurs. The lower-priority sound could be muted, mixed with the higher-priority sound, or paused until the higher-priority access is complete.

For more information regarding priorities and priority preferences you should consult the documentation in a phone-specific SDK.

7.7   Using the ICL

Section 7.2.5 gave an overview of the ICL architecture, so this section explains the main APIs of the ICL (image conversion library). It chiefly replaces the (now deprecated) `CMdaImageUtility` API and is used for three main tasks, which are the decoding and encoding of images, and the transformation of bitmaps.

Due to the encapsulated way in which the ICL deals with encoded images, there is no need to know the details of how they are stored. However, an uncoded image – regardless of whether it is a source to be encoded, the target for decoding, or the source or target of a bitmap transformation – is always encapsulated as a bitmap, in the `CFbsBitmap` class.

Since the ICL subsystem is much smaller and less complicated than the multimedia framework, the description will be illustrated here with code fragments, rather than with a complete example application.

7.7.1   Active Object-Based Image Processing

In contrast to other parts of the multimedia framework, the ICL makes explicit use of active objects when performing potentially extended operations such as a conversion or a transformation of an image. (See Chapter 1 for a basic discussion of active objects and their use.)

Before starting a conversion or transformation, the application has to create an active object and pass a pointer to the active object's `iStatus` member to the relevant function. The active object is signaled upon completion and its `RunL()` function should initiate any necessary response to the completion.

The requirement that the client application must supply an active object means that the application's structure will depend on how the client is using the ICL, for example on how many images are to be held open at any given time. The simplest arrangement for a basic application that opens a single image at any given time is for a 'watcher' active object's `RunL()` to call code in either the application's AppUI or one of its views in order to update the application when an operation is complete. A more complicated application, that has many images open at a given time, would require some form of list or array of images. Typically these list objects could either be active objects themselves, or have the active objects encapsulated within their structure.

7.7.2   Decoding Images

The image decoder class, `CImageDecoder`, is defined in `imageconversion.h` and provides the ability to decode images. The source image data may be in a variety of formats, and can be stored in either a file or a descriptor.

Unlike with the MMF utility classes, a new instance of the image decoder class must be created for each image that is to be decoded. There are four suitable `XxxNewL()` functions, two of which are shown below:

```
static CImageDecoder* FileNewL(RFs& aFs,
              const TDesC& aSourceFilename,
              const TOptions aOptions = EOptionsNone,
              const TUid aImageType = KNullUid,
              const TUid aImageSubType = KNullUid,
              const TUid aDecoderUid = KNullUid);

static CImageDecoder* DataNewL(RFs& aFs,
              const TDesC8& aSourceData,
              const TOptions aOptions = EOptionsNone,
              const TUid aImageType = KNullUid,
              const TUid aImageSubType = KNullUid,
              const TUid aDecoderUid = KNullUid);
```

The only parameter that differs between the two is the one that specifies the source data. For a file-based image it is passed as a reference to a `TDesC` that contains the file name, whereas for a descriptor-based image it is a reference to the `TDesC8` descriptor that contains the image data.

The first parameter is a reference to the file server session that the image decoder should use, and is the only essential parameter other than the data itself. If just these two parameters are specified then the image decoder will attempt to locate a suitable decoder plugin by examining the supplied data.

If you wish, you can specify the UID of a decoder plugin, which will force the image decoder to use that specific plugin. If it is known, you can also specify the image type. This is specified as a main type and a subtype (the subtype is normally set to KNullUid), using the values defined in ImageCodecData.h. You can obtain a list of the supported types and subtypes by calling GetImageTypesL() and GetImage-SubtypesL().

The remaining parameter allows you to specify any ored combination of the following options:

EOptionsNone	This is the default value and instructs the image decoder not to use any extra options.
EOptionNoDither	By default the image decoder uses error-diffusion dithering when decoding to a bitmap display mode that differs from the one that is recommended. This option turns off dithering and can be useful, for example, if the image is to be rescaled, so that the dithering can be applied after rescaling rather than at the decoding stage.
EOptionAlwaysThread	This value instructs the image decoder to perform the decoding operation in a separate thread, rather than just relying on the active object approach. Its main use is when processing many images, in which case the decoder active object might make the application unresponsive for short periods. However, the usage of threads requires extra overhead – and possibly extra memory and decoding time. If you specify this option when the decoder plugin itself has already requested that a separate thread be used, then setting the option has no effect.
EOptionAllowZeroFrameOpen	This setting allows the opening stage of the decoding process to complete with KErrNone, even if there is less than a whole frame available. This is useful for streaming, to allow decoding to be attempted whilst the source image is downloading.
EAllowGeneratedMask	An image decoder will normally only report that transparency is possible if mask data is encoded together with the image. However, some decoders have the ability to automatically generate a mask. Setting this flag instructs the image decoder to enable this operation, if it is supported.

The remaining two functions are overloads of `FileNewL()` and `DataNewL()` that differ from the ones described above by taking an extra MIME type parameter to specify the format of the image. The image decoder uses this information to identify a suitable decoder plugin. Some image formats, for example WBMP` ` and OTA images, lack sufficient header data to enable the image decoder to determine which format or plugin to use. In these cases the format and plugin must be explicitly supplied.

If a suitable decoder plugin can not be found, the `XxxNewL()` function will leave, reporting an error of `KErrNotFound`. Once a decoder plugin has been selected, the image decoder uses it to scan the image looking for headers. These are used to set internal properties that support the returning of frame counts and frame information. If the plugin is unable to decode the image, `XxxNewL()` will leave with the error `KErrCorrupt`. However, if there is more than one decoder plugin that can process a given source image, then in the event that the first decoder fails due to `KErrCorrupt`, the image decoder will try to open the second decoder, and so on. Failure is reported only if *all* of the decoders were unable to process the source image.

Once an instance of the image decoder class has been successfully created, you can find the number of frames in the source image by calling `FrameCount`. You can find out more about any of the source image frames by calling the `FrameInfo()` and `FrameData()` functions. `FrameInfo()` supplies basic information, such as the color depth and frame size, stored in a `TFrameInfo`, and `FrameData()` specifies more advanced settings, such as image quality and color settings, stored in an instance of `CFrameImageData`. The frame information also contains parameters such as whether transparency and scaling are supported for the current frame, and the delay that should be observed between the current frame and the next frame if they are part of an animated sequence. You can find the full definition of both classes in `ImageData.h`.

Before proceeding, it is generally necessary to set up the destination bitmap in accordance with the dimensions and display mode that are specified in the frame information:

```
CFbsBitmap* frame = new(ELeave) CFbsBitmap;
CleanupStack::PushL(frame)
const TFrameInfo* theFrameInfo = &iImageDecoder->FrameInfo();

frame->Create(theFrameInfo->iOverallSizeInPixels,
     theFrameInfo->iFrameDisplayMode ));
```

This code fragment assumes that an instance of `CImageDecoder` has already been created and is stored in the `iImageDecoder` member data of some other class.

All decoders support 'thumbnail' decodes to bitmaps that are one half, one quarter or one eighth of the full image dimensions. Decoders that set the EFullyScaleable flag in the frame information support decoding to arbitrarily sized bitmaps. If the ECanDither flag is set then the decoder also supports dithering, and can decode to any display mode. If dithering is supported it is usually best to set the destination type to be the same as the current window's display mode, since this will result in faster drawing. Finally, it is worth mentioning that any mask bitmap must be the same size as the main decoded image, but the display mode can (and usually does) differ.

Once the destination bitmaps have been created and set up correctly, you can start decoding one or more frames. The standard approach is to then use one of the asynchronous Convert() function overloads, the most basic of which is:

```
void Convert(TRequestStatus* aRequestStatus,
    CFbsBitmap& aDestination,
    TInt aFrameNumber = 0);
```

You pass this function a pointer to the request status of the active object being used to monitor the conversion, a reference to the destination bitmap, and the frame number of the frame to be decoded (which defaults to zero). You are free to decode frames in any order and to repeat the decoding of any given frame.

When the conversion is complete, the active object will be signaled and an error code will be present in its iStatus member data. If the conversion was successful, then the error code will be KErrNone and the bitmap can then be displayed. It is possible (though rare) that a KErrCorrupt error could be reported, if there was an error in the image that XxxNewL() did not identify. More likely errors are KErrNoMemory and KErrInUse. If the error code is KErrUnderflow, then this is not strictly an error condition, as is explained later in this section.

7.7.2.1 Using Masks

If a source image contains a mask – which is indicated by the presence of the ETransparencyPossible flag in the iFlag member of the frame information – then you need to decode it using a different overload of Convert(), whose prototype is:

```
void Convert(TRequestStatus* aRequestStatus,
    CFbsBitmap& aDestination,
    CFbsBitmap& aDestinationMask,
    TInt aFrameNumber = 0);
```

In this case you need to supply references to two bitmaps – one for the main image and one for the mask. Two types of mask are supported. If the EAlphaChannel flag is set, then the mask is deemed to be an eight-bit alpha blend, and the supplied mask bitmap must be of display type EGray256. If the EAlphaChannel flag is not set, then there will be a two-bit mask present and you can supply a mask bitmap that is either of type EGray2 or of type EGray256. In general, the decoding of a mask is optional and you can choose to decode only the main image even in an image where mask data is present.

Although most image formats combine image and mask data, Symbian OS normally keeps them in two separate bitmaps.

7.7.2.2 Frame Animation

The image decoder provides support for GIF animation through the values that a call to FrameInfo() writes to the returned TFrameInfo. The most significant information is the value of the iDelay member, which specifies the delay in microseconds that should be observed before displaying the next frame in the sequence. In addition, the following flag values, in the iFlags member, are relevant to animation:

ELeaveInPlace	Specifies that the current frame should be left in place when displaying the next frame. Thus any pixels from the current frame which are not covered by the next frame will continue to show.
ERestoreToBackground	Specifies that the current frame should be overlaid on top of the current background tile or color, which will therefore be visible in any transparent regions within the frame.
ERestoreToPrevious	Specifies that the current frame should be overlaid on top of the most recent frame set to ELeaveInPlace, which will therefore be visible in any transparent regions within the frame.

Although the image decoder can supply the correct animation settings, it is the application's responsibility to display the sequence of bitmaps associated with the animation. It must, for example, call the image decoder to decode the images, and display successive images at intervals specified by the TFrameInfo's iDelay member.

The architecture required in an application to support the displaying of animated images will very much depend on the requirements of the application itself. However, it is likely to need an internal state machine and one or more timer objects. In the interests of efficiency, the application will probably also use a number of bitmaps to hold pre-decoded frames from the animation.

7.7.2.3 Additional Frame Information

As previously mentioned, you can use the `FrameInfo()` function, and the more advanced `FrameData()` function, to access a variety of data items associated with images, including the color palette, lookup tables, copyright messages and other strings. To a large extent, much of this information is not designed to be used by application writers, but is included either for completeness or to provide plugin writers with a mechanism for providing additional features.

> Bear in mind that the data in a `TFrameInfo` or a `CFrameImage-Data` are valid only for the lifetime of the `CImageDecoder` class from which they were derived.

One type of data that applications frequently access is comment strings, which can provide useful extra information about a given image or frame. The functions to access them are built directly into the `CImageDecoder` class itself. You can find the number of such comments by using the `NumberOfImageComments()` and `NumberOfFrameComments()` functions. Any particular string can then be accessed by means of the `ImageCommentL()` and `FrameCommentL()` functions. Both of these functions return a pointer to an `HBufC` that contains the given string.

The image decoder's `FrameInfoStringsL()` function (also supplied in a `FrameInfoStringsLC()` variant) returns a pointer to a `CFrameInfoStrings` object that can provide information such as the format, dimensions and decoder plugin description, all in readable text strings. You can use this information, for example, when preparing a frame or image properties dialog.

7.7.2.4 Streamed or Progressive Decoding

The ICL provides facilities to support simultaneous decoding and displaying images as they are being downloaded. When opening a partial image, the framework will attempt to do as much as possible with the data available. The general steps involved are shown below.

1. If there is insufficient data to identify the image format, then the `XxxNewL()` function will leave with `KErrUnderflow`. This will not occur if you have explicitly specified the format or the plug-in – in this case the image decoder will attempt to open a suitable plugin regardless.

2. If there is enough data in an image, the image decoder will open a suitable plugin and scan the image for frame headers. While this is occurring, an internal flag is set so that `IsHeaderProcessing-Complete()` will return `EFalse`. It is generally the case that the frame headers will be stored before the actual frame data and thus the frame count of the image will be updated as soon as each header is parsed – without having to download every frame first.

3. Normally, the decoder creation will leave with `KErrUnderflow` if the current frame count is less than one, but this can result in inefficiencies, such as having to recreate the decoder multiple times before it is successfully opened. To counter this, you can supply the `EOptionAllowZeroFrameOpen` flag in the options passed to `XxxNewL()`. This allows the image to be opened, even if the frame count is zero. Of course, the creation can still leave if there is insufficient data to identify the image type, or if some other error condition occurs. Regardless of whether you have set the `EOptionAllowZeroFrameOpen` flag, you should not call either of the `FrameInfo()` or `Convert()` functions until the frame count is at least one, or the ICL will generate a panic.

4. The frame count will be updated when a header has been fully parsed. If you wish to decode a frame number that is greater than or equal to the current frame count, and `IsHeaderProcessingComplete()` returns `EFalse`, then you must wait until more data has been received. While downloading is in progress, it is the client application's responsibility to supply more data. When it has appended more data to the source descriptor (or file), it should then call `ContinueProcessingHeadersL()`, which will scan the data for any additional headers. This may increase the frame count and may cause `IsHeaderProcessingComplete()` to return `ETrue`.

5. When the required frame number is less than the frame count, you can start decoding the frame by calling the relevant `Convert()` function. This will perform as much decoding as possible. A successful decode will complete with `KErrNone` as normal, but if the whole frame is not yet present, the call will complete with the error code `KErrUnderflow`. In general, it is possible to display a partial frame if required. The exception to this is when the `EPartialDecodeInvalid` flag has been set in the frame information, indicating that, for the current format, only complete frames can be displayed.

6. If the previous decode completed with `KErrUnderflow` then, when more data has been appended, you should call `ContinueConvert()` to continue the decoding process from where the previous data ended, without repeating the processing that was performed in any previous conversion stage. For obvious reasons, you must therefore pass a reference to the same bitmaps that were supplied to the

original call to `Convert()`. The ICL will generate a panic if it detects that the target bitmaps have been changed between calls. If the `ContinueConvert()` function completes with `KErrUnderflow` then you should call it again when more data has been appended.

Probably the best way to run this kind of process is to implement a state machine that calls the various decoder functions as appropriate. The ICL itself has no means of determining whether all the source data has been downloaded, so the application will need to manage this. It will also need to handle the case where the image is incomplete, but no more data will be downloaded, say, if the source file at the remote site has been truncated. The state machine must not be permitted to enter an infinite loop in such a situation, but whether it is considered to be an error will depend on the exact nature of the application.

7.7.2.5 Buffered Decoding

The ICL provides another decoder class, `CBufferedImageDecoder`, which encapsulates both a `CImageDecoder` and the descriptor used to store the image. Unlike the `CImageDecoder` class, a buffered decoder can always be created, even when there is no source data. Internally, the `CImageDecoder` instance is not created until a valid image source has been supplied and the appropriate decoder has been found.

The `CBufferedImageDecoder` class is instantiated using a `NewL()` function which takes only one parameter – a reference to the file session to use. Source images are opened using one of the two `Open()` functions, prototyped as:

```
void OpenL(const TDesC8& aSourceData,
    const TDesC8& aMIMEType,
    const CImageDecoder::TOptions aOptions=CImageDecoder::EOptionNone);

void OpenL(const TDesC8& aSourceData,
    const CImageDecoder::TOptions aOptions=CImageDecoder::EOptionNone,
    const TUid aImageType=KNullUid,
    const TUid aImageSubType=KNullUid,
    const TUid aDecoderUid=KNullUid);
```

The parameters supplied to these functions are identical (with the exception of the file session) to those of the `CImageDecoder`'s `DataNewL()` functions, and the behavior for whole images is the same. When streaming, however, the following differences are present:

- When additional data becomes available, you supply it to the decoder using the `AppendDataL()` function, and process it using the `ContinueOpenL()` function. This stage copies the data and attempts to identify a suitable plugin, and should be repeated until `ValidDe-coder()` returns `ETrue`.

- The process for decoding the image headers and frames is exactly the same as for the `CImageDecoder` class, except that data is appended using the `AppendDataL()` function, rather than appending it to the source descriptor.

The calling application can discard any data once it has been passed to `AppendDataL()`, since the buffered decoder keeps a local copy.

You can, if you wish, reuse a buffered decoder. A call to its `Reset()` function destroys the internal buffer and decoder and you can then restart it by calling `Open()` again.

7.7.3 Encoding Images

The image encoder class `CImageEncoder` is defined in `imageconversion.h`, and provides the ability to encode `CFbsBitmap` encapsulated bitmaps into images stored in a variety of formats. The destination can be a file or a descriptor.

The functionality of `CImageEncoder` basically mirrors the decoder class but is much more limited in its use. It is designed to perform only two tasks: the saving of a bitmap from screen and the saving of a photo. Unlike the image decoder, `CImageEncoder` does not support streaming or buffered operation. Apart from that, the two classes are very similar, so we'll only discuss the differences in the way they are used.

As with the image decoder, you need to create a new instance of the `CImageEncoder` class for each image that you want to convert. The four encoder `XxxNewL()` functions are virtually identical to those of the decoder and, for comparison, the prototypes for the two standard functions are:

```
static CImageEncoder* FileNewL(RFs& aFs,
              const TDesC& aDestinationFilename,
              const TOptions aOptions=EOptionNone,
              const TUid aImageType=KNullUid,
              const TUid aImageSubType=KNullUid,
              const TUid aEncoderUid=KNullUid);

static CImageEncoder* DataNewL(HBufC8*& aDestinationData,
              const TOptions aOptions=EOptionNone,
              const TUid aImageType=KNullUid,
              const TUid aImageSubType=KNullUid,
              const TUid aEncoderUid=KNullUid);
```

As with the decoder class, there is an overload of each function that takes an explicit MIME type instead of the image types. The file-based versions need little additional discussion, but the `DataNewL()` function

is worthy of attention. Always pass a reference to a NULL pointer of type
HBufC8* for the aDestinationData parameter. The buffer will be
allocated during the encoding process. Ensuring that you do not need to
know the buffer's size prior to image creation, it also emphasizes that the
ownership of the buffer remains with the image encoder.

The other parameters supplied to these functions have the same
meanings as for the CImageDecoder class, with the exception that
the only flag values that can be passed in the aOptions parameter
are EOptionNone (the default) and EOptionAlwaysThread, which
specifies threaded encoding. As with the decoder class, if no image type
is supplied then the image decoder will choose one. However, it has little
or no information to guide its choice, so you would generally be advised
to specify one yourself.

Once the image encoder class has been instantiated, convert the image
using the Convert() function, prototyped as:

```
void Convert(TRequestStatus* aRequestStatus,
      const CFbsBitmap& aSource,
      const CFrameImageData* aFrameImageData=NULL);
```

As you can see, the only mandatory parameters are the request status
of an observing active object and a reference to the CFbsBitmap that
contains the image to be encoded. Upon completion, the active object
will be signaled, with an error code stored in its iStatus. An error code
of KErrNone indicates that the conversion completed successfully and
that there is a valid encoded image in the destination that was passed to
XxxNewL().

The last parameter to the Convert() function is optional and allows
you to 'prime' the encoder with format-specific data. The data that
you can supply, in an instance of the CFrameImageData class, is
linked to the capabilities of the plugin being used, and is different for
every plugin. You append data – contained in classes derived from TIm-
ageDataBlock and TFrameDataBlock – by calling CFrameImage-
Data's AppendImageData() and AppendFrameData() functions,
respectively. The following example demonstrates this process, using a
TJpegImageData class supplied for use with JPEG processing:

```
TJpegImageData data = TJpegImageData;
data.iSampleScheme = TJpegImageData::EColor444;
data.iQualityFactor = 95;

iFrameImageData = CFrameImageData::NewL();
User::LeaveIfError(iFrameImageData->AppendImageData(data));
```

In general, the destination image's size and display mode are taken to be
the same as those of the bitmap to be encoded.

As with the image decoder, the `CImageEncoder` class can leave during instantiation, or return an error through the `iStatus` of the active object observer. The error is unlikely to be `KErrCorrupt` unless there is a problem with the encoder itself, and will never be `KErrUnderflow`. It could, however, be any of the system-wide error codes – such as `KErrNotFound` if the required encoder plugin could not be found, or `KErrNoMemory` if the system has run out of memory.

7.7.4 Static Functions

The `CImageDecoder` and `CImageEncoder` classes supply a number of static utility functions that can be used to aid the decoding and encoding of images.

`GetMimeTypeFileL()`	Retrieves the MIME type associated with a particular image file. Supported by `CImageDecoder`.
`GetMimeTypeDataL()`	Retrieves the MIME type associated with a particular image descriptor. Supported by `CImageDecoder`.
`GetImageTypesL()`	Retrieves a list of primary supported image formats. Each entry contains a text description and an image type UID. Supported by `CImageDecoder` and `CImageEncoder`.
`GetImageSubTypesL()`	Retrieves a list of image subtypes supported for a given primary image type. Each entry contains a text description and an image subtype UID. Supported by `CImageDecoder` and `CImageEncoder`.
`GetFileTypesL()`	Retrieves a list of supported MIME types and file extensions. Supported by `CImageDecoder` and `CImageEncoder`.

In addition, `CImageDecoder` provides `GetImplementationInformationL()`, which returns, in an instance of `CImplementationInformation`, the ECom implementation information for a specified decoder plugin.

Since all of these functions are static, there is no need for a decoder or encoder to exist in order to use them. You can use them, for example, to compile a list of supported decode and/or encode formats so that a user could select a desired format from a list of those that are supported.

7.7.5 Bitmap Transformation

The bitmap transforms library is defined in `BitmapTransforms.h` and provides the ability to rotate and scale bitmaps. Both the rotation class and the scaling class are AO driven in their internal processing and signaling, and are therefore conceptually similar in use to the encoder and decoder `Convert()` functions. As with the image decoder and encoder, both transformation classes can be interrupted using `Cancel()`.

Unlike the decoding and encoding utilities, both transformation classes can be used repeatedly, to perform asynchronous transformations on more than one source bitmap.

7.7.5.1   Bitmap Rotation

The bitmap rotation functionality is provided by the `CBitmapRotator` class, which is instantiated using its parameterless `NewL()` function. You can then use the instance to rotate source bitmaps, using either of the `Rotate()` function overloads shown below:

```
void Rotate(TRequestStatus* aRequestStatus,
    CFbsBitmap& aSrcBitmap,
    CFbsBitmap& aTgtBitmap,
    TRotationAngle aAngle);

void Rotate(TRequestStatus* aRequestStatus,
    CFbsBitmap& aBitmap,
    TRotationAngle aAngle);
```

In addition to a pointer to the active object's request status, both overloads take a 'rotation' parameter, which must be one of the following enumerated values:

```
enum TRotationAngle
  {
  ERotation90DegreesClockwise,
  ERotation180DegreesClockwise,
  ERotation270DegreesClockwise,
  EMirrorHorizontalAxis,
  EMirrorVerticalAxis
  };
```

The first `Rotate()` function takes references to both a source and a target bitmap. It outputs the rotated image to the target bitmap, leaving the source bitmap unchanged. The second overload takes a reference to a single bitmap and performs an in-place rotation. In either case the rotation is performed asynchronously and the active object is signaled when the process is complete. A successful outcome is indicated by a value of `KErrNone` in the active object's request status (`iStatus`) data member.

7.7.5.2   Bitmap Scaling

The bitmap scaling functionality is provided by the `CBitmapScaler` class, which is, like the rotation class, instantiated using its parameterless

`NewL()` function. You can then use the instance to scale source bitmaps by calling either of the `Scale()` functions shown below:

```
void Scale(TRequestStatus* aRequestStatus,
     CFbsBitmap& aSrcBitmap,
     CFbsBitmap& aTgtBitmap,
     TBool aMaintainAspectRatio=ETrue);

void Scale(TRequestStatus* aRequestStatus,
     CFbsBitmap& aBitmap,
     const TSize& aDestinationSize,
     TBool aMaintainAspectRatio=ETrue);
```

Both overloads take the request status of an active object and can be passed an optional boolean value, which dictates whether the aspect ratio of the source bitmap should be preserved.

The first of these two functions outputs the scaled image to the specified target bitmap, leaving the source image unchanged. It determines the scale factor to be applied from the relative dimensions of the source and target bitmaps. The second function calculates the scale factor from the dimensions of the bitmap and the specified destination size. It performs the scaling in-place, adjusting the bitmap's dimensions accordingly.

As with the bitmap rotator, the scaling is performed asynchronously and the active object is signaled when the process is complete. A successful outcome is indicated by a value of `KErrNone` in the active object's request status (`iStatus`) data member.

7.8 Using ECam

As mentioned previously, the ECam Onboard Camera API is an optional Multimedia subsystem, which a mobile phone manufacturer can choose to implement. In Symbian OS v7.0s, ECam is shipped as a header file (and appropriate DEF files) only.

This section provides a brief overview of how you might use the ECam architecture to perform a few simple tasks.

The ECam API, defined in `ECam.h`, provides the functionality to allow an application to access and control any camera hardware that is attached to a phone. It supplies functions to query the status of the camera, adjust camera settings, control the camera viewfinder and capture still images. It also contains video capture APIs, although you would normally access this functionality through the `VideoRecorderUtility` class.

In order to use ECam, you must derive an observer from the camera observer interface class, `MCameraObserver`, whose class definition is:

```
class MCameraObserver
  {
public:
  virtual void ReserveComplete(TInt aError)=0;
  virtual void PowerOnComplete(TInt aError)=0;
  virtual void ViewFinderFrameReady(CFbsBitmap& aFrame)=0;
  virtual void ImageReady(CFbsBitmap* aBitmap,HBufC8*aData,TInt aError)=0;
  virtual void FrameBufferReady(MFrameBuffer* aFrameBuffer,TInt aError)=0;
  };
```

The interface to ECam is through the CCamera base class, which encapsulates a camera device. Since it has to encapsulate a large number of possible operations and variables, this class is very complicated and a full description is beyond the scope of this book. However, performing basic tasks, such as creating a camera, displaying a viewfinder, or taking a picture, is relatively straightforward.

Since Symbian OS does not implement ECam, CCamera is defined as a base class that can not be instantiated, and most of its member functions are declared as pure virtual functions. Mobile phone manufacturers supply a concrete implementation of the three static functions defined in CCamera, one of which is the NewL() below.

Creating a Camera Device

The first step is to create an instance of the CCamera class to represent the camera that you want to use. You do this by calling the NewL() function, which is prototyped as:

```
static CCamera* NewL(MCameraObserver& aObserver,TInt aCameraIndex);
```

Note that the NewL() will instantiate a derived class, but this normally should be accessed through the CCamera API.

The aObserver parameter is a reference to the camera observer class, derived from MCameraObserver. You must ensure that the value of aCameraIndex represents a valid camera. Fortunately, CCamera defines the static function CamerasAvailable() which returns the number of cameras available for the phone's use. You can then set aCameraIndex to any value in the range from 0 to CamerasAvailable() −1 inclusive.

A simple instantiation might take the form:

```
if (!CCamera::CamerasAvailable())
  {
  User::Leave(KErrNotSupported);
  }
iCamera = CCamera::NewL(*this,0); // derived from MCameraObserver
```

Once a camera has been created, you need to request exclusive access, so that the application can have access to the viewfinder and image capture capabilities of the camera. You do this by calling the `Reserve()` function, for example:

```
iCamera->Reserve();  // request exclusive access
```

The observer's `ReserveComplete()` function will be called when the reservation process is complete and, if the reservation was successful, the passed error code will be `KErrNone`. Possible errors include `KErrInUse`.

Finally, the camera should be switched on using the `PowerOn()` function:

```
iCamera->PowerOn();
```

This results in the observer's `PowerOnComplete()` function being called. If the passed error code is `KErrNone`, then the initialization has been successful and the camera is ready for use.

Displaying the Viewfinder

The ECam API supports two methods for displaying the viewfinder. The first of these is direct screen access, where the application specifies which region of the screen it wishes to use and the camera draws the current view directly to this area. In the other supported viewfinder option, the camera provides a series of bitmaps and the application has to handle drawing them.

A given camera device may support either or both of these options, so you first have to determine which are available. You do this by calling the `CameraInfo()` function, which fills a `TCameraInfo` with the entire set of parameters for a given camera device. The `iOptionsSupported` data member contains a combination of `TOptions` flags, a subset of which is shown below:

```
enum TOptions
  {
  /** Viewfinder display direct-to-screen flag */
  EViewFinderDirectSupported     = 0x0001,
  /** Viewfinder bitmap generation flag */
  EViewFinderBitmapsSupported    = 0x0002,
  /** Still image capture flag */
  EImageCaptureSupported         = 0x0004,
  /** Video capture flag */
  EVideoCaptureSupported         = 0x0008,
  /** Viewfinder display mirroring flag */
  EViewFinderMirrorSupported     = 0x0010,
  ...
  };
```

Thus, in order to check the viewfinder capabilities, you might use code such as:

```
TCameraInfo info
iCamera->CameraInfo(info) ;

// see if direct screen access is supported first
if (info.iOptionsSupported & EViewFinderDirectSupported)
  {
  ViewFinderDirectScreenAccess()
  }
else if (info.iOptionsSupported & EViewFinderBitmapsSupported)
  {
  ViewFinderBitmaps();
  }
```

If both options are available, it is usually more desirable to use a direct screen access viewfinder, which is why the example code checks for this option first and uses bitmaps only if direct access is not supported.

Once the viewfinder mode has been established, you have to set up the parameters for your chosen method. For direct screen access you do this using the StartViewFinderDirectL() function, whose prototype is:

```
virtual void StartViewFinderDirectL(RWsSession& aWs,
              CWsScreenDevice& aScreenDevice,
              RWindowBase& aWindow,
              TRect& aScreenRect)=0;
```

As illustrated below, you pass this function a reference to the current window server session, a reference to a screen device, a reference to a window and the required screen rectangle (in screen coordinates).

```
CExampleCameraApp::ViewFinderDirectScreenAccessL()
  {
  TRect screenRect = *iAppView->Rect();
  screenRect.Shrink(10,10 );

  iCamera->StartViewFinderDirectL(iCoeEnv->WsSession(),
              *iCoeEnv->ScreenDevice(),
              *iAppView->DrawableWindow(),
              screenRect);
  }
```

There is also an overloaded version of StartViewFinderDirectL() that additionally allows you to specify a clip rectangle.

With direct screen access, the viewfinder will start displaying images immediately and can be stopped by calling StopViewFinder().

If, instead of direct screen access, you are using the bitmap display mode, you need to call the StartViewFinderBitmapsL() function, prototyped as:

```
virtual void StartViewFinderBitmapsL(TSize& aSize)=0;
```

The only parameter to this function is a `TSize` that specifies the region in which the bitmap should be displayed. In general, it is best to choose a standard capture size for the camera, from those that can be found by means of calls to the `EnumerateCaptureSizes()` function. This function, whose prototype is shown below, writes a supported capture size to the passed `TSize` reference, given a size index and an image format.

```
virtual void EnumerateCaptureSizes(TSize& aSize,
                TInt aSizeIndex,
                TFormat aFormat) const=0;
```

The size index is a value between `0` and `TCameraInfo::iNumImage-SizesSupported-1`, and the image format is any one of the `CCamera::TFormat` enumerated flags that are set in `TCameraInfo::iImageFormatsSupported`.

This process is illustrated in the following example code. Of course, in a real application, you would iterate through the available size indexes and choose a suitable size for your application, rather than simply selecting the first size index. You would also need to select a supported format from those listed in `TCameraInfo::iImageFormatsSupported`.

```
CExampleCameraApp::ViewFinderBitmapsL()
  {
  TSize imageSize;
  iCamera->EnumerateCaptureSizes(imageSize, 0,
    CCamera::EFormatFbsBitmapColor16M);

  iCamera->StartViewFinderBitmapsL(imageSize);
  }
```

If you are using the bitmap display mode, ECam calls the observer's `ViewFinderFrameReady()` function whenever a new viewfinder frame is ready to be displayed and it is the application's responsibility whether or not to display the latest frame.

Image Capture

Capturing a still image is reasonably straightforward. The first task is to select a suitable image format and then to find a suitable size, using `EnumerateCaptureSizes()`, as described in the previous section. Once you have chosen an image index for a given image format, you then need to prepare the camera for image capture by calling `PrepareImageCaptureL()`.

These processes are outlined in the following example code.

```
// Get the camera info
TCameraInfo info
iCamera->CameraInfo(info) ;
```

```
// Select a format
CCamera::TFormat format;
if (info.iImageFormatsSupported & CCamera::EFormatFbsBitmapColor16M)
  {
  format = CCamera::EFormatFbsBitmapColor16M;
  }
else
  {
  User::Leave(KErrNotSupported);
  }

// Choose the index for the required image size.
TInt index = 0 ;
TSize size;
iCamera->EnumerateCaptureSizes(size,index,format);

// Set up, and perform, the image capture
iCamera->PrepareImageCaptureL(format,index);
iCamera->CaptureImage()
```

This example is slightly more realistic than that used to set up a viewfinder, in that it checks that the desired format is supported, and leaves if it is not. However, it still arbitrarily selects an image index of zero and accepts the resulting image size. A real selection would be highly dependent on the nature of the application.

If `PrepareImageCaptureL()` does not leave, then the image preparation has been successful and you can proceed to capture an image by calling the `CaptureImage()` function. The image is captured asynchronously and, on completion, is passed to the application by means of a call to the observer's `ImageReady()` function. The image is passed as either a pointer to a `CFbsBitmap` or a pointer to an `HBufC8` descriptor, depending on the format specified in the call to `PrepareCaptureImageL()`. An error code is also passed to `ImageReady()`, with a value of `KErrNone` indicating whether the capture was successful.

You can cancel an image capture at any time using the `CancelCaptureImage()` function.

8

Comms and Messaging

This chapter provides an overview of the comms facilities that Symbian OS provides, and explores in detail the APIs for messaging, in particular those used for sending and receiving multimedia messages (MMS).

8.1 Introduction

Comms development on Symbian OS can be considered to fall into three main areas:

- Developing drivers to talk to communications hardware, such as serial ports, or telephony hardware. Development of this type is undertaken by the phone developer when a phone is produced. Symbian supplies drivers that work with some specified reference hardware, such as the Intel Lubbock board, which can be used while the real phone hardware and drivers are under development.

- Developing protocol implementations, such as implementations of HTTP for accessing the web, or infrared for infrared protocols. Phone developers and third parties can develop protocol implementations to add to those already supplied as part of Symbian OS.

- Developing applications that use the available protocols: the phone developer supplies the phone with basic communication applications, such as telephony and messaging. However, applications in many areas can benefit from having integrated communications. These include:

  - games, which can be made interactive by multi-player competition over local communications protocols such as Bluetooth, or over packet data on 2.5G telephone networks

Symbian OS C++ for Mobile Phones, Volume 2.   Edited by Richard Harrison
© 2004 Symbian Software Ltd   ISBN: 0-470-87108-3

– enterprise applications, to communicate with specialist application servers.

8.1.1 Communications Components

Among the currently available communications components are:

- serial communications framework
- sockets framework
- telephony framework
- TCP/IP stack
- Bluetooth stack
- infrared stack
- SMS and EMS stack
- WAP stack
- HTTP transport framework
- Telnet and FTP engines
- messaging protocol support, including for MMS, SMTP, POP3 and IMAP4.

These components offer APIs that your applications can access.

8.1.2 Comms and Platforms

As in other areas of Symbian OS development, when doing comms-related work you will need to understand what is provided by Symbian OS and what is added by the UI platforms, such as Series 60 and UIQ.

The first principle, as you might expect, is that Symbian OS provides engine components that implement particular communications protocols and publish APIs to these engines; UI platforms provide applications that use these engines. For example, Symbian OS provides components that implement Internet email protocols, while UIQ and Series 60 provide messaging applications that allow users to send and receive email.

There are, however, ways in which this straightforward picture can be complicated:

- New comms protocols are being developed, and quite often the mobile phone manufacturers are leaders in this development. They naturally wish to put these protocols into their phones, and this can occur ahead of when an implementation has been produced for Symbian OS. For example, Series 60, which first appeared based on Symbian OS v6.1, provides its own implementation of multimedia messaging, which is different from the Symbian OS multimedia messaging that was added in Symbian OS v7.0.

- Comms is, of course, not just about software: it requires hardware too. Protocols such as Bluetooth and infrared require hardware that may not be present on all phones.

- For telephony, and telephony-related protocols such as WAP, the availability of certain features may also depend on the network operator, or on the user's account with their operator.

- Comms standards typically specify some features as mandatory and others as optional. Phone developers can modify the Symbian OS protocol implementations to add optional features if they are not already present.

- For comms features that are closely tied to a user interface, a licensee may add additional APIs that are preferable to use than the equivalent Symbian OS APIs. For example, for integrating simple message sending into applications, both Series 60 and UIQ provide APIs that would usually be preferred to the generic Symbian OS equivalent.

Because of these complexities, most of the comms support takes the form of frameworks into which modules that implement particular protocols are loaded as plugins at runtime when they are needed. For example, there is a framework for accessing protocols through sockets; protocols that can be accessed in this way, such as TCP, are implemented as plugins to this framework.

Generally, when you're using such an API, you'll need to know about both the general API provided by a framework, and also the particular parameter values or classes that relate to the individual protocol. To continue the sockets example, the sockets API provides a general class, `TSockAddr`, that encapsulates a socket address; the TCP/IP support provides a class derived from this, `TInetAddr`, that encapsulates an IP address.

The use of frameworks means that application code written for use with one protocol can be modified relatively simply to use a different protocol that has been written for the same framework. In some cases, flexible applications can be written that can query for the available protocols and, say, allow the user to choose which one to use. The messaging application is usually written in this way: it queries the messaging framework for the available messaging protocols, and dynamically modifies its user interface to offer the user commands relating to those protocols.

Unfortunately, this combination of frameworks and plugins can mean that the APIs are sometimes not the easiest to use. This difficulty has been reduced, for messaging and telephony, by the provision of simpler wrapper APIs that encapsulate basic functionality.

8.2   Overview of Symbian OS Comms Architecture

As mentioned above, to understand Symbian OS comms, you'll need to understand both the general frameworks provided and the implementations of particular protocols that plug in to these frameworks. This section will look first at the frameworks and then go on to the protocols.

Each of the key frameworks uses the Symbian OS client/server architecture. In this architecture, a program running in the background (the server) offers services to multiple other programs (the clients). This arrangement is chosen when there is some common resource which multiple client programs on a phone may want to access. It's the server's job to control access to the resource. For the lower-level comms servers, the resource in question may be a hardware resource, such as a serial port. A resource can also be shared data, such as the store for messages.

Although a framework usually has elements other than just the server itself, such as libraries of utility classes, a whole framework is often referred to, as a shorthand, as the server.

In some cases, the APIs make it obvious that the client/server architecture is being used. For example, the primary task when using the telephony functionality is to create an RTelServer object, which provides an initial connection to the telephony server. (The Symbian OS convention is that API classes, such as RTelServer, which are used to access a server begin with 'R'.) Other APIs, such as messaging, provide extensive client-side classes that hide the direct use of the client/server interface from the client program.

The core servers are as follows:

- Sockets: provide communications between addressable end-points, using protocols such as TCP/IP. This has been part of Symbian OS since its first version; v7.0s adds a new API for creating and managing connections.

- Serial comms: provide communications over simple serial connections, such as for handling RS232.

- Messaging: provide sending, retrieving, and storage of messages, using such protocols as Internet email and SMS.

- Telephony: provide control of telephone calls and services, and of the configuration of the telephony functionality.

8.2.1   Sockets

The idea of sockets first appeared in the Berkeley Software Distribution (BSD) of Unix from the University of California at Berkeley, with an API in the C language. Since then, sockets have become common across many operating systems and languages.

A socket represents the logical end of a communication 'channel'. It is a combination of a physical machine network address and a logical port number to which another socket can transmit data.

Because a socket is identified by a machine address and a port number, each socket is uniquely identified on a particular computer network. This allows an application to uniquely identify another location on the network with which to communicate.

The traditional use of sockets is to communicate over a network running the Internet Protocol (IP). In that case, the machine address would be an IP address, and the port would specify some Internet application, such as the Web or FTP.

Compared with sockets from another operating system, there are two main differences with the Symbian OS implementation of sockets:

- Sockets can be used to access a number of protocols, not just TCP/IP. These include Bluetooth protocols L2CAP and RFCOMM, and the infrared protocols IrDA, IrTinyTP and IrMUX.

- The APIs are in C++ and are not identical to the traditional BSD C API. If you consider it essential to use the C API, for example when porting code designed for another operating system, you can consider using the C API available in the Symbian OS implementation of the C Standard Library (STDLIB). For details of this, and of the issues that may arise, see the Developer Library's guide section on the C Standard Library.

The Symbian OS C++ API to use for sockets is the sockets client API, published in the header es_sock.h and in the library esock.dll. The client interface to the socket server is provided by RSocketServ, while a socket itself is encapsulated by RSocket.

The socket's client APIs make asynchronous calls to a sockets server, which coordinates client access to socket services and manages communications with protocol modules that provide support for the particular networking protocols. The protocol modules are plugin DLLs that the server loads and unloads as required.

In addition to connecting to sockets, and reading and writing data, the API provides access to other facilities:

- Hostname resolution (RHostResolver): some network types have the ability to convert between symbolic machine addresses that are suitable for displaying to end users, and the numeric addresses used internally within the protocols. In the case of TCP/IP, the hostname resolution service is the Domain Name Service (DNS). For Bluetooth and infrared, the resolution interface can be used to discover which other devices are in range and available for communication using those protocols. Queries made through RHostResolver objects are packaged in `TNameEntry` descriptors, which hold `TNameRecord` objects that contain the host name and address.

- Protocol information (`TProtocolDesc`): you can query which socket protocols are supported on the phone, and get information for each, such as the protocol name and flags indicating its capabilities.

The sockets API also offers the following, which are less likely to be useful to you:

- Network database access (`RNetDatabase`): this is intended for access to databases about devices. For infrared, there is such a service, the IrDA Information Access Service (IAS). It has no relevance for TCP/IP or Bluetooth.

- Service resolution (`RServiceResolver`): this is intended for queries about the capabilities of a remote device, that is, what services the device can offer over the relevant protocol. This is not implemented for any of TCP/IP, Bluetooth, or infrared. The Bluetooth standard does have such a service, the Bluetooth Service Discovery Protocol (SDP), but the Symbian OS Bluetooth implementation does not use the sockets API for this purpose, as it has its own dedicated SDP API.

Before Symbian OS v7.0s, how a network connection was made in order to fulfill a socket request was not the concern of the caller of the socket's client API. A connection was made implicitly, for example, as follows.

An application requests, for example, a TCP socket to a certain remote address. The Symbian OS component concerned with managing network interfaces (NIFMAN) checks that no network connection already exists. It reads from a database of communication settings (CommDb) how to make a connection. For example, the settings might specify to dial up an ISP. Other components would then be called that could perform the dial-out and connect to the ISP using a suitable protocol (e.g. PPP). The required settings to establish the connection, such as the ISP's phone number and log-on information, are also stored in the communications database.

Technologies such as W-CDMA and later releases of GPRS are capable of establishing multiple subconnections within a connection. This is supported in v7.0s by the connection management interface `RConnection`. It provides clients with functionality to create, configure, and monitor connections and subconnections.

8.2.2 Serial Communications

Serial communications are simpler than sockets. Instead of having many possible devices, and services thereon, to connect to, data is simply read to and written from a port on the phone. Traditionally, this has been used when the Symbian OS device is connected to a PC for synchronization, by a cable or by infrared, or to an external modem.

As with sockets, the Symbian OS serial comms implementation uses a server that can load plugin modules to handle particular communication protocols. These plugin modules, known as CSY modules, are loaded by the Serial Comms server and are not directly accessed by client applications.

A Symbian OS phone may include a number of CSY modules as standard, such as for handling RS232 and infrared serial communications. The Serial Protocol Module API allows new CSY modules to be developed.

You can discover the available serial ports on a phone and their protocols through the server session class RCommServ. Once you have selected a port to use, you access it through RComm, the serial port interface. This class is used to read, write, configure, set break conditions, and get port state information.

Typically, before using a port, you will set its configuration. This includes settings such as the data rate, parity type, and handshaking control. These settings are held in a serial port configuration block of type TCommConfigV01. You can check the capabilities of a serial port before configuring it, to ensure that the desired configuration is possible. The capabilities are encapsulated by an object of type TCommCapsV01.

8.2.3   Messaging

The messaging components of Symbian OS provide a framework for multi-protocol messaging, and support for particular messaging protocols.

Messaging offers opportunities to build highly featured message client applications and to create plugin modules to support individual messaging protocols. The set of components that make up such a plugin module is called a Message Type Module (MTM). All interaction with lower-level communication protocols, such as TCP/IP, is performed by the MTMs.

The Messaging architecture provides base classes which define the component interfaces for MTM implementations. Using the base class interfaces allows client applications to discover and use dynamically the available messaging protocols.

Protocol providers develop new libraries for their MTM implementations. These implementations access lower-level communications libraries as needed by the protocol.

8.2.3.1   Message Server

The Message server controls access to message data and delegates protocol-specific requests to server-side MTMs.

Each message that is stored by the server has an integer identifier, of type TMsvId. The state of a message (for instance, whether it is read or unread) and some generic properties common to most messages, such as date and subject, are held in a header data object, accessed through the class TMsvEntry. A message can also have:

- a file store that holds the message body text, and protocol-specific data; this is encapsulated by the CMsvStore class

- a directory in which associated files (e.g. message attachments) can be stored.

A message as a whole, from which its ID, header, store, and file directory, can be accessed and manipulated, is encapsulated by CMsvEntry.

Aside from messages themselves, the server also stores entries that represent services and folders. A service is a useful abstraction that collects settings information. For example, for SMTP, a service would specify settings for a mail account. Folders hold groups of entries (messages and other folders). Some of these folders, for example Inbox and Drafts, are always present and are created by the message server on its first startup. Users can also create their own folders.

Messages are held in a tree, similar in form to a file system directory tree. Each entry in the tree represents a service, folder of messages or message part. An example is given in Figure 8.1.

The tree is broken down into three levels:

- The root entry: this is present just to tie the tree structure together.

- Service entries: there is one local service under which local folders and messages are stored and zero or more remote services. Remote services represent message accounts.

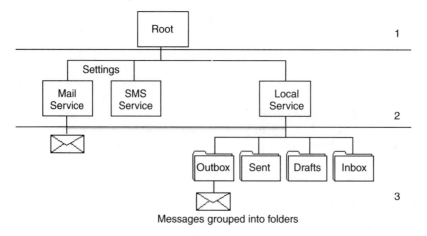

Messages grouped into folders

Figure 8.1 Message storage

- Messages and folder entries: messages and folders under the local service represent messages stored on the device. Messages and folders under remote services represent a local copy of messages that are present on a remote server. For example, under a POP3 email service you would have copies of all the messages present on the POP3 email server; under an SMS service you might find SMS messages that are stored on a SIM.

8.2.3.2 MTM Base Classes

MTM base classes are subclassed to provide support for a messaging protocol. There are four classes to consider.

The User Interface MTM (CBaseMtmUi) offers user interface operations. These include the following:

- Create: launches a message editor with a new message.

- Edit: if the entry is a message this launches the editor; if an entry is a service, it edits the settings.

- View: launches a viewer for the message.

- Displaying the progress of an operation.

The client-side MTM (CBaseMtm) provides a generic interface to operating on message data. The class defines functions to:

- create a message

- reply to a message

- forward a message

- add/remove addressees

- add/remove body text

- add/remove subject

- add/remove attachments.

The UI Data MTM (CBaseMtmUiData) provides access to certain UI MTM-specific resources. These include:

- MTM-specific icons for Message Server entries

- user interface text, e.g. for menus, for MTM-specific operations

- information functions for checking whether MTM functions are suitable for use on an entry.

Finally, the server-side MTM (`CBaseServerMtm`) provides message transport over the relevant communications protocol to remote services. Its functions include:

- moving or copying an entry that is currently under a remote service

- moving or copying from the local service to a destination under a remote service

- creating, deleting, or changing remote entries, if the protocol allows messages on a remote server to be manipulated

- implementing MTM-specific commands. For example, such a command might be to synchronize the entries with the messages on a remote server.

8.2.3.3 Registry

MTMs must be registered with the message server. This allows clients to query what MTMs are present, and the server to know which DLLs to load to create a given MTM component. Registration is done by providing a resource file for each MTM.

The registry classes allow MTM components to be identified and instantiated. Key classes are `CClientMtmRegistry` and `CMtmUi-DataRegistry`.

8.2.3.4 SendAs

This interface is simple to use and allows applications to create outgoing messages. Note that despite the name of this interface, SendAs provides an API only to create, but not to send a message.

When SendAs is used, it first provides to the caller a list of all the registered MTMs that support sending messages. The application can then add more constraints on the list of MTMs: for example, it can insist that the MTM supports attachments, or a certain message size. SendAs queries each of the MTMs about these additional requirements and removes any MTMs from the list that do not meet them. When the application has finished adding requirements, it selects one of the MTMs left in the list and uses that to compose the message, adding addresses, a subject, body text and attachments.

The basic interface is provided by `CSendAs` and `MSendAsObserver`. These are, however, 'engine' APIs: they do not provide a user interface for creating and sending messages that can be simply added to applications. However, the user interface platforms do provide wrapper classes to achieve this. These are:

- UIQ API: `CQikSendAsDialog`
- Series 60 API: `CSendAppUi`.

8.2.3.5 Scheduled Sending

The Schedule Send functionality provides a client with the ability to schedule messages to be sent at a later time, instead of immediately. It allows scheduling messages, deleting their schedule, rescheduling them and checking their schedule. It also provides meaningful status information, such as whether a message is currently scheduled, sending, failed, etc.

MTMs can choose whether or not to provide support for scheduling. To support scheduling, the server MTM must derive from `CSchedule-BaseServerMtm`. The server MTM uses another Symbian OS component called the Task Scheduler that can handle initiating specified actions at specified times. The interface to the task scheduler is encapsulated by `CMsvScheduleSend`.

8.2.4 Telephony

Mobile telephony is a complex area that includes many standards for phone services, networks, and hardware. The telephony APIs and frameworks that Symbian OS offers are intended to reduce this complexity by providing common interfaces to the phone functionality, whatever the underlying hardware or network.

The API uses the Symbian OS client/server framework. The API provides `R` classes that send requests to a telephony server. The server in turn passes requests to an appropriate telephony driver plugin that handles the physical device.

Telephony APIs are used by the phone manufacturer to provide a telephony application on the phone that allows the user to make calls and to set up service preferences. Apart from this purpose, telephony APIs can be considered low-level and are used mainly by other, higher-level comms components. For example, the sockets section explained how an application's request to connect a socket can result ultimately in a telephone call being made to an ISP.

If you do need to use telephony directly from an application, you may find a difficulty in that licensees often choose not to expose the more advanced telephony APIs in their SDKs. So unless you are a partner of the licensee or of Symbian, your applications may not be able to access all the phone's telephony functionality.

SDKs do, however, expose two APIs of general interest to application programmers: ETel Core and Third Party Telephony.

ETel Core defines a core set of functions that are supported by almost all telephony devices and services. Applications can use it directly to access general telephony devices. The API breaks down into four main classes, which encapsulate a session to the telephony server, a phone, a line, and a call, respectively.

- The session class, RTelServer, provides access to system telephony information, in particular the available phones and TSYs.

- The phone class, RPhone, abstracts a particular telephony device. It allows the client to access the status and capabilities of the device and to be notified if these change.

- The line class, RLine, represents a line belonging to a phone. Clients can access the status and capabilities of a line and can be notified if these change.

- Finally, lines can have zero or more active calls, which are encapsulated by the RCall class. A call has the functionality to dial a number, to wait for an incoming call, and to hang up. As before, a client can get status and capabilities information and can be notified of changes.

As you can see, ETel Core has a fairly straightforward and logical structure, but a simpler encapsulation of its functionality, the Third Party Telephony API, is also available. This provides a high-level interface for telephony data calls. It allows the API's user to make an outgoing telephone call, wait for an incoming call, or check the status of the line. The API has a single class, CTelephony. The caller passes an RComm (serial port) object to the CTelephony object, and when the call is established it uses that object to read and write data over the connection.

8.2.4.1 Advanced APIs

Modern mobile telephony standards offer much more than simply the ability to make and receive calls. Symbian OS offers extensions to the basic ETel Core API to provide access to these advanced capabilities. However, as mentioned, these extensions may not be available on SDKs.

The principal advanced API is ETel Multimode, which extends the ETel Core API to offer access to common mobile telephony services for a number of air-interface cellular standards (or modes): at v7.0s, the modes are GSM, GPRS, EDGE, CDMA (IS-95), 3GPP2 cdma2000 1x (Release A), and 3GPP W-CDMA. The API also offers access to specialized services for particular standards, such as GSM.

Other advanced telephony APIs are:

- ETel Multimode Packet Data, which provides a common API for telephony packet data standards such as those in GPRS, CDMAOne and CDMA2000.

- SIM Application Toolkit, which provides an API to access functionality offered by the phone's SIM card.

To complete the survey of telephony APIs, there are a few less important APIs to mention:

- Dial: provides utilities related to dialing, such as phone number manipulation.

- PhoneBook Synchronizer: synchronizes the two stores of address information on a Symbian OS phone, the phonebooks on the phone's ICC (on GSM phones called the SIM card), and the Contacts Database.

Finally, fax support is also closely related to the telephony APIs:

- The ETel fax client is used to send and receive faxes to or from a Symbian OS fax store file.

- The Faxio API provides libraries for fax encoding and compression.

- The Fax store API is used to access stored faxes.

With the telephony framework, we've completed a survey of the most important components in the Symbian OS comms architecture. We can now go on to look at the many concrete comms protocols that have been developed for this architecture.

8.3   Protocol Support

The protocols that Symbian OS supports are discussed in this section:

- networking: TCP/IP, the basis for all Internet communications

- Internet application protocols: HTTP, FTP, and Telnet

- short range communications: Bluetooth and IrDA infrared

- WAP

- telephony: SMS and EMS

- messaging: BIO messaging, email, SMS and MMS.

8.3.1   TCP/IP

TCP/IP is a family of protocols that provide communications over the Internet and have also been adopted for local networks. Symbian OS provides support for both IPv4, the version of the Internet protocol commonly used at present, and the next generation IPv6 which offers, among other enhancements, a far larger number of possible addresses (each address is 128 bits rather than 32 bits) than IPv4. This large address space is expected to be required as many more devices, such as phones, require IP addresses. IP-based Symbian OS clients such as email, HTTP, and SSL can use IPv6 addressing as well as IPv4 addressing.

Symbian OS implements TCP/IP as a plugin to the socket server. The main features are as follows:

- Sockets are available for the IP, ICMP, TCP, and UDP protocols. Socket services are provided through the generic interface RSocket.

- Domain name services (DNS) are available through the generic host name resolution interface RHostResolver.

- A socket address, specifying an IP address and port, is encapsulated in a TInetAddr object. It stores either an IPv4 (32-bit integer) or an IPv6 address (encapsulated in the class TIp6Addr).

The TCP/IP plugin also provides the security protocols for secure electronic commerce, the Transport Layer Security (TLS) and Secure Sockets Layer (SSL). Secure sockets are accessed through the CSecureSocket class.

Also related to security is support for IPSec. This IP layer protocol is used to secure host-to-host or firewall-to-firewall communication. IPSec is a plugin module to the IP stack, and provides tunneling, authentication and encryption for both IPv4 and IPv6.

8.3.2 HTTP

Symbian OS does not have its own web browser, leaving browser development to licensees or to specialists such as Opera. It does, however, implement the transport protocol underlying the web, HTTP.

The HTTP 1.1-compliant stack enables applications such as SyncML, GPRS, OCSP and streaming multimedia to operate over TCP/IP. Third-party browsers can use this stack as well.

HTTP is provided as a plugin to a framework that has not been mentioned so far, the Transport Framework. This provides a generalized mechanism for HTTP-like protocols that operate over various transports. The framework provides a unified, high-level API that is independent of particular data formats, commands, and transports. Apart from HTTP itself, the HTTP-like WAP protocol, WSP, can be used through the framework.

The user of the framework first sets up a session, an RHTTPSession object. This defines the HTTP protocol, encoding and transport that are used. Once a session is established, any number of transactions, encapsulated by RHTTPTransaction, can be carried out between the client and the remote server. Normally a transaction consists of a single exchange of messages between client and server (a client request and a server response).

Both the request and response portion of a transaction are composed of a header and an optional body. The header, accessed through RHTTP-Headers, has fields that are used to convey information about the transaction, the message, or the client and server themselves. The body portion of requests and responses is represented in the API as an abstract interface, `MHTTPDataSupplier`, which allows the real implementation of the classes that generate body data to be hidden.

The framework allows a protocol to be extended through the use of filters. These are add-on modules that provide additional behaviors to a session: for example, caching received data to improve communications efficiency. Filter classes implement the abstract interface `MHTTPFilter`.

8.3.3 Telnet

Telnet is a well-established Internet protocol that is typically used to connect to remote machines, in order to execute commands and receive command responses. Symbian OS has a Telnet engine to implement the protocol. The main class, `CTelnetSession`, provides functions that connect to, read from, write to and disconnect from a Telnet server. It also offers functions that get and set the Telnet configuration.

8.3.4 File Transfer Protocol (FTP)

The Internet File Transfer Protocol is used to transfer files to and from a remote server. As with Telnet, Symbian OS has an engine component to implement the protocol. A class `CFtpProtocol` allows a client to connect to an FTP server and to access individual FTP commands. Another class, `CFTPSession`, provides a simplified interface to FTP, by wrapping the most common FTP operations in simple-to-use functions.

8.3.5 Bluetooth

Bluetooth is a short-range radio communications technology, standardized by the Bluetooth SIG. Symbian OS has a Bluetooth stack that is fully compliant with the Bluetooth v1.1 specifications.

The services that a Bluetooth implementation can offer are characterized by the Bluetooth standards as a number of profiles. The Symbian OS Bluetooth stack fully implements the Generic Access Profile, the Serial Port Profile and the General Object Exchange Protocol. All other Bluetooth profiles are dependent on these three core profiles.

A Bluetooth stack theoretically contains the following layers:

- RFCOMM: allows an application to treat a Bluetooth link in a similar way as if it were communicating over a serial port.

- Logical Link Control And Adaptation Protocol (L2CAP): allows finer-grained control of the link. It controls how multiple users of the link are multiplexed together, handles packet segmentation and reassembly, and conveys quality of service information. It is the usual choice for applications.

- Service Discovery Protocol (SDP): used for locating and describing services provided by or available through a Bluetooth device. Applications typically use it when they are setting up communications to another Bluetooth device.

- Host Controller Interface (HCI) driver: packages the higher level components to communicate with the hardware.

The Symbian OS implementation of this stack is made up of a Bluetooth sockets server module, a Bluetooth security manager, a Bluetooth communications server module and a Service Discover Protocol server module. Together these provides six interfaces that allow access to the different levels of Bluetooth functionality.

A Bluetooth serial communications API is provided by a plugin to the serial communications server. It provides a serial port emulation over Bluetooth, though only for outgoing connections; incoming serial connections must be implemented using RFCOMM sockets. The API is intended for use by legacy applications that already use the serial communications API, and for the Serial Port Profile.

A Bluetooth sockets API is provided by a plugin to the sockets server. The API provides support for both the Logical Link Control and Adaptation Protocol (L2CAP) and the RFComm protocol layers. The Bluetooth sockets API is the preferred means of managing Bluetooth connections, as it offers more control and flexibility than the serial communications API. As a plugin to the sockets server, access to the API is through the sockets classes, such as `RSocket`, that we have already seen being used for TCP/IP. Bluetooth provides a specialized address class, `TBTSockAddr`, which adds a Bluetooth device address field, encapsulated by `TBTDevAddr`. You can also discover which remote Bluetooth devices are available to communicate with through the sockets API, using the class `RHostResolver`. A specialist Bluetooth sockets address class, `TInquirySockAddr`, which encapsulates Bluetooth address, Inquiry Access Code, and service and device classes, is provided for use with such inquiries.

The Bluetooth manager provides a highly configurable means of maintaining the security of a Symbian OS phone while using Bluetooth. It consists of a system server, the Security Manager and a Registry module to store information about phones and services. It allows the implementation of flexible security access policies for the range of services supported by the phone. For example, high security can be given for services that require authorization and authentication, medium for services that require authentication only, and low for services open to all devices. Access to the manager is provided by two main classes, `RBTMan` and `RBTSecuritySettings`.

The Bluetooth Service Discovery Protocol (SDP) API allows services to be discovered and to be registered. The Service Discovery Agent API enables you to discover the Bluetooth services, and the attributes of those services, that are available on a remote device. The central class of this API is CSdpAgent. The Service Discovery Database allows a local service to enter its properties into a local Bluetooth service database, which enables remote Bluetooth services to discover that the service is available. To access the database, a client must create and connect a session, RSdp, and a subsession, RSdpDatabase. A subsession allows service records and their attributes to be added, deleted, and updated.

The Host Controller Interface (HCI) module lives at the bottom of the Bluetooth stack, and communicates to standard HCI-compliant Bluetooth hardware. Although there is no direct Symbian OS interface to the HCI, HCI-related commands can be given through asynchronous I/O control (RSocket::Ioctl()) commands on an L2CAP or RFCOMM socket.

A final way to access Bluetooth is through the Object Exchange (OBEX) protocol, which provides a means to exchange objects such as appointments and business cards between devices. Symbian OS implements OBEX v1.2, and provides an API to access OBEX over Bluetooth RFComm sockets as well as over infrared TinyTP sockets. The CObex-Client class provides functions to connect an OBEX session to a remote machine, running over an IrDA or Bluetooth socket, and to perform commands to get or receive objects. The CObexServer class provides a framework for servicing OBEX requests from remote clients.

8.3.6 Infrared

Infrared provides a means for short-range communications between devices, with standards defined by the Infrared Data Association (IrDA). Applications typically use device discovery to find other IR devices in range, service queries to discover whether the found device supports the required service, and then use either a reliable or an unreliable data protocol to transfer data. Symbian OS supports both slow infrared (SIR), at speeds of 9.6 kbps to 115.2 kbps, and fast infrared (FIR), at speeds of 0.576 Mbps to 4 Mbps.

Like many other comms protocols, infrared can be thought of as a stack of layers providing different levels of functionality. The layers of interest are as follows:

- Link Management Multiplexer (LM-MUX): provides infrared communications equivalent to an unreliable datagram service.

- IrDA Tiny TP: provides infrared communications equivalent to a reliable packet service.

- Link Management Information Access Service (LM-IAS): gets information about a remote device's capabilities.

- Link Access Protocol (IrLAP): provides low-level control of the infrared link, such as baud rate.

The IrDA stack is implemented by Symbian OS in a plugin to the sockets server. The layers of the stack are accessed through the various abstractions that the sockets API provides:

- IrDA, IrTinyTP and IrMUX can be accessed through the generic socket interface `RSocket`.

- IrLAP options can be set through options on such sockets.

- LM-IAS is accessed through the sockets `RNetDatabase` class.

Above these basic sockets layers, further higher-level infrared services are available:

- IrOBEX v1.2 (IrDA object exchange): OBEX, a protocol to exchange objects such as business cards, has already been mentioned in connection with Bluetooth. The same APIs, `CObexClient` and `CObexServer`, can be used with infrared as the transport.

- IrTRANP v1.0: this is a protocol for transferring pictures from a digital camera over infrared. The `CTranpSession` class encapsulates the behavior for receiving a picture from a peer device.

- IrCOMM v1.0 provides an emulation of a serial port over infrared. It is implemented as a plugin to the serial communications server, and is accessed through the generic serial interfaces, such as `RComm`.

8.3.7 WAP

The Wireless Application Protocol (WAP) is family of protocols, developed by the WAP Forum, and then by The Open Mobile Alliance (***http://www.openmobilealliance.org***), which provides data communications and browsing services over mobile telephony.

Symbian OS provides a WAP stack with support for WAP 1.2.1 (WAP June 2000), push functionality and GPRS as a bearer. The stack has the following layers:

- WSP, session protocol for WAP

- WTP, transaction protocol for WAP

- WTLS, transport layer security protocol for WAP

- WDP, datagram protocol for WAP, client and server.

The key interfaces to this stack are `RWAPConn`, for connections over WSP, WTP, and WDP; `RWSPCLConn` for WSP-specific functionality; and `RWDPConn` for WDP-specific functionality.

WAP, however, is an area where Symbian OS licensees have taken the lead and chosen to provide their own WAP stack implementations. For this reason, client applications are recommended not to use WAP stack APIs directly, but to use one of two higher-level APIs that are provided to wrap the lower-level stack implementations. These APIs are as follows:

- The Transport Framework, which we've mentioned in the context of HTTP. WSP is similar to HTTP in form, and can be also accessed through this API.

- The WAP Messaging API, which provides a datagram service, encapsulating WAP Push and WDP. Its classes are:

  - `CWapBoundCLPushService`: listens for WAP Push messages from any sender

  - `CWapBoundDatagramService`: sends and receives datagrams over WDP

  - `CWapFullySpecCLPushService`: listens for WAP Push messages from a single, named remote host

  - `CWapFullySpecDatagramService`: sends and receives datagrams over WDP using a specified local port and a single, named remote host.

8.3.8   SMS and EMS

The Short Messaging System (SMS) is used to send and receive short text messages between mobile phones. It is defined as part of the ETSI's GSM telephony standards (03.40).

Symbian OS implements an SMS stack that offers functionality to:

- send and receive SMS messages, including concatenated messages

- enumerate, read, write and delete access to the SMS storage areas of the phone and SIM

- receive messages that match a specified text.

Seven-bit SMS alphabet, 8-bit SMS alphabet and the UCS2 data coding scheme are supported.

Like some of the other comms protocols we've looked at, the SMS stack is implemented as a plugin to the sockets server. Sending an SMS message involves:

- opening a socket, an `RSocket` object, for the SMS protocol

- creating an object, of class `CSmsMessage`, to hold the message text and settings such as the destination address. The object owns a `CSmsBuffer` object that encapsulates the text

- writing this object to the socket. A specialized stream class, `RSms-SocketWriteStream`, is provided for this purpose

- calling an ioctl socket command (`KIoctlSendSmsMessage`) to start sending.

Receiving SMS messages is also done through a socket. The messages intended for a particular application, rather than general messages that should be displayed to the user through the messaging application, are usually recognized by matching a distinctive string at the start of a message. An alternative is to use SMS port numbers. A socket can be set to receive suitable messages by setting the match criteria in a `TSmsAddr` object, and binding the socket to this object. An ioctl command can then be given to the socket to asynchronously wait until a matching message is received. Once a message is received, it can be read from the socket, using a stream class `RSmsSocketReadStream`, into a `CSmsMessage` object. From there, an application can easily access the message text for appropriate processing.

There are difficulties in using SMS and EMS APIs, as Symbian OS licensees generally do not publish the ETel multimode header in their SDKs, and as the SMS headers depend on the multimode header, programs that use the SMS headers will not compile. For UIQ developers, the Symbian Developer Network publishes a workaround that supplies enough of the ETel multimode header to allow the SMS headers to compile. This can be obtained at ***http://www.symbian.com/developer/development/utilapi .html***. The same location provides some useful wrapper classes that offer a simple API to send and retrieve SMS messages. For Series 60, Nokia have chosen not to make SMS headers available in their kits.

SMS is a text-only medium. The Extended Messaging Service (EMS) is a telephony standard (3GPP release 4, TS 23.040) that extends SMS to offer support for graphic and sound elements. Its features include:

- pictures at various resolutions

- animations, predefined and user-defined

- sounds

- formatting text in specified sizes, styles, and alignments.

The API for EMS is a simple extension to the SMS API. Each extended element, such as a picture, is encapsulated as an information element class that can be added to a `CSmsMessage`. For example, to add a picture, create a `CEmsPictureIE` object that specifies it, and add the object to your `CSmsMessage`.

8.3.9 Messaging Protocols

The messaging protocols include BIO Messaging, Internet Email (SMTP, POP3, IMAP4), Multimedia Service (MMS), Short-Message Service (SMS), and OBEX protocol support.

8.3.9.1 BIO Messaging

The BIO Messaging component provides a means of handling incoming messages that are intended for processing by the device as opposed to being displayed to a user. Examples of these types of messages include:

- Internet Access configuration: automatic updates of Internet settings

- email settings: update or creation of email accounts

- vCards: electronic business cards

- vCals: electronic calendar entries.

Standards for these types of messages are defined by the proprietary Nokia Smart Messaging standard, and the Nokia/Ericsson Over The Air Settings Specification.

BIO stands for Bearer-Independent Object: this is a reminder that the handling of such a message does not depend on the type of transport (e.g. SMS or email) over which it was received.

The BIO Messaging component can be split into two parts: the BIO framework and the BIO parsers. BIO parsers provide functionality to parse and process particular types of message. It is possible to add new BIO parsers to provide support for different types of BIO message. All parsers derive from a base class `CBaseScriptParser`.

The framework manages a database of BIO message information that can be used to determine the type of the incoming message. A component called the BIO watcher is responsible for watching for new messages that fit one of these types.

Once a BIO message has been received, the BIO MTM is invoked to parse and process it. The BIO MTM routes the message data to the appropriate BIO parser.

The parse and process operations may perform different functions depending on the type of message. Sometimes the message payload may be saved as a file for some other component to process; sometimes the data is processed immediately.

8.3.9.2 *Email*

The email MTMs implement a number of different email protocols:

- SMTP: to create and send new messages, reply to and forward email messages

- POP3: to download messages, or optionally just message headers

- IMAP4: to download messages, or optionally just message headers, or just the header and body text. Messages can also be managed (moved, etc.) on a remote Internet IMAP4 server.

Applications can use the client MTM interfaces for these MTMs, together with various utility classes, to perform email operations programmatically.

8.3.9.3 *SMTP Client MTM*

The SMTP Client MTM, `CSmtpClientMtm`, provides an API for creation of outgoing email messages. It also provides an SMTP-specific operation, to enable automatic sending of waiting messages either immediately or whenever a dial-up connection is made. Some generic messaging operations, such as receiving messages, are not supported by the MTM.

Settings for SMTP connections, such as server address and email address, are defined for each SMTP service entry. Encapsulation of service settings is provided by `CImSmtpSettings`.

For SMTP operations, progress information includes such things as connection state and number of messages sent. Progress information is provided by `TSmtpProgress`.

As mentioned in the earlier overview of the message server, the server keeps an index entry for each message with fields that describe its properties. Email messages have some extra email specific properties: a class `TMsvEmailEntry` is provided to read these.

8.3.9.4 *POP3 Client MTM*

The POP3 Client MTM, `CPop3ClientMtm`, provides MTM-specific operations for POP3 connections and mail retrieval. Some generic messaging operations, such as creating and sending messages, are not appropriate for POP3 and are not supported by the MTM.

In addition, a helper class is provided by `CImPOP3GetMail`, which encapsulates all the operations required to connect and retrieve mail. Options are available for moving or copying mail, and for getting all mail or selected mail.

A POP3 service's settings define email server and user log-on details. They are encapsulated by `CImPop3Settings`.

8.3.9.5 IMAP4 Client MTM

The IMAP4 Client MTM, `CImap4ClientMtm`, provides IMAP-specific operations, the most important of which are synchronizing with a remote server, or a folder on a remote server. Some generic messaging operations, such as sending messages (for which you should use SMTP), are not supported by the MTM. Extra functions are provided for obtaining and setting IMAP4 service settings.

As for POP3, a helper class is available that wraps up many individual IMAP operations into a single call. A large number of options are available, which fall into the following groups:

- get mail when already connected

- connect, get mail and then disconnect

- connect, get mail and then stay online.

The get mail helper class is `CImImap4GetMail`.

IMAP4 service settings are encapsulated by `CImImap4Settings`.

8.3.9.6 SMS

We've seen earlier that the SMS stack provides access to SMS functionality. Messaging's support for SMS provides a further level that integrates SMS sending, receiving, and storage into the messaging subsystem. Note that where SDKs do not allow the use of the lower-level SMS APIs, the messaging SMS APIs are not available either.

The SMS Client MTM, `CSMSMtmClient`, provides MTM-specific operations for SMS, including accessing a telephone handset's service center information, and scheduled sending of SMS messages. SMS receiving is normally done automatically by system telephony and messaging components, with received SMS messages placed in the Inbox.

Settings for SMS connections, such as service center addresses, are encapsulated by `CSmsSettings`.

For SMS operations, progress information includes such things as type of operation and number of messages processed. Progress information is provided by `TSmsProgress`.

8.3.9.7 OBEX

We've seen that APIs exist for OBEX over infrared and Bluetooth. An OBEX MTM also exists to integrate OBEX functionality into the messaging framework, so that, for example, an object sent to a phone shows up as an attachment to a message in the user's inbox. The classes `CIrClientMtm` and `CBtClientMtm` provide client-side functionality for OBEX messaging over infrared and Bluetooth respectively. Headers for such messages can be accessed through `CIrHeader` (infrared) and `CBtHeader` (Bluetooth).

This concludes the overview of the comms protocols. The next section discusses in more detail the remaining messaging protocol, the multimedia messaging service, MMS.

8.4   MMS

The Multimedia Message Service (MMS) is a standard defined by The Open Mobile Alliance (originally by the WAP Forum) for sending and receiving messages with rich content, from user to server to user, over mobile telephony networks.

MMS operates over dial-up (CSD) or GPRS connections and uses WAP or HTTP for message transports and notifications. MMS messages themselves are similar to MIME email messages. They contain headers that define the message properties, and a MIME body content that, in theory, could be any media type. In practice, the expected format for MMS messages is MIME Content-Type: Multipart/Related; i.e. the message contains a number of media objects, such as pictures, sounds and text. The first object is expected to be a Synchronized Multimedia Language (SMIL) file, an XML document that provides instructions to the client about how to present the other media objects in the message.

As we've seen, comms is not the simplest area for compatibility between Symbian OS devices, and the MMS implementations differ between generic Symbian OS, used in UIQ, and Series 60. The APIs described here are the Symbian versions. For the Nokia versions, see the document 'Designing MMS Client Applications for the Series 60 Platform' on the Series 60 SDK.

8.4.1   MTMs and APIs

The MMS implementation is supplied by plugins to the messaging framework, and some supporting utility libraries:

- The MMS Client MTM provides interfaces for MMS message creation, retrieval, sending, and forwarding and reply operations.

- The MMS Server MTM calls lower-level comms components to send and receive MMS messages. It does not have client interfaces.

- The MMS utilities library provides classes to encapsulate an MMS message, and its parts, such as message headers and media objects.

These libraries don't supply any functionality for rendering or editing MMS messages, or any other user interface operations. There is, however, a library to parse SMIL documents. In fact, to complicate matters, there are two! An original library, called SMIL Parser and Composer (`smiltranslator.lib`), was produced for v7.0. v7.0s substantially modified this approach to make it more flexible, with the ability to process XML documents for any specified DTD. This API is called the Messaging XML Support API. The v7.0 API is still available, but if you are writing new code you should prefer to use the new API.

We'll now work through the tasks involved in using the messaging framework and the MMS APIs to send and receive MMS messages. These tasks are:

- connecting to the message server
- getting the client MTM
- creating a message
- setting the message content and properties
- sending the message
- receiving messages.

8.4.2   Server Session

The messaging framework is based around a server program, to which a connection (a session) must be made before anything else with messaging can be done. The session class is called `CMsvSession`, and message client applications typically create an instance of this class upon startup. Instances of Client-side MTMs, User Interface MTMs, and high-level client library classes maintain a reference to the message client application's session object, so that they can make requests to the server as needed.

```
// Create message server session
iMsvSession = CMsvSession::OpenSyncL(*this);
```

The `OpenSyncL()` method takes a parameter of type `MMsvSession-Observer`, so, for this code to work, the containing class must implement this observer interface. It enables the server to call the client to inform it of significant events, such as new MTMs being registered.

8.4.3   Client MTM

An application cannot directly create an object for the MTM that it wishes to use. Instead, it must ask the message server to create the object for it. A record of all the MTM components installed on a machine is held in a dedicated registry file managed by the message server. Registry classes use this registration data to allow MTM components to be identified and instantiated. For example, the `CClientMtmRegistry` class has a member function to create a Client-side MTM object:

- A message client application calls this function, passing in the UID of the MTM component that it wants.

- `CClientMtmRegistry` searches the registry for a record for the required MTM component, and obtains from it the library name and ordinal of the library's exported factory function that can create an MTM object.

- `CClientMtmRegistry` loads the DLL and calls the factory function, passing the new object back to the caller.

A similar process is used for User Interface and UI Data MTMs. Such MTM objects are owned by the clients that create them. Clients are also responsible for deleting these objects.

In the current case, you want the MMS client MTM, so you use `CClientMtmRegistry`:

```
// Create client registry object
iClientMtmRegistry = CClientMtmRegistry::NewL(*iMsvSession);
// Request MMS client MTM object
iMmsClientMtm = (CMmsClientMtm *)iClientMtmRegistry->
   NewMtmL(KUidMsgTypeMMS);
```

The registry `NewMtmL()` function returns a pointer to the client MTM base class, `CBaseMtm`, so you cast it to the MMS client class, `CMms-ClientMtm`.

8.4.4   Message Creation and Deletion

Before creating a new message, you need to decide where you want to store it. The usual place in which to create a new entry is the Drafts folder. The Client-side MTM has the notion of a current entry, or context, on which operations take place: you therefore need to set this current entry to the Drafts folder. You may remember from the earlier discussion of the messaging architecture that every entry has an identifier by which it can be referred to. The identifier for the Drafts folder is published in the API, and is `KMsvDraftEntryId`. The following call sets the current entry to this folder:

```
// Set context to Drafts folder
iMmsClientMtm->SwitchCurrentEntryL(KMsvDraftEntryId);
```

Also before creating a message, you need to set to which service it belongs. You can programmatically discover which services are available.

The following code uses the message entry `ChildrenWithMtmL()`
function to ask for a list of entries, of the MMS type, under the message
root entry. This will get a list of MMS services, as services are stored
under the root.

```
// get the root entry of the message server
CMsvEntry* root=CMsvEntry::NewL(*iMsvSession, KMsvRootIndexEntryId,
        TMsvSelectionOrdering(KMsvNoGrouping,EMsvSortByDescription));
CleanupStack::PushL(root);
// get the IDs of all the children of the root, of the MMS type
CMsvEntrySelection* services=root->ChildrenWithMtmL(KUidMsgTypeMMS);
CleanupStack::PushL(services);
// Leave if no MMS service
if (services -> Count() == 0) User::Leave(KErrNotFound);
```

The code gets the list of IDs of MMS services, or leaves if no MMS service
is available.

Messaging also has the concept of default services, which are to be
used unless the user specifies otherwise. Default service settings can be
found through the class `CMsvDefaultServices`. They are stored in the
message store associated with the root entry. The following code reads
the default service settings from this store, and gets the ID of the default
MMS service in `serviceId`.

```
CMsvDefaultServices* services = new(ELeave)CMsvDefaultServices;
CleanupStack::PushL(services);
TMsvId serviceId;

if (root->HasStoreL())
    {
    // if the root has a store, restore the default services
    CMsvStore* store = root->ReadStoreL();
    CleanupStack::PushL(store);
    services->RestoreL(*store);
    CleanupStack::PopAndDestroy(); // store
    }

TInt error = services->DefaultService(KUidMsgTypeMMS, serviceId);
CleanupStack::PopAndDestroy(); // services
```

After these preliminaries, creating the message is straightforward:

```
// Create new message entry
iMmsClientMtm->CreateMessageL(serviceId);
```

You can also create new messages by forwarding or replying to an
existing MMS message, through `CMmsClientMtm::ForwardL()` and
`CMmsClientMtm::ReplyL()` respectively. In these cases, the new
message is created for you, with the appropriate content and headers
copied into it from the original message.

8.4.5 Setting the Message Content

The next task is to set the message content and headers. The client MTM base class `CBaseMtm` defines some functions that you would use for this task with most MTMs:

- `SetSubjectL()`: to set the message subject

- `Body()`: to get and set body text

- `AddAddresseeL()`: to set recipients

- `CreateAttachmentL()`: to add attachments.

However, the properties and contents of an MMS message are more complex than can be specified using the above generic functions. For this reason, the API provides a full encapsulation of an MMS message in its class `CMmsClientMessage` and its base class `CMmsMessage`. The message class owns an object that encapsulates the message headers, `CMmsHeaders`, and a list of `CMmsMediaObject` objects, each of which describes a multimedia object, such as a SMIL or image file.

To get a `CMmsClientMessage` object for the message you've just created, you use the following call:

```
// Set new entry to be an MMS message
iMmsClientMessage = CMmsClientMessage::NewL(*iMsvSession,
    iMmsClientMtm->Entry().EntryId());
```

The second parameter of `NewL()` here is the ID of the MMS message. The sequence `iMmsClientMtm->Entry().EntryId()` asks the client MTM to return the ID of its current entry: as the `CreateMessageL()` function sets the current entry to be the new message, the ID of the new message is returned.

The first task to undertake with the `CMmsClientMessage` object is to set the message headers. The MMS specification defines a large number of fields that can be set in an MMS message header. Some of these are compulsory, such as message type, transaction ID, MMS version number, from-address, content-type, and at least one recipient address field (To, Cc, or Bcc).

Fortunately, the MMS server-MTM gives appropriate values to all the mandatory fields except for recipients, which the client must do itself. This is straightforward: the following sets a To recipient for the message (`aNewRecipient` being a descriptor that specifies the address).

```
iMmsClientMessage->Headers().AddRecipientL(CMmsHeaders::ETo,
    aNewRecipient);
```

You can also set the optional fields. This sets the message subject:

```
iMmsClientMessage->Headers().SetSubjectL(aSubject);
```

8.4.6   Media Objects

The media content of the message is defined through adding one or more media objects. A media object can be added through `CMmsMessage::CreateMediaObjectL()`.

`CreateMediaObjectL()` returns a `CMmsMediaObject` object, another of the MMS utility classes. The object specifies a filename, to which the media data (e.g. image or SMIL file) for the object should be copied. When a media object is added, the caller can specify that the object is the message's *root* object: this is typically the SMIL file that contains the presentation of the other objects. The following code adds a root SMIL file with a file name specified in `smilFile`.

```
// Request the MO to be created, passing the root option, MIME type, and
// file extension
// File's MIME type
_LIT(KSMILMIME, "application/smil");
// File's extension
_LIT(KSMILExt, ".smil");
CMmsMediaObject& newMO = iMmsClientMessage->CreateMediaObjectL
(ETrue, KSMILMIME, KSMILExt);
// Copy the data into a target file specified by the media object
err = BaflUtils::CopyFile(iFs, smilFile, newMO.Filename());
```

Each object also has an ID, a `TMmsMediaObjectId`, that can act as a handle. The caller can recover this with the `Id()` method and store it for later reference.

```
TMsvId smilId = newMO.Id();
```

When you have added all the media objects, you are nearly ready to send the message. The final task before that is to tell the client MTM that you've finished making changes to the message. This ensures that any changes that the client MTM has kept locally are saved to the server's message store.

```
// save the modified or new message to the (server-side) message store
iMmsClientMtm->SaveMessageL();
```

8.4.7   Sending the Message

Perhaps surprisingly, the client MTM base class `CBaseMtm` does not provide a method of sending a message. It does, however, provide a standard means for MTMs to make MTM-specific functionality available. To do this, the MTM implements either or both of the virtual `InvokeAsync-FunctionL()` (for asynchronous operations) and `InvokeSyncFunctionL()` (for synchronous operations) methods. Many MTMs use these functions to offer a command to send a message, and MMS does so through `InvokeAsyncFunctionL()`.

The function takes:

- a parameter specifying the command to perform. The enum `TMmsMtmCmds` defines the IDs for MMS MTM-specific operations; `KMmsMtmSendMessage` is the ID for the send command

- the ID(s) of the message(s) to perform the command on

- a buffer (not used here) that can hold optional parameters

- the usual request status parameter for asynchronous operations.

The following code initiates the sending of the messages listed in the array `iIdToSend`, which is of type `CMsvEntrySelection`.

```
TBuf8<1> param; // unused
// send the message asynchronously
CMsvOperation* msvOperation = iMmsClientMtm->InvokeAsyncFunctionL(
KMmsMtmSendMessage, *iIdToSend, param, aCompletionStatus);
```

A feature of the message APIs is that functions, such as this, that perform asynchronous operations return a `CMsvOperation` object. This allows the caller to get progress information about the operation (for example, that two out of four messages have so far been sent), and to cancel the operation.

An alternative way to send a message is to copy the message from its local folder to the entry for the remote service using `CMsvEntry::CopyL()`.

To conclude the discussion of MMS, I'll look at how an application can process incoming messages.

8.4.8 Incoming Messages

The Messaging Architecture declares a pair of abstract interfaces, which define session observers and entry observers. A session observer, which must implement the interface `MMsvSessionObserver`, is notified of events such as the arrival of new messages, and the registration of new MTMs. An entry observer, which implements interface `MMsvEntryObserver`, is notified when an individual entry changes or is accessed.

To handle inbound MMS messages, you need a class that is a session observer. The interface has just one function to implement:

```
void HandleSessionEventL(TMsvSessionEvent aEvent, TAny* aArg1,
    TAny* aArg2, TAny* aArg3)=0;
```

The `TMsvSessionEvent` parameter will tell you what type of event occurred; the other parameters can supply relevant data for the event, depending on the event type. For inbound MMS messages, two event types need handling:

- `EMsvEntriesCreated` which occurs when notification of a new MMS message arrives on the phone. `aArg1` is a `CMsvEntrySelection` containing the new entry, while `aArg2` is the `TMsvId` of the parent entry.

- `EMsvEntriesChanged` which occurs when the message body is fetched and available to be played or displayed. The event parameters are as for `EMsvEntriesCreated`.

There are two events because an MMS message is not necessarily fetched from the server completely. Instead, the remote MMS server first sends the phone a notification that a new message is available for it. At this point, the Symbian OS MMS implementation creates a new entry in the message store, and sets those header fields which are available from the notification (the others are left null). The phone then has the option as to whether to get the message body (its media objects) or not.

Fetching can happen automatically or on request. An MMS service can be set to fetch automatically the complete message, using the method `CMmsSettings::SetAutomaticFetch()`. Where automatic fetching is not set, an application can explicitly request the fetch using a `KMmsMtmFetchMessage` or `KMmsMtmBackgroundFetchMessage` command through `CMmsClientMtm::InvokeAsyncFunctionL()`. You can always check whether a message has been completely fetched by checking whether its `TMsvEntry::Complete()` flag is true.

Another complication to be aware of is that your observer will receive notifications for all types of message, not just MMS. So your MMS-handling code should make sure to check first that the event relates to an MMS message (or messages). The following code does this, assuming that `entries` is the `CMsvEntrySelection` containing the ID of the new entry.

```
TInt count = entries->Count();
while (count-- != 0)
    {
    const TMsvId id = (*entries)[count];
    CMsvEntry* msvEntry = iMsvSession->GetEntryL(id);
    CleanupStack::PushL(msvEntry);
    if (msvEntry->Entry().iMtm == KUidMsgTypeMMS) ProcessMyMmsL(id);
    CleanupStack::PopAndDestroy(msvEntry);
    }
```

If your application wants to handle only some MMS messages, but not all, you will need to read the message and check some aspect of it, such as the subject, to see whether you should process it. This can be done by instantiating a new `CMmsClientMessage` for the entry and accessing the required header fields.

If you determine that your application should process the message, you can prevent the user seeing the message in the Inbox of the standard messaging application. To do this, set the message entry's visible flag to false, using `TMsvEntry::SetVisible()`.

8.5 Summary

This chapter has:

- summarized the main frameworks used in Symbian OS communications programming

- shown how these frameworks are used to implement, in a flexible and extensible way, a large number of comms protocols

- described how to use the messaging framework with the MMS protocol.

9

Testing on Symbian OS

The ability to test Symbian OS and to gain full confidence in its robustness and reliability is a necessity. Testing the operating system is not a straightforward task, due to the very different types of functionality that are available, as we have seen through the previous chapters. Each area demands that its very own specific nature be validated.

Tools to support two types of test approach are described:

- a more usual approach that provides the framework and the tools common to all software engineering activity regardless of the nature of the software being produced, i.e., code coverage, unit testing, etc.
- a more surgical approach that targets, through concrete applications and tools, very specific functionality that is not met by more mainstream methodologies.

In this chapter, we will be guiding you through some of the test infrastructure and tools that are currently available for Symbian OS, in order to provide the testing granularity that makes our phones more reliable.

Please note that most of the discussions in this chapter assume that you will have installed the TechView Technology Kit available on the CD provided with this book.

We do not intend to provide full details on how to use these tools, and we will refer you to the Test Tools section of the Symbian OS developer library, available on the CD or on the Symbian developer website. Unless stated otherwise, tools and examples in this chapter can all be found by visiting *www.symbian.com/books*.

9.1 Code Coverage Analysis

9.1.1 Overview

This tool allows you to make sure that the code you are writing is structurally valid, using a methodology better known as 'white box testing'.

Symbian OS C++ for Mobile Phones, Volume 2. Edited by Richard Harrison
© 2004 Symbian Software Ltd   ISBN: 0-470-87108-3

An external tool provided by Bullseye Testing Technology is being used as the de facto dynamic analysis tool on most Symbian OS subsystems.

BullseyeCoverage allows, through code instrumentation, feedback on function, branch and statement coverage. Using this tool will allow you to measure any deficiency of your test script in terms of coverage of the API it aims to test.

As it is not working straight out of the box, a Symbian OS wrapper needs to be installed for supporting WINS as well as WINSCW variants.

The tool is downloadable from ***http://www.bullseye.com***, and you will need to purchase a license to use it. The version currently supported by Symbian is 6.4.6.

A wrapper that will install the right environment in order to link and compile with the right development tools must be installed (for example, support for WINSCW emulator built with Metrowerks compilation tools).

This wrapper, `SymbianCoverUpdateV1.1-Install.exe`, is available at ***www.symbian.com/books***.

9.1.2   Rationale Behind the Wrapper

The wrapper is needed in order to save coverage data for the different emulator builds, i.e. WINS and WINSCW.

BullseyeCoverage calls the Win32 API function `LoadLibrary()` to prevent the DLL from being unloaded after the Symbian OS has unloaded all references to it, for the time needed to save coverage data. Otherwise, the DLL would be unloaded by the Symbian OS DLL loader before the coverage data can be written out.

This function is not compatible with Symbian OS polymorphic DLLs, as their entry points are not recognized.

The wrappers provide the framework that will enable BullseyeCoverage to save the information.

9.1.3   Installing the Tool

Install BullseyeCoverage v6.4.6 on the C drive of your computer. You must choose the `c:\program files` directory that is suggested as default.

Then, run `SymbianCoverUpdateV1.1-Install.exe`.

This will have overwritten the linker and compiler proxies that were installed previously in the BullseyeCoverage directory.

9.1.4   How to Use the Tool

In order to gather coverage information, you will need to build your code with BullseyeCoverage compilation enabled from the command line.

You should first gain familiarity with the build process which is explained in *Symbian OS C++ for Mobile Phones Volume 1*, Chapter 1, Sections 1.2.4 to 1.2.6.

The installation of BullseyeCoverage will have copied on to your PC the environment necessary to instrument your code at build time.

- Type 'covbuild on' to enable building with BullseyeCoverage.

- Build the component using 'abld build wins'.

- Type 'covbuild off' to disable building under CCover.

- Build the test using 'abld test build wins'.

- Run the test.

- Open the function coverage file from `C:\test.cov` to check the coverage. This will trigger the BullseyeCoverage GUI in which you will be able to visualize the coverage rates per component, file, and function.

9.1.5 Graphical Examples

Figure 9.1 is an example of coverage for the app-engine subsystem (as structured under `src\common\generic\app-engines` in the Tech-View installation folder).

You can see the functional and condition coverage for each of the instrumented components when running test code on a WINS emulator.

For example, 85% of the functions and 65% of the conditions and decisions defined in `cntmodel` have been covered when running a certain set of unit tests.

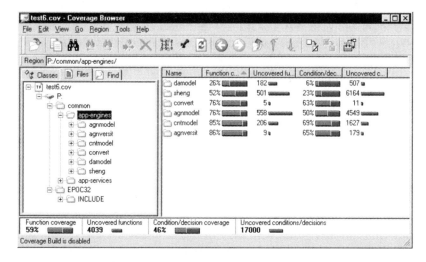

Figure 9.1 Coverage Browser – component view

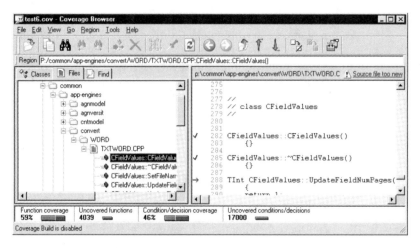

Figure 9.2   Coverage Browser – function view

Of course, you can refine the view, and zoom down to the functions defined in a particular source file. Figure 9.2 provides an example.

The `CFieldValues` function in `txtword.cpp` has been covered, while `UpdateFieldNumPages` has not. You are then in a position to know where to refine your test code.

All the coverage information is saved by default in a `test.cov` file, and depending on the number of tests you are running, this file can become very big.

In order to make the information more accessible, it is possible to dump the full `test.cov` file and format the content into html or comma-separated values files, using the `covfn` and `covbr` functions provided with the tool.

Typically, you will generate a one-line-per-function `csv` file when typing:

```
>covfn --csv >coveragefile.csv
```

A one-line extract of this file will show respectively the `"class::func-tion()"`, `"SourcePath"`, line and coverage information:

```
"CShgWorkSheet::DoPasteActionDataL","P:/common/appengines/
    sheng/source/SHGWKCPY.CPP",422,0,0,116,65%
```

In this case:

- `CShgWorkSheet::DoPasteActionDataL` shows the class and function names.

- `P:\common\app-engines\sheng\source\SHGWKCPY.CPP` is the local path to the file where this function is called.

- `422` is the line number and `0` is the function coverage (in this case not covered).

9.1.6 Conclusion

We have seen how code coverage can be obtained for Symbian OS using the BullseyeCoverage tool when developing your code; this will help you write the best unit tests, hence ensure your software has been thoroughly verified.

It is beyond the scope of this chapter to provide all possible use cases for this tool, although we encourage you to find out more on this tool from the BullseyeCoverage website, ***http://www.bullseye.com***.

9.2 Binary Compatibility Tool

9.2.1 Overview of the Functionality

When developing your application for a new version of Symbian OS, one of the first questions you will be asking yourself is: how portable is the previous version of your application to this newer system? Basically, you will want to check for any binary break that might prevent your component from being used directly without having to rebuild it.

The BC Comparator allows you to check for binary compatibility breaks between two different releases of Symbian OS. Any break that would occur, and on which your component is dependent, will be an issue for you. Being able to address it as early and as accurately as possible will make your work easier.

The BC (Binary Compatibility) Comparator comprises a set of tools that allows you to assess the binary compatibility between two Symbian OS releases following certain characteristics such as the differences in ordinal exports of `def` files, class sizes, and `vtables` (ordering and layout of virtual functions).

The tools are installed as part of the Symbian OS Kit and can be found in the `epoc32\tools\` directory once you have installed the CD provided with this book.

In order to compare two given Symbian OS builds, you will need to run successively four tools that will each play a specific role in the reporting of any possible break between the two versions of the platform.

You will need access to ROM images in order to use these tools, so this section is of most benefit to Devkit license holders. You can find out more about licensing at ***www.symbian.com/partners***.

9.2.2   Running the Tool

It would be very lengthy to detail here the full suite of actions necessary to obtain the list of binary breaks between two releases of the operating system.

If you browse the developer library, available either on the CD or on the Symbian developer network website (***http://www.symbian.com/developer***), under the Test Tools section, you will be able to find all the information necessary to run the tool efficiently.

9.3   Test Driver

Considering the number of unit tests that need to be run on a device to ensure the quality of any production code being released, you will need to give yourself the means to achieve this as smoothly and as swiftly as possible.

Running test code on a device and getting the results of the tests fed back to the developer so that he or she can process and use them can be a mammoth task if undertaken manually.

In this particular case, automation is the solution and Test Driver provides the environment to do exactly this.

Running this tool will allow a tester to set up his or her PC to execute tests for all variants of the operating system, on target or on emulator.

To achieve this, the tests will need to be described by a set of XML files that will serve as input for the process.

In this section we will go through all the necessary steps to set up the environment needed to run this tool:

- configure Test Driver
- build tests and test suites into a repository on your machine
- run tests on a target device, or on the emulator
- view test logs and test results.

9.3.1   Preliminary Information

The Test Driver tool is provided on the CD and can be found in a zip format in the `bin\epoc32\tools\TestDriver` directory.

Your PC will need to have the following minimum specifications:

- PII 500 MHz
- Microsoft Windows NT4/2000
- 256 MB RAM
- 1 GB free disk space.

You will also need the following to run your tests on a device (not necessary if running your tests on emulator):

- Assabet/Lubbock development board (provided by Intel)
- 1 × NULL modem cable
- 1 × Compact Flash (CF) card, 32 MB.

If running the tool on a Symbian OS emulator, you will need to launch `bin\techview\epoc32\release\winscw\urel\epoc32.exe`.

9.3.2 Installing the Tool

The following section will assist you in setting up the tool:

- Firstly, unzip the file `bin\techview\epoc32\tools\testdriver\testdriver.zip` to any location. This will unzip all the files needed into a subdirectory called `testdriver`, within the directory from which the file was unzipped.
- Then, from the command line type:

```
> testdriver install
```

This will create a registry entry and set up the XML DTD files that will be necessary to describe your tests.

9.3.3 Configuring Test Driver

The first time you install Test Driver, you are prompted to set up your environment and define the following information:

- The EPOC drive, i.e. the drive where the EPOC32 tree has been installed
- The XML Root, i.e. the root path to your XML tree that defines your test suite hierarchy
- The Repository location, i.e. the root path where all the binaries and related test files are stored
- The Result location, i.e. the root path where the test logs are stored.

An example of installation is shown below:

```
D:\testdriver\bin>testdriver install
Please enter the drive (eg. j:) of your epoc32 tree:
   d:\symbian\<OS_build_no>\bin\techview\
```

```
Please enter the path to your test XML: d:\TEST\xml -tree
Please enter the path to where you want your test repository: d:\test
Please enter the path to where you want to store the results: d:\test
Installation complete!
D:\testdriver\bin>
```

The `config` option is used to set, view and modify the parameters, as shown below:

```
>testdriver config
```

This is used to display the current configurations. Applied to the example shown above:

```
D:\testdriver\bin>testdriver config
epoc drive= d:\symbian<OS_build_no>\bin\techview\
xml root= d:\TEST\xml -tree
repository root= d:\test
result root= d:\test
```

You can of course reconfigure some of these values, using the appropriate flags. For example:

- `-e` will reset the EPOC drive
- `-x` will reset the path to the root of your XML structure.

9.3.4 Defining a Hierarchy for the Set of Tests that will be Run

The Test Driver tool requires you to organize your tests into a test suite hierarchy on your PC. Suites can contain tests, other suites or a combination of both.

XML files describe the structure that mirrors the hierarchy of the set of tests you wish to run.

An example of test hierarchy and the associated structure is shown in Figure 9.3.

Figure 9.4 mentions the `RTest` and `TestExecute` terms. It is beyond the scope of this chapter to delve into these concepts in full detail, but we will briefly introduce them.

`RTest` is basically a class defined within `E32Test` that will create a console window to which test results are logged. It provides quite a lot of functions necessary for testing, and you will find its definition in `e32test.h` (`bin\epoc32\include\`).

A simple use case reporting success is provided below:

```
RTest test(_L("My particular test"));
test.Start(_L("Test Whatever"));
```

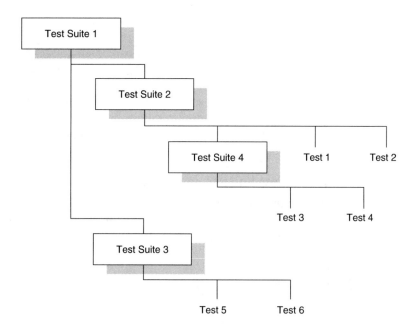

Figure 9.3 Test structure

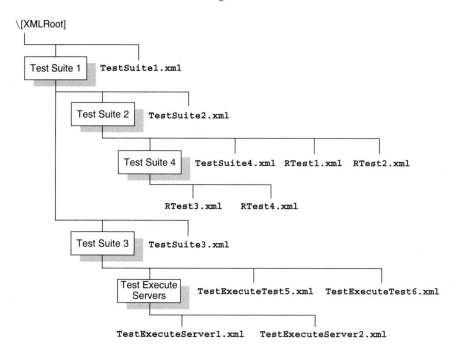

Figure 9.4 XML definitions

```
test.Title();
test(iAllTestsOK);
test.End();
test.Close();
```

A `TestExecute` test is a server that enables you to optimize the running of your tests, through a series of steps executed either sequentially or simultaneously. Each step is defined by using the `CTestStep` class, and will contain the calls to the APIs you are willing to test.

9.3.5 Defining a Test Suite Using XML

The test suite hierarchy is implemented using XML files based on specific, supplied DTDs. Test suites can contain three types of tests: `RTests`, `TestExecute` tests and `CommandLine` tests.

Let us first examine the structure of a simple `CommandLine` test.

9.3.5.1 Defining `CommandLine` Tests

A generic test harness called `CommandLineTest` is available. This enables any test harness to be executed as it would be from command line.

Below is an example of how to define such a command line test, based on the `commandLineTest.dtd` provided when the tool is installed:

```
<commandLineTest>
<name>HelloWorld</name>

<commandLine>c:\system\programs\helloworld.exe symbianuk.script
   -i</commandLine>

<logFile>c:\HelloWorld\log.txt</logFile>

<timeout>60</timeout>

<dependencies>
<data>
<hostPath>epoc32\data\Test.data</hostPath>
<devicePath>c:\HelloWorld\Test.data</devicePath>
</data>

<build type="test">
<hostPath>cinidata.dll</hostPath>
<devicePath>c:\system\libs\cinidata</devicePath>
<bldInfPath>development\cinidata\group\bld.inf</bldInfPath>
<mmpFile>cinidata.mmp</mmpFile>
</build>

<retrieve>
<devicePath>c:\system\programs\HelloWorld\Retrieve.me</devicePath>
<hostPath>HelloWorld\Retrieve.me</hostPath>
</retrieve>
```

```
<delete>
<devicePath>c:\system\programs\HelloWorld\Delete.me</devicePath>
</delete>
</dependencies>

</commandLineTest>
```

Let us go through the meaning of the XML flags encountered so far (the optional ones are shown in *italic*) :

`<name>`	States the name of the application under test.
`<commandLine>`	Describes the full path to the exe to be started on the device.
`<timeout>`	States the time in seconds for how long the test should be run; taking longer than what is described here would result in the killing of this particular test.
`<logFile>`	Set to the path of the test log file.
`<dependencies>`	Used to list blocks of description for any dependencies for this test such as:

- *`<data>`* defining the `<hostPath>` and `<devicePath>`

- *`<build>`* defining buildable dependencies.

9.3.5.2 *Defining an `RTest` Type Hierarchy*

Let us now have a look at the directory structure of a test suite comprising `RTest` unit tests validating the `app-engines` subsystem:

```
C:\TESTDRIVER\XML -TREE
|    root.xml
|    testSuite.dtd
|
|____root
|    |    CoreApps.XML
|    |
|    |___coreapps
|        |    RTestRom.xml
|        |
|        |____RTestRom
|                AGT_FILE.xml
|                AGT_GENERAL.xml
|                T_VIEWSOOM.xml
|                T_VIEWSORT.xml
|                T_VIEWSORTERROR2.xml
|                T_WKBK.xml
|                T_WRONGFIELDMATCH.XML
```

Let us have a look at `root.xml`. It includes all tests and suites:

```
<?xml version="1.0"?>
<!DOCTYPE testSuite SYSTEM "file:///c:/testdriver/xml/testSuite.dtd" [ ]>
<testSuite>
    <name>root</name>
        <!-- List all tests and suites here !! -->
    <testItems>
        <suite>CoreApps</suite>
    </testItems>
</testSuite>
```

The `CoreApps` suite will be defined in the root folder under `CoreApps.xml`:

```
<?xml version="1.0"?>
<!DOCTYPE testSuite SYSTEM "file:///c:/program files/common
                        files/symbian/testSuite.dtd" [ ]>
<testSuite>
        <!-- Name of Suite, can be any arbitrary name -->
        <name>CoreApps</name>
        <!-- Test Items found in Suite:  Tests and Suites -->
    <testItems>
        <suite>RTestRom</suite>
    </testItems>
</testSuite>
```

Finally, the `RTestRom.xml` file will list the `xml` description of all unit tests.

For example, let us have a look at the fourth one in the list: `TViewSort.xml`:

```
<?xml version="1.0"?>
<!DOCTYPE RTestRom SYSTEM "file:///c:/program files/common
                        files/symbian/RTestRom.dtd" [ ]>
<RTestRom>
<name>T_VIEWSORT</name>
<devicePath>
Z:\System\Programs\CntmodelTest\T_VIEWSORT.exe
</devicePath>
<timeout>
900
</timeout>
</RTestRom>
```

There are two important tags defined here:

- `<devicePath>` defines where in the ROM the test is located. Note that in this case we have put the executable in the Z drive.

- `<timeout>` defines the maximum time to be spent on this particular test in seconds.

9.3.6 Building a Test Suite

Now we have seen what these suites and hierarchies should look like, let us see a concrete example of how to build these test suites.

Typically, type the following from the command line:

```
> testdriver build -p <platform> -b <build> -s suite [-a architecture]
```

where the mandatory flags are as follows:

- -p is the platform
- -b is the variant: udeb or urel
- -s specifies the test suite to build.

For example:

```
testdriver build -p arm4 -b urel -s testsuite1.testsuite4
```

You will then need to add all the built executables corresponding to these test suites on the device as defined in your XML description.

9.3.7 Running Test Suites

Tests can be run both on the emulator and on a device, from a PC. The boards tested so far are Assabet and Lubbock. When tests are run, the tests and their dependencies are copied to the C drive of the device (or emulator). Tests are then executed on the device and the logs are collected and transferred back to the PC.

The command line for running a test suite is:

```
> testdriver run -p <platform> -b <build> -s <suite> [-t transport]
    [-a architecture] [-1 rdebug|console] [-c collection path]
```

Taking our test suite example defined in Section 9.3.5.2, we would type:

```
Testdriver run -p arm4 -b urel -s root.coreapps -t serial2
```

9.3.8 Connecting to a Device

At present, Test Driver supports three different methods to allow communication between your PC and the device under test: TCPIP, Bluetooth and serial connection. Here again, we refer you to the Symbian Test Tools entry on Test Driver to learn how to connect your device to your host PC.

9.3.9   Analyzing the Results

Test Driver will have created a result file in the results directory that was defined during the configuration of the tool.

Let us take a look at the result provided when running `t_viewsort.exe` defined in Section 9.3.5.2:

```
ATRTEST: Level  001
Next test - T_ViewSort
RTEST: Level  002
Next test - Test the view sort order
RTEST: Level  003
Next test - Starting Test 1
" a" "" ""
" a" " y" ""
" a" "x" ""
...
"h" "b" "a"
RTEST: Level  004
Next test - Starting Test 2
"ibbb" "" ""
"Ibbb" "" ""
RTEST: Level  005
Next test - Starting Test 3
"_Sunshine" "" ""
" Symbian" "" ""
"S_unshine" "" ""
...
"Symmetrical" "" ""
RTEST: Level  006
Next test - Starting Test 4
" .ally" "" ""
".alli" "" ""
...
"ally" "" ""
RTEST: SUCCESS : T_ViewSort test completed O.K.
```

Have a look at the source for that particular test under `tsrc\t_viewsort.cpp` which you'll find at ***www.symbian.com/books***.

You will be able to trace the calls detailed in the different test levels shown in the extract above, in particular calls to `CTestResources::CreateTestThreeContactsL()` and `CTestResources::CreateTestOneContactL()`.

9.4   Network Emulator

The *Network Emulator for Symbian OS* is designed to provide a complete, controllable, test environment for network applications and services. No additional hardware, such as a handset, is required to use the *Network*

Emulator. It allows all Symbian OS networking software, from data link protocol to applications that take advantage of Symbian OS networking features, to be tested in a controllable, repeatable way, thus improving software testing.

The *Network Emulator for Symbian OS* can significantly improve the coverage and efficiency of developing, testing, and debugging networking software for Symbian OS.

Testing can begin earlier in the project, since the *Network Emulator* emulates both the phone and the network, so there is no need to wait for the hardware to be available.

Coverage is increased, since the *Network Emulator* can test fault and error conditions that are difficult or impossible to generate on real networks (e.g. setting a particular level of quality of service). It can also test services that have not yet been deployed in live networks such as 3G, CDMA and Mobile IP.

9.4.1 Overview

The *Network Emulator for Symbian OS* comprises three components (Figure 9.5).

- The *Emulation Environment* is a set of PCs that emulate the network environment external to the Symbian OS device, thus playing the role of the operator's network and attached data networks. This

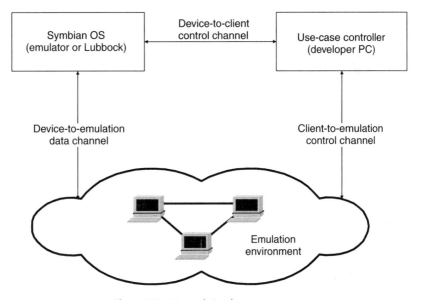

Figure 9.5 Network Emulator components

environment can emulate the behavior of the air-interface on data traffic, receive and initiate PPP, TCP/IP, and MobileIP connections, and control other services that run on top of these protocols.

- The *Use Case Controller* is a client software that runs on a developer's PC and controls the Emulation Environment. It configures the environment before, and during, the test. It can also collect status information from services to assist debugging. The Use Case Controller can be controlled interactively by a developer, or automatically via a remote connection from the Symbian device under test.

- *Symbian OS* is the device under test. This may be either the emulator or a hardware reference board. The device can control the Use Case Controller via a remote connection, allowing it to synchronize its test code with the configuration of the Emulation Environment (ideal for automated testing). It also interacts directly with the Emulation Environment via a network connection to perform the actual test.

9.4.2 Supported Services

The Emulation Environment currently includes support for testing the following scenarios:

- Packet filtering to emulate bandwidth limitation, latency, data duplication, and error-rate across the air-interface

- PPP connection establishment, negotiation, and data-flow

- Standard TCP/IP connection establishment and data-flow

- Mobile IP connection establishment, negotiation, data-flow, and hand-off.

9.4.3 Application Servers

Application servers that run over TCP/IP can easily be added to the Emulation Environment and controlled from the Use Case Controller. This means that users can simply augment the *Network Emulator* with their own application servers (e.g. HTTP), expanding the capabilities of the Emulation Environment, allowing them to test an infinite variety of Symbian OS network application software.

External IP networks can also easily be attached to the Emulation Environment. This allows users to access their existing test application servers, in a controlled way, from the Symbian OS device under test, preventing duplication of their test environment.

9.4.4 Availability

You will find binaries and documentation on the Symbian Developer Network website under Test Tools (NetworkEmulator.zip).

9.5 Sampling Profiler

9.5.1 Overview

A profiler is a tool that will allow you to analyze the performance of the applications you run. Tracing the information you get back to the actual code will help you in assessing where the system is spending time.

You will be able to measure how long a routine takes to run, and how much of total CPU time at some spot is spent executing that routine. Used appropriately, it constitutes a must-have feature during the development or testing phases of your projects.

The profiler provided with the Symbian OS platform interrupts the CPU regularly (the sample rate is fixed at 1 kHz, typically generating 3 kB of data per second of processor activity) to record the currently executed instruction for the application it is profiling.

It consists of a program running on the target device which periodically samples the program counter and current thread ID, saving the resulting trace data to a file. This program comprises:

- a kernel-dependent device driver; currently only ARM targets are supported (the logical device driver has to be built for the ASSP of the device being used)

- an executable which can be controlled via command line, programmatically, or through a console interface.

9.5.2 Installing the Profiler

To generate the profiling trace, the profile program must be installed on the Symbian OS device (e.g. by copying it to a CF card). Again, you will need to be a Symbian OS Devkit licensee to have access to this tool. Two parts of the program need to be installed on the device as follows:

- `Profiler.exe` from `\epoc32\release\<ABI>\urel\` to `\system\programs`

- `Sampler.ldd` from `\epoc32\release\<ASSP>\urel` to `\system\libs`.

To load the profiler, the `Profiler.exe` program must be executed on the target device; it then remains loaded until asked to exit. As it is a text mode executable, it will run on any UI family.

9.5.3 Profiler Commands

The Profiler understands the following four instructions:

- Start, to start sampling
- Stop, to stop sampling
- Close, to close the trace data file
- Unload, to unload the profiler and save all the trace data.

9.5.4 Command Line Control

The Profiler can be controlled from the command line, by specifying the action as a parameter to the `Profiler.exe` program. If the profiler is already loaded, this will just pass on the control request to the running program; otherwise it will load the profiler ready for further commands.

For example, use the following command in `eshell` (the Symbian OS command line interpreter):

```
cd system\libs
start profiler start
```

`eshell.exe` can be started from the `z:\system\samples` directory on your device. You will need to have a file manager on that device.

Once `eshell` is started, type at the prompt shown above. If you then move back to the root of the C drive and type **dir**, you will notice that a trace file (`profiler.dat`) has been generated.

9.5.5 Programmatic Control

The profiler can also be controlled from within another program on the device by using the API defined in `profiler.h`. This defines the Profiler class, which has static functions corresponding to each profiler command.

Note that `Profiler::Start()` will not load the profiler if it is not currently loaded; it will only cause an already loaded profiler to start sampling. It is possible to check for this possibility and use the command line to run the profiler using the following code:

```
if (Profiler::Start() == KErrNotFound)
    {
    _LIT(KProfiler,"profiler");
    _LIT(KStart,"start");
```

```
RProcess p;
if (p.Create(KProfiler,KStart) == KErrNone)
    {
    p.Resume();
    p.Close();
    }
}
```

9.5.6 Analyzing the Results

In order to visualize the results, Symbian OS provides a tool that formats the `profiler.dat` file into a more readable format.

You will first need to transfer the trace data file to your PC, and then use `analyse.exe` (available in `bin\techview\epoc32\tools`).

9.5.6.1 Generating a Full Listing of the Trace

The full trace recorded in the trace file can be output in text form using the `-listing` (`-l`) option:

```
analyse profiler.dat -listing
```

A more interesting trace can be generated if a ROM symbol file matching the device ROM is available. This is done using the `-rom` (`-r`) command option:

```
analyse profiler.dat -rom \epoc32\rom\myrom.symbol -listing
```

9.5.6.2 Example Using the ROM Image Available in a Devkit

- Load the `la_001.techview.img` onto your Lubbock development board (this can be purchased from Intel).

- Start the Profiler as explained in Section 9.5.4.

- Collect the `profiler.dat` file back to your PC.

- From the prompt, type:

```
analyse profiler.dat -rom \..\epoc32\rom\la_001.techview.symbol
    -listing
```

An extract of the analysis is provided below:

```
0    0    MatchesHereFoldedWithPrefixTest(TCharIteratorCanonicalOrder)+ fc
1    0    TCharIteratorCanonicalOrder::AtEnd(void) + 1c
```

```
2     1    TGulBorder::OuterRect(TRect const &) const + 52
3     1    __ArmVectorSwi(void) + 130
4     2    Kern::MessageComplete(int, int) + 4
5     2    CActiveScheduler::DoStart(void) + 5c
6     1    TDes8::Zero(void) + 4
7     2    CScreenBackup::ClientDestroyed(CWsClient const *) + 46
8     1    CActiveScheduler::DoStart(void) + 5c
.............................  . .
11882   9    flushDataCache(void) + 2c
12664   2    TParameterManager::TParameterManager(void) + 68
12679   3    flushDataCache(void) + 44
12836   2    TSwapScanLineIntersections::Swap(int, int) const + c
12898  17    TScheduler::ReSchedule(void) + c4
13410   2    CCaptureKeys::ProcessCaptureKeys(TKeyData &) const + 1e
13411   0    TInt64::GetTInt(void) const + 0

Threads:
 0  EFile::FileServer
 1  Profiler::Profiler
 2  EwSrv::Wserv
 3  System::System
 4  EikSrvs::AppArcServerThread
 5  EKern::Supervisor
......................... .
18  EFile::LoaderThread
19  FLogSvr::FLogger server
20  EComServer::ecomserver
```

9.5.6.3 *Finding Out Where Threads Spend Most of their Time*

It is possible to find out the amount of time spent by each thread in the ROM:

```
analyse profiler.dat -rom \epoc32\rom\myrom.symbol -byfunction
```

You can eliminate from your report any thread that would spend less than a certain amount of time:

```
analyse profiler.dat -r \epoc32\rom\myrom.symbol -byfunction -cutoff0.5
```

This produces the same results as last time, but does not report on any threads or locations with less than 0.5% of the overall samples.

If we take our previous `profiler.dat` and run the following command:

```
analyse d:\PROFILER.DAT -rom la_001.techview.symbol -byfunction -cutoff1.5
```

we would obtain all threads that spend more than 1.5% of the time:

```
EPOC Profile Analyser   Version 1. 0(build 000)
Copyright (c) Symbian Limited 2000. All rights reserved.
```

```
Profile distribution

Samples: 13417
Active: 638   (4.76%)
Counted: 638   (4.76%)


ID   Thread name
A    EwSrv::Wserv
B    EDbsrv::DBMS server
C    C32exe::EtelServer
D    EFile::FileServer
E    EKern::Supervisor
F    System::System
G    EwSrv::MMFControllerProxyServerde8e07d3
H    Profiler::Profiler
A       B       C     D     E      F      G      H     Total
45.30% 16.77%  7.99% 7.37% 6.43%  3.76%  3.45%  2.82% 100.00% Total
0.63%   1.57%  0.16% 0.16%        0.63%  0.31%  0.31%   4.55% TScheduler::ReSchedule()
1.72%   0.63%  0.78% 0.47% 0.16%         0.16%  0.47%   4.39% Mem::Copy()
1.41%   0.16%  0.31% 0.31%        0.31%         0.31%   3.61% __ArmVectorSwi()
3.13%                                                   3.13% etext=.
2.66%                             0.16%                 2.82% CFontStore::UpdateSupportListL()
2.51%                                                   2.51% CPdrModelList::ListModelsL()
1.57%          0.16% 0.16%                             1.88% Mem::Fill()
1.72%                                                   1.72% COpenFont::Rasterize()
1.10%                       0.16%                       1.41% TUnicode::GetData()
0.31%          0.16%        0.16%                0.31% 1.41% CActiveScheduler::DoStart()
        0.31%  0.16% 0.47% 0.16% 0.31%               1.41% TCharIterator::SetCurrent()
```

9.5.6.4 *Filtering by Thread Name, DLL Name, Function Name, or Sample Number*

The samples for a specific thread can be analyzed using the `-thread` (`-t`) option; this is followed by a basic wildcard pattern for the thread name to analyze. (The `*` character can represent zero or more of any character.)

For example, to analyze just the threads in the file server the following commands could be used:

```
analyse profiler.dat -r \epoc32\rom\la_001.techview.symbol -bf -thread efile*

Samples: 13417
 Active: 638   (4.76%)
Counted: 51   (0.38%)
ID   Thread name
A    EFile::FileServer
B    EFile::LoaderThread


  A       B      total
92.16%   7.84%  100.00%  total
 3.92%   1.96%    5.88%  __ArmVectorSwi(void)
 5.88%            5.88%  TCharIterator::SetCurrentCharacterAndWidth(void)
 5.88%            5.88%  Mem::Copy(void *, void const *, int)
 3.92%            3.92%  TParseBase::Set(TDesC16 const *, TDesC16 const *
 1.96%   1.96%    3.92%  TCharIterator::AtCombiningChar(void)
 3.92%            3.92%  Fault(TFsFault)
```

Analysis for the thread/location distribution can be restricted to DLLs matching a provided pattern using the `-dll` (`-d`) option, e.g. to restrict analysis to `C32.dll` the following command could be used:

```
analyse profiler.dat -r \epoc32\rom\la_001.techview.symbol -cutoff1 -bf -dll
    c32.dll

ID  Thread name

A   EwSrv::Wserv
B   EDbsrv::DBMS server
C   C32exe::EtelServer
D   EFile::FileServer
E   EKern::Supervisor
F   System::System
G   EwSrv::MMFControllerProxyServerde8e07d3
H   Profiler::Profiler
I   WAPSTKSRV::PosSock-54
J   EikSrvs::AppArcServerThread

A       B       C       D       E       F       G       H       I       J       Total
45.30%  16.77%  7.99%   7.37%   6.43%   3.76%   3.45%   2.82%   1.25%   1.10%   100.00%
45.30%  16.77%  7.99%   7.37%   6.43%   3.76%   3.29%   2.82%   1.25%   1.10%   99.84%
```

9.6   Countloc – Measuring Source Code Size

9.6.1   Overview

Countloc is a tool that counts lines of code (LOC). It currently supports C, C++ and Java.

Countloc is a Symbian tool, which is actually a wrapper around a tool called CodeCount from the University of Southern California, Center for Software Engineering.

In addition to counting lines of code, Countloc can measure other aspects of the code, such as the occurrence of specific keywords, and the percentage of comments.

9.6.2   What Does Countloc Measure?

Countloc reports logical or physical lines of code (LOC) within source files. Comments and blank lines are excluded from the line count.

It is independent of programming language syntax, which enables it to collect other useful information such as comments, blank lines, and overall size, all independent of information content. The logical LOC definitions will vary depending on the programming language, due to language-specific syntax.

The logical LOC definitions are compatible with the SEI's Code Counting Standard. Basically, the physical line count counts the number of physical lines, and the logical line count looks for delimiters like ';' or '{ }' in C/C++ and counts the number of lines based on that.

If you need to compare sizes of source files written in different languages, you should use the physical LOC measure.

9.6.3 Using Countloc

Countloc is currently available in two 'flavors':

- `c_countloc.exe`, for C and C++ code
- `java_countloc.exe`, for Java code.

An environment file determines settings to be used, for example whether to use the physical or logical line counting rules. A default working environment file is supplied with the tools.

To run the tool, simply type:

```
>c_countloc <location>
```

where `<location>` is the path to the file(s) to measure.

For example,

```
c_countloc g:\mydir
```

will measure all the supported files in `g:\mydir`.

The following command line options are available:

c_countloc *pattern* [*options*]
java_countloc *pattern* [*options*]

pattern may be one of the following:

- File – a full or relative path to the file(s) to measure.
  For example, `c_countloc g:\mydir\*.cpp` measures LOC in all `.cpp` files in `g:\mydir`.

- Directory – a full or relative directory. All files found in the specified directory that match one of the supported files are measured.
  For example, `java_countloc g:\mydir` measures all supported files in the `g:\mydir` directory.

- MMP file – a full or relative path to an MMP file (not supported by `java_countloc`).
  For example, `c_countloc myfile.mmp` measures all supported files in `myfile.mmp`.

options may be one of the following:

- **/recurse** – measures all files or directories from the specified location and below.

- **/e=<file>** – specifies a specific input file.

- **/out_file=<file>** – specifies an alternative output file.
- **/h** – displays the help screen.

9.6.4 Output File

This text file (c_outfile.dat or java_outfile.dat) contains the resulting sizes of the files measured. It is created in your working directory.

This is an example of an output file:

```
C/C++ SOURCE LINES OF CODE COUNTING PROGRAM
( c ) Copyright 1998 - 2000 University of Southern California,
CodeCount (TM)

          app-engines/agenda/agvapp

Total Blank|   Comments    |Compiler Data  Exec.| Number |     Module
Lines Lines|Whole Embedded|Direct   Decl  Instr|of Files|SLOC Name
--------------------------------------------------------------------
1620   152 |  152      75 |  45      74   1197 |   1316 |CODE  AGVAPP.CPP
  27     6 |    7       0 |   2       1     11 |     14 |CODE  AGVDLL.CPP
1085   133 |  163      82 |  22      47    720 |    789 |CODE  AGVDOC.CPP
 186    11 |   27       8 |  11       1    136 |    148 |CODE  AGVENV.CPP
  30     5 |    4       1 |   2       1     18 |     21 |CODE  AGVSCNCL.CPP

 University of Southern California retains ownership of this copy of
software.  It is licensed to you. Use, duplication, or sale of this
product, except as described in the CodeCount License Agreement, is
strictly prohibited.  This License and your right to use the software
automatically terminate if you fail to comply with any provisions of the
License Agreement.  Violators may be prosecuted.

This product is licensed to : USC CSE and COCOMO II Affiliates
```

9.6.5 Availability

You will find the Symbian OS wrappers on the Symbian developer website (*http://www.symbian.com/developer*) under Tools. These will work only if you have purchased and installed the codecount and javacount executables provided by the University of Southern California. You can sign up for a license at *http://sunset.usc.edu/research/codecount/*.

9.7 Summary

In this chapter we have provided different solutions for enhancing the quality of the software you intend to develop, in harmony with the Symbian OS platform characteristics you are targeting.

Some of the methodology and tools presented are fairly common to the software engineering activity (backward compatibility checking, code coverage analysis), yet their applicability to a wireless platform is novel.

Others, such as the network emulator or the performance profiler, are more specific to a functionality that enables Symbian OS to be more accepted as the standard platform for mobile phones.

Each of these tools brings a unique perspective to the testing of the operating system, as it addresses specific types of validation, ranging from code quality to system performance and robustness, and from repeatability to traceability.

Armed with this test infrastructure for Symbian OS development, your software development process will be able to detect faults more acutely and to prevent you from releasing them.

We encourage you to regularly check the Symbian developer website (***http://www.symbian.com/developer***) under the Tools section for updates and latest tools available for Symbian OS.

This is only the beginning of the quest for better test tools. Symbian is committed to providing an even more comprehensive test platform: stay tuned for the next releases.

Appendix 1

Example Projects

This appendix lists the projects described throughout the book.

Source code is available from *http://www.symbian.com/books*. You can download the examples to any location you choose, but the text of this book assumes that they are located in subdirectories of a \scmp2 top-level directory of the drive into which the UIQ SDK is installed.

The example subdirectories contain one project each. If there are any specific instructions, or additional information about a project, they will be found in a readme.html file in that project's top-level directory.

Programs have been tested under the emulator on Windows NT and Windows 2000 Professional, on a Nokia 6600 phone and, where appropriate, on a Sony Ericsson P900 phone.

The Projects

Example	Description	Chapter
active	Demonstrates the use of active objects	1
audio1	Multimedia example that opens a .wav file and plays it	7
audio2	Multimedia example that records a .wav file	7
audio3	Multimedia example to demonstrate streaming audio data	7
audio4	Multimedia example to illustrate tone generation	7
cleanup	Illustrates error-resistant class instantiation and destruction	1
conslauncher	A simple application to launch console (.exe) applications	4
helloblank	A version of 'Hello World' that uses a blank control for its view	4
oandx	A game of noughts and crosses (tic tac toe)	4, 6
viewex	An illustration of the use of views and the view architecture	5

Symbian OS C++ for Mobile Phones, Volume 2. Edited by Richard Harrison
© 2004 Symbian Software Ltd   ISBN: 0-470-87108-3

Appendix 2

Symbian OS System Model

The System Model is a high-level representation of the Symbian OS architecture, showing a layered view of Symbian OS components. Please refer to the color diagram bound into the back of this book.

The **five Symbian OS layers** are:

- UI Framework

- Application Services

- OS Services

- Base Services

- Kernel Services and Hardware Interface.

Java J2ME is shown as a single block spanning the two highest Symbian OS layers.

The System Model follows these organizing principles:

- Each layer abstracts the complexity of the layer beneath and provides services to the layer above.

- Within each layer, components are encapsulated (grouped into collections) according to functionality.

- The OS Services layer is further encapsulated by grouping collections into broader blocks or suites of like technologies, for example Comms Services is the logical group for *communications external* to Symbian OS.

Encapsulation provides manageable granularity to the OS architecture. However, some degree of dependency (i.e. break of encapsulation) inevitably exists between adjacent blocks and components.

Similarly, while the layer model works well for Symbian's purposes, some relaxing of the layers is necessary to cope with the efficiency demands placed on the operating system. For example, some lower-layer functionality is accessible to all layers above.

The System Model should therefore be regarded as a useful approximation rather than a rigid specification of architectural relationships.

Components shown in the System Model have a concrete interpretation in the source system, and correspond approximately to parts of the source tree controlled by a single build file.

Tools and test components are not shown in the System Model. Key reference components (hatched) and plugin components (bold outline) are identified.

The System Model assumes that a Symbian OS licensee-supplied UI sits over the UI Framework.

To find out more about the Symbian OS System Model, including versions for different releases of Symbian OS, see ***www.symbian.com/ developer***.

Appendix 3

Writing Good Symbian OS Code

When writing applications for Symbian OS there are many considerations – beginning with design, all the way through to the final finishing touches of releasing your .SIS file. All contribute to the quality and robustness of your application. This appendix gathers useful hints and tips which you can use to produce the most reliable Symbian OS applications possible.

A3.1 General Tips

1. The **Symbian Developer Network website** (*www.symbian.com/ developer*) hosts valuable information to help you write applications. Visit it regularly to obtain the latest SDKs, technical information, code examples and white papers. You should also sign up to the Symbian Developer Network newsletter – this is distributed monthly by email and is the best way to keep up to date with Symbian developer news.

2. Symbian's licensees also run developer-related programs. You should register at their sites to get access to the latest phone-specific information and tips.

3. Hosted on the Symbian Developer Network website, the **Symbian OS FAQ database** (*http://www.symbian.com/developer/techlib/faq.html*) is an invaluable source of information for developers, covering the most frequently asked design and coding issues. The **Symbian OS C++ Coding Standards** are also hosted there. In the course of developing Symbian OS itself, Symbian has defined several important coding idioms and styles. The coding idioms paper explains them for third-party developers. By using these tried and tested conventions, you will benefit from all of the experience of Symbian's engineers while developing Symbian OS.

A3.2   Design Tips

1. The most important design tip for Symbian OS is to separate out your 'engine' and UI code into different modules. Symbian OS itself is designed in this way and it aids porting between different UI systems.

All non-UI related code should be put into a separate engine .DLL file. Your Application UI can then link to this .DLL for access to engine functionality. For example, code that opens up a connection to the Agenda Server (RAgendaServ) uses no UI components and is therefore clearly engine code. The Agenda Server is common to all Symbian OS releases since v6.0, so your engine code would work on all phones without modification.

By coding in this way, you ease the porting burden for new UI platforms. Pure engine code often runs unaltered on any UI platform. This means that all you have to port and optimize for a new UI is your own, separate UI layer.

2. Always design with support for localization in mind. Never hard-code strings or literals into your source files – always use the resource file mechanism Symbian OS provides to store strings.

3. Take care to stick to using APIs which are documented and supported for given SDK and Symbian OS releases. The use of unsupported or deprecated APIs can lead to future problems for your application – for example, Symbian reserves the right to change or remove APIs which are not intended to be used by external developers.

A3.3   Coding Tips

The following is a collection of general tips which you should bear in mind when actually writing your code.

1. Ensure your application responds to system shutdown events. It is vital that you respond to `EEikCmdExit` (and any platform-specific events, for example `EAknSoftkeyBack` and `EAknCmdExit` on Series 60) in your `AppUi::HandleCommandL()` method.

2. Respond to incoming system events. Remember that your application is operating on a multitasking phone system. You need to pay careful attention to focus gained/lost events, etc., to ensure you respond correctly when the user receives a high priority notification. For example, ensure you save your state and data when there is an incoming phone call which will cause your application to lose focus (i.e. you need to appropriately act on standard 'to background' events – see the SDK for more details).

3. Memory handling on Symbian OS is a key consideration. Note that behavior on the phone can sometimes differ from that on the emulator. It is *vital* that you test your application on a real phone.

4. KERN-EXEC 3 crashes are often symptomatic of stack corruption/overflow – as per the Symbian SDK recommendations, prefer the heap to the stack.

5. Under low memory situations, failing gracefully is important – an application panic indicates a real bug in your code. Here are some key, common errors:

- Failure to have properly added a non-member, heap-allocated variable to the Cleanup Stack.

- The 'double delete' – e.g. failure to correctly `Pop()` an already destroyed item from the Cleanup Stack, causing the stack to try to delete it again at a later time.

- Accessing functions in variables which may not exist in your destructor; e.g.

```
CMyClass::~CMyClass()
    {
    iSomeServer->Close();
    delete iSomeServer;
    }
```

should always be coded as

```
CMyClass::~CMyClass()
    {
    if (iSomeServer)
        {
        iSomeServer->Close();
        delete iSomeServer;
        }
    }
```

- Putting member variables on the Cleanup Stack – never do this; just delete them in your destructor as normal.

6. Always use `CleanupClosePushL()` with `R` classes which have a `Close()` method. This will ensure they are properly cleaned up if a leave occurs. For example:

```
RFile file;
User::LeaveIfError(file.Open(...));
CleanupClosePushL(file);
...
CleanupStack::PopAndDestroy(); // file
```

7. In addition, remember that the Cleanup Stack is an extensible mechanism that can be used for cleaning up *anything* when there is a leave. If you have a more complex situation to deal with, don't just ignore proper cleanup. See the SDK documentation on `TCleanupItem` for more information. This can be used to ensure proper cleanup of other R classes which do not have `Close()` methods (for example, ones which have `Release()` or `Destroy()` instead).

8. Always set member `HBufC` variables to `NULL` after deleting them. Since `HBufC` allocation (or reallocation) can potentially leave, you could find yourself in a situation where your destructor attempts to delete an `HBufC` which no longer exists.

9. If you have cause to use a `TRAP` of your own, do not ignore all errors. A common coding mistake is:

```
TPAPD(err, DoSomethingL());
if (err == KErrNone || err == KErrNotFound)
    {
    // Do something else
    }
```

This means all other error codes are ignored. If you must have a pattern like the above, leave for other errors:

```
TPAPD(err, DoSomethingL());
if (err == KErrNone || err == KErrNotFound)
    {
    // Do something else
    }
else
    User::Leave(err);
```

10. Do not wait to `PushL()` things on to the Cleanup Stack. Any newly allocated object (except member variables) should be added to the stack immediately. For example, the following is wrong:

```
void doExampleL()
    {
    CSomeObject* myObject1=new (ELeave) CSomeObject;
    CSomeObject* myObject2=new (ELeave) CSomeObject;
    }
```

because the allocation of `myObject2` could fail, leaving `myObject1` 'dangling' with no method of cleanup. The above should be:

```
void doExampleL()
    {
    CSomeObject* myObject1=new (ELeave) CSomeObject;
    CleanupStack::PushL(myObject1);
    CSomeObject* myObject2=new (ELeave) CSomeObject;
    CleanupStack::PushL(myObject2);
    }
```

11. Note that functions with a trailing C on their name *automatically* put the object on the Cleanup Stack. You should *not* push these objects onto the stack yourself, or they will be present twice. The trailing C functions are useful when you are allocating non-member variables.

12. Two-phase construction is a key part of Symbian OS memory management. The basic rule is that a Symbian OS constructor or destructor must never leave. If a C++ constructor leaves, there is no way to clean up the partially constructed object, because there is no pointer to it. For this reason, Symbian OS constructors simply instantiate the object, which then provides a ConstructL() method where member data can be instantiated. If ConstructL() leaves, the standard destructor will be called to destroy any objects which have been successfully allocated to that point. It is essential that you mirror this design pattern to avoid memory leaks in your code. For each line of code you write, a good question to ask yourself is 'Can this line leave?'. If the answer is 'Yes', then think: 'Will all resources be freed?'.

13. Do not use the _L() macro in your code. This functionality has been deprecated since Symbian OS v5 – you should prefer _LIT() instead. The problem with _L() is that it calls the TPtrC(const TText*) constructor, which has to call a strlen() function to work out the length of the string. While this doesn't cost extra RAM, it does cost CPU cycles at runtime. By contrast, the _LIT() macro directly constructs a structure which is fully initialized at compile time, so it saves the CPU overhead of constructing the TPtrC.

14. When using descriptors as function parameters, use the base class by default. In most cases, pass descriptors around as a const TDesC&. For modifiable descriptors use TDes&.

15. Active Objects (AOs) are a key piece of Symbian OS functionality. You should carefully study the SDK documentation and Symbian Developer Network white papers to get a good understanding of the way these work. Here are a few useful tips:

- There is no need to call TRAP() inside RunL(). The Active Scheduler itself already TRAPs RunL() and calls CActive::RunError() after a leave.

- To this end, you should implement your own RunError() function to handle leaves from RunL().

- Keep RunL() operations short and quick. A long-running operation will block other AOs from running.

- *Always* implement a DoCancel() function and *always* call Cancel() in the AO destructor.

16. You should make use of the Active Object framework wherever possible. Tight polling in a loop is highly inappropriate on a battery-powered device and can lead to significant power drain. Pay particular attention to this when writing games – there are technical papers on the Symbian Developer Network website with more detailed discussions on this subject.

17. ViewSrv 11 panics are a hazard when writing busy applications, for instance games. They occur when the ViewSrv Active Object in any application does not respond to the view server in time. Typically 10–20 seconds is the maximum allowed response time. See FAQ-0900 for a more detailed explanation and FAQ-0920 for practical tips to avoid this problem.

18. You don't need to use `HBufC::Des()` to get into an `HBufC`. All you have to do is dereference the `HBufC` with the `*` operator – this is particularly relevant, for example, when passing an `HBufC` as an argument to a method which takes a `TDesC&` parameter (as recommended above).

19. When making use of the standard application .INI file functionality (i.e. by using the `Application()->OpenIniFileLC();` API in your Application UI class), be sure to write into streams with version number information. This allows you to create new streams for future versions of your application and means that if an end user installs a new version of your product in the future, an 'old' .INI file will not cause the new version to panic if it cannot find the right settings or stream.

A3.4   Testing Tips

1. The most important testing tip is to exit your application under the emulator, *not* just to close the entire emulator. In debug mode, there is memory and handle checking code surrounding the application framework shutdown functions. If you exit your application, this code will be invoked and you will be able to see whether you have leaked memory or left a handle (e.g. an R object) open. For UIQ applications it is customary to provide an Exit menu item *in debug mode only* for this purpose.

2. Another vital tip is to ensure that the correct platform dependency information is included in your .PKG file prior to deployment. More details of the dependency string you require should be in your platform-specific SDK. FAQ-0853 on the Symbian OS FAQs also provides useful information.

3. When writing your .PKG file, also ensure you use the `!:\` syntax where appropriate. In general, your application should install and run from any drive on the end user's phone. Very few things (e.g. .INI files) should be put on `C:\` only.

A3.5 Debugging Tips

1. When writing and debugging a new control class, put `iEikonEnv ->WsSession().SetAutoFlush(ETrue);` in the `ConstructL()` function of your AppUi. This means that `gc draw` commands will show up in the emulator immediately, rather than when the window server client buffer is next flushed. This means you can step through the draw code and see the effect each line is having. However, you must ensure that this line does not make it into released software, as it has efficiency implications.

2. If your application panics on shutdown due to a memory leak, casting the leaked address to `CBase*` will often give you the type of the leaked object.

3. An important recent addition to the functionality available for Symbian OS developers is on-target debugging. While this is not currently available for all SDK and tool variants, most of the newest SDK and IDE releases do support it. Where available, you are advised to use it prior to releasing your application to help track down any phone-specific defects. See your SDK and/or IDE documentation for more information.

Appendix 4

Developer Resources

A4.1 Symbian OS Software Development Kits (SDKs)

SDKs are built based on a particular reference platform (sometimes known as a 'reference design') for Symbian OS. A reference platform provides a distinct UI and an associated set of system applications for such tasks as messaging, browsing, telephony, multimedia and contact/calendar management. These applications typically make use of generic application engines provided by Symbian OS. Reference platforms intended to support the installation of third-party applications written in native C++ have to be supported by an SDK which defines that reference platform, or at least a particular version of it. Since Symbian OS v6.0, four such reference platforms have been introduced, resulting in four 'flavors' of SDK which can be found at the websites listed here:

- **UIQ** (*www.symbian.com/developer*)

- **Nokia Series 90** (*www.forum.nokia.com*)

- **Nokia Series 60** (*www.forum.nokia.com*)

- **Nokia Series 80** (*www.forum.nokia.com*)

Prior to this, SDKs were targeted at specific devices, such as the Psion netPad. Symbian no longer supports these legacy SDKs, but they are still available from Psion Teklogix at *www.psionteklogix.com*.

For the independent software developer, the most important thing to know in targeting a particular phone is its associated reference platform. Then you need to know the Symbian OS version the phone is based on. This knowledge defines to a large degree the target phone as a platform for independent software development. You can then decide which SDK you need to obtain. In most cases you will be able to target – with a single version of your application – all phones based on the same reference platform and Symbian OS version working with this SDK. The Symbian OS System Definition papers give further details of possible differences between phones based on a single SDK.

- **Symbian OS System Definition**
  www.symbian.com/developer/techlib/papers/SymbOS_def/ symbian_os_sysdef.pdf
- **Symbian OS System Definition (in detail, incl. Symbian OS v8.0)**
  www.symbian.com/developer/techlib/papers/SymbOS_cat/ SymbianOS_cat.html

A4.2   Getting a UID for Your Application

A UID is a 32-bit number, which you get as you need from Symbian.

Every Uikon application should have its own UID. This allows Symbian OS to distinguish files associated with that application from files associated with other applications. UIDs are also used in other circumstances, such as to identify streams within a store, and to identify one or more of an application's views.

Getting a UID is simple. Just send an email to ***uid@symbiandevnet.com***, titled 'UID request', and requesting clearly how many UIDs you want – 10 is a reasonable first request. Assuming your email includes your name and return email address, that's all the information Symbian needs. Within 24 hours, you'll have your UIDs.

If you're impatient, or you want to do some experimentation before using real UIDs, you can allocate your own UIDs from a range that Symbian has reserved for this purpose: `0x01000000–0x0fffffff`. However, you should never release any programs with UIDs in this range.

> **Don't build different Symbian OS applications with the same application UID – even the same test UID – on your emulator or Symbian OS machine. If you do, the system will recognize only one of them, and you won't be able to launch any of the others.**

A4.3   Symbian OS Developer Tools

As well as the following tools offerings from Symbian DevNet partners, Symbian DevNet provides a number of free and open source tools:

www.symbian.com/developer/downloads/tools.html

AppForge

Develop Symbian applications using Visual Basic and AppForge. App-Forge development software integrates directly into Microsoft Visual Basic, enabling you to immediately begin writing multi-platform applications using the Visual Basic development language, debugging tools and interface you already know.

www.appforge.com

Borland

Borland offers C++BuilderX Mobile Edition and JBuilder Mobile Edition as well as the more recent Borland Mobile Studio for developers who want to develop rapidly on Symbian OS using C++, Java or both. These multi-platform IDEs offer target debugging, GUI RAD and a unifying IDE for Symbian OS SDKs and compilers.

www.borland.com

Forum Nokia

In addition to a wide range of SDKs, Forum Nokia also offers various development tools to download, including the Nokia Developer Suite for J2ME, which plugs in to Borland's JBuilder MobileSet or Sun's Sun One Studio integrated development environment.

www.forum.nokia.com

Metrowerks

Metrowerks offer the following products supporting Symbian OS development:

- CodeWarrior Development Tools for Symbian OS Professional Edition
- CodeWarrior Development Tools for Symbian OS Personal Edition
- CodeWarrior Wireless Developer Kits for Symbian OS.

www.metrowerks.com

Sun Microsystems

Sun provides a range of tools for developing Java 2 Micro Edition applications including the J2ME Wireless Toolkit and Sun One Studio Mobile Edition.

http://java.sun.com

Texas Instruments

Development tools for the OMAP platform easy-to-use software development environments are available today for OMAP application developers and OMAP Media Engine developers, as well as device manufacturers.

Tool suites that include familiar third-party tools and TI's own industry leading eXpressDSP DSP tools are available, allowing developers to easily develop software across the entire family of OMAP processors.

http://focus.ti.com

Symbian DevNet Tools
Symbian DevNet offers the following tools as an unsupported resource to all developers:

- **Symbian OS SDK add-ons**
  www.symbian.com/developer/downloads/tools.html

- **Symbian OS v5 SDK patches and tools archive**
  www.symbian.com/developer/downloads/archive.html

A4.4   Support Forums

Symbian DevNet offers two types of support forum:

- **Support newsgroups**
  www.symbian.com/developer/public/index.html

- **Support forum archive**
  www.symbian.com/developer/prof/index.html

Symbian DevNet partners also offer support for developers:

Sony Ericsson Developer World
As well as tools and SDKs, Sony Ericsson Developer World provides a range of services including newsletters and support packages for developers working with the latest Sony Ericsson products such as the Symbian OS powered P900.

http://developer.sonyericsson.com

Forum Nokia
As well as tools and SDKs, Forum Nokia provides newsletters, the Knowledge Network, fee-based case-solving, a Knowledge Base of resolved support cases, discussion archives and a wide range of C++ and Java-based technical papers of relevance to developers targeting Symbian OS.

forum.nokia.com/main.html

Sun Microsystems Developer Services
In addition to providing a range of tools and SDKs, Sun also provides a wide variety of developer support services including free forums, newsletters, and a choice of fee-based support programs.

- **Forums**
  http://forum.java.sun.com

- **Support and newsletters**
  http://developer.java.sun.com/subscription

A4.5 Symbian OS Developer Training

Symbian's Technical Training team and Training Partners offer public and on-site developer courses around the globe. For course dates and availability, see ***www.symbian.com/developer/training***.

Early bird discount: Symbian normally offers a 20% discount on all bookings confirmed up to 1 month before the start of any course. This discount cannot be used in conjunction with any other discounts.

Course	Level	Language
Symbian OS essentials	Introductory	C++
Java on Symbian OS	Introductory	Java
Symbian OS: Application engine development	Intermediate	C++
Symbian OS: Application UI development	Intermediate	C++
Symbian OS: Internals	Advanced	C++
Symbian OS: UI system creation	Advanced	C++

Please note

Intermediate and advanced courses require previous attendance of OS Essentials. UI system creation course also requires previous attendance of Application UI course.

A4.6 Developer Community Links

These community websites offer news, reviews, features and forums, and represent a rich alternative source of information that complements the Symbian Development Network and the development tools publishers. They are good places to keep abreast of new software, and of course to announce the latest releases of your own applications.

My-Symbian

My-Symbian is a Poland-based website dedicated to news and information about Symbian OS phones. This site presents descriptions of new software for Symbian OS classified by UI. It also features discussion forums and an online shop.

http://my-symbian.com

All About Symbian

All About Symbian is a UK-based website dedicated to news and information about Symbian OS phones. The site features news, reviews, software directories and discussion forums. It has strong OPL coverage.

www.allaboutsymbian.com

SymbianOne

SymbianOne features news, in-depth articles, case studies, employment opportunities and event information all focused on Symbian OS. A weekly newsletter provides up-to-date coverage of developments affecting the Symbian OS ecosystem. This initiative is a joint venture with offices in Canada and New Zealand.

www.symbianone.com

NewLC

NewLC is a French-based collaborative website dedicated to Symbian OS C++ development. It aims to be initially valuable to developers just starting writing C++ applications for Symbian OS, and with time will cover more advanced topics.

www.newlc.com

infoSync World

infoSync World is a Norway-based site providing features, news, reviews, comments and a wealth of other content related to mobile information devices. It features a section dedicated to Symbian OS, covering new phones, software and services – mixed with strong opinions that infoSync is not afraid to share.

symbian.infosyncworld.com

Your Symbian

Your Symbian (YS) is a fortnightly magazine distributed exclusively by email. YS takes a lighthearted look at the Symbian OS world. Major news is covered in its editorial and it includes a software round-up. To sign up, browse the archives, or get in touch with the editorial team.

www.yoursymbian.com

TodoSymbian (Spanish)

TodoSymbian is a Spain-based website for everyone wanting to read in Spanish about Symbian OS. It provides news, reviews, software directories, discussion forums, tutorials and a developers' section.

www.todosymbian.com

A4.7 Symbian OS Books

Symbian OS C++ for Mobile Phones, Vol. 1
Richard Harrison *et al.*
John Wiley & Sons, ISBN 0470856114

Programming the Java 2 Micro Edition for Symbian OS
Martin de Jode *et al.*
John Wiley & Sons, ISBN 047092238

Programming for the Series 60 Platform and Symbian OS
Digia, Inc.
John Wiley & Sons, ISBN 0470849487

Symbian OS Communications Programming
Mike Jipping
John Wiley & Sons, ISBN 0470844302

Wireless Java for Symbian Devices
Jonathan Allin *et al.*
John Wiley & Sons, ISBN 0471486841

Developing Series 60 Applications
Edwards, Barker
Addison Wesley, ISBN 032126875X

A4.8 Open Source Projects

Many open source projects are happening on Symbian OS. They are a rich source of partially or fully functional code which should prove useful to learn about use of APIs you're not yet familiar with. Please also consider contributing to any project that you have an interest in.

Repository websites

- **SymbianOS.org**, *http://symbianos.org*
  Community website dedicated to the development of open source programs for Symbian OS. Hosted projects include Vim, Rijndael encryption algorithm, MakeSis package for Debian GNU/Linux, etc.

- **Symbian open source**, *http://www.symbianopensource.com/*
  Repository for Symbian OS open-source software development. It provides free services to developers who wish to create, or have created, open-source projects.

- **Open Source for EPOC32**, *http://www.edmund.roland.org/osfe.html*
  Website of Alfred Heggestad where he maintains a list of open source projects for Symbian OS.

Appendix 5

Build Process Overview

A5.1   Compilers and IDEs

At the time of writing, support for building Symbian OS C++ projects is provided by the Borland, Metrowerks and Microsoft C++ compilers and their associated IDEs, which are:

- Borland C++Builder 6 Mobile Edition, and Borland C++BuilderX Mobile

- Metrowerks CodeWarrior

- Microsoft Visual C++ version 6.

You can use any of the three compilers to build projects to run on a PC, under the emulator that is supplied with each SDK. However, because of incompatibilities between the outputs of the different compilers, you must make sure that you obtain a version of the SDK that is appropriate to your chosen compiler.

Regardless of which of these compilers you use, builds for an ARM target device are made using the GNU C++ compiler that is supplied with each SDK.

The remainder of this appendix gives a brief overview of the build process. You can find more complete information in the SDK's Developer Library documentation.

A5.2   Command Line Builds

Builds can be run from the command line, or from within an IDE. All builds can be derived from the application's **component description file**

(`bld.inf`) and the one or more **project definition files** (with a `.mmp`
extension) that it references.

To perform a command line build, you first need to run the command

```
bldmake bldfiles
```

from the directory that contains your project's `bld.inf` file. This generates the project's makefiles, and an `abld.bat` batch file that you use to
run the rest of the build.

The build is carried out in six stages (`export`, `makefile`, `library`,
`resource`, `target` and `final`) and you can, if you wish, use the
`abld.bat` batch file to run each individual stage. However, the most
straightforward method is to run the entire build by means of a command
of the form

```
abld build <platform> <buildtype>
```

The `<buildtype>` parameter can be `udeb`, to perform a debug build,
`urel`, to perform a release (non-debug) build, or `all` to perform both
debug and release builds.

The `<platform>` parameter depends on both the target processor and
the compiler that you use. If you are building to run under the emulator,
for example, the different compilers use the following values:

C++ compiler	Parameter
Borland	`WINSB`
Metrowerks	`WINSCW`
Microsoft	`WINS`

You use exactly the same procedure to run a command line build for
a real target phone, except that the possible values for the `<platform>`
parameter are:

- `ARM4`, to compile for the ARMv4 instruction set

- `ARMI`, to compile for the THUMB (16-bit) mode subset of the ARMv4T
  instruction set[1]

- `THUMB`, to compile for the ARM mode subset of the ARMv4T instruction set.

---

[1] For technical reasons, the Borland IDEs, used with any Series 60 version 1.x SDK, use
the term `ARMIB` to refer to the `ARMI` platform.

It is recommended that third-party applications should be built using the `ARMI` platform, since such code will work with code compiled for any of the three platforms.

A5.3 Using an IDE

To build from within an IDE, you first need to create the project file appropriate to the IDE you are using. Most IDEs allow you to create a suitable Symbian OS project, together with the necessary skeleton source files, but it is often simpler to start with a basic application (such as one of the examples used in this book, or the Hello World example that is supplied in all SDKs). All the IDEs mentioned above allow you to create a project in this way, but the details vary, depending on the particular IDE that you are using.

For Microsoft Visual C++ version 6, you create the project file with the aid of the `abld.bat` batch file described earlier. Move to the directory containing the project's `abld.bat` file and, at the command line, type:

```
abld makefile vc6
```

This generates the project's `.dsw` workspace file, which you can then open from the Microsoft Visual C++ IDE.

The Borland and Metrowerks IDEs allow you to import the project from within the IDE. Both Borland IDEs provide a means, under their **File| New** menu options, of importing a project from its `bld.inf` component description file. In contrast, the Metrowerks IDE imports a project's `.mmp` project definition file, by means of the **File|Import Project From .mmp File** menu option.

Unlike the other IDEs, Microsoft Visual C++ version 6 does not support builds for any ARM platform so, when using this IDE, you have to run such builds from the command line.

Appendix 6

Specifications of Symbian OS Phones

Additional technical information on a range of phones can be found at *www.symbian.com/phones*.

Please note that this is a quick guide to Symbian OS phones. For full specifications, C++ developers can retrieve extended information using HAL APIs or check the manufacturer's website.

Nokia 9210i

OS Version	Symbian OS v6.0
UI/Category	Series 80
Memory available to user	40 MB
Storage media	Yes
Screen	640 × 200; 4096 colors
Pointing device	No
Camera	No

GSM/HSCSD/GPRS/3G

GSM 900	Yes
GSM 1800	Yes
GSM 1900	No (GSM 900/1900 on 9290)
HSCSD	Yes
GPRS	No
3G	No

Connectivity

Infrared	Yes
Bluetooth	No
USB	No
Serial	Yes

Java APIs	CLDC 1.0
	MIDP 1.0
	PersonalJava 1.1.1
	JavaPhone

Browsing

WAP	WAP 1.1
XHTML (MP)	Yes
Browser available	Yes (built-in and third-party)

Nokia 7650

OS Version	Symbian OS v6.1
UI/Category	Series 60
Memory available to user	4 MB NOR flash user data storage
Storage media	No
Screen	176 × 208; 4096 colors
Pointing device	No
Camera	Yes; 640 × 480 resolution
GSM/HSCSD/GPRS/3G	
GSM 900	Yes
GSM 1800	Yes
GSM 1900	No
HSCSD	Yes
GPRS	Yes (2 + 2, 3 + 1, class B and C)
3G	No
Connectivity	
Infrared	Yes
Bluetooth	Yes
USB	No
Serial	No
Java APIs	MIDP 1.0
	CLDC 1.0
	Nokia UI API
Browsing	
WAP	WAP 1.2.1
XHTML (MP)	No
Browser available	Yes (third-party)

Nokia 3600/3650

OS Version	Symbian OS v6.1
UI/Category	Series 60 (v1)
Memory available to user	3.4 MB
Storage media	Yes; MMC
Screen	176×208; 4096/65 536 colors
Pointing device	No
Camera	Yes; 640×480 resolution
GSM/HSCSD/GPRS/3G	
GSM 900	Yes
GSM 1800	Yes
GSM 1900	Yes
HSCSD	Yes
GPRS	Yes (2 + 2, 3 + 1, class B and C)
3G	No
Connectivity	
Infrared	Yes
Bluetooth	Yes
USB	No
Serial	No
Java APIs	MIDP 1.0
	CLDC 1.0
	Nokia UI API
	Wireless Media API
	Mobile Media API
Browsing	
WAP	WAP 1.2.1
XHTML (MP)	Yes
Browser available	Yes (third-party)

Nokia 3620/3660

OS Version	Symbian OS v6.1
UI/Category	Series 60 (v1)
Memory available to user	4 MB
Storage media	Yes; MMC
Screen	176 × 208; 4096/65 536 colors
Pointing device	No
Camera	Yes; 640×480 resolution

GSM/HSCSD/GPRS/3G

GSM 850	Yes
GSM 1800	No
GSM 1900	Yes
HSCSD	Yes
GPRS	Yes
3G	No

Connectivity

Infrared	Yes
Bluetooth	Yes
USB	No
Serial	No

Java APIs

	MIDP 1.0
	CLDC 1.0
	Nokia UI API
	Mobile Media API
	Wireless Media API

Browsing

WAP	WAP 1.2.1
XHTML (MP)	Yes
Browser available	Yes

Siemens SX1

OS Version	Symbian OS v6.1
UI/Category	Series 60
Storage media	Yes; MMC
Screen	176 × 208; 65 536 TFT
Pointing device	No
Camera	Yes; 640 × 480 and 160 × 120 resolution
GSM/HSCSD/GPRS/3G	
GSM 900	Yes
GSM 1800	Yes
GSM 1900	Yes
HSCSD	Yes
GPRS	Yes (class 10, B (2Tx, 4Rx))
3G	No
Connectivity	
Infrared	Yes
Bluetooth	Yes
USB	Yes
Serial	No
Java APIs	MIDP 1.0
	Wireless Media API
	Mobile Media API
Browsing	
WAP	WAP 2.0
XHTML (MP)	Yes
Browser available	Yes (third-party)

Nokia N-Gage

OS Version	Symbian OS v6.1
UI/Category	Series 60
Memory available to user	4 MB NOR flash user data storage
Storage media	Yes; MMC
Screen	176 × 208; 4096 colors
Pointing device	No
Camera	No

GSM/HSCSD/GPRS/3G

GSM 900	Yes
GSM 1800	Yes
GSM 1900	Yes
HSCSD	Yes
GPRS	Yes (2 + 2, 3 + 1, class B and C)
3G	No

Connectivity

Infrared	No
Bluetooth	Yes
USB	Yes
Serial	No

Java APIs	MIDP 1.0
	CLDC 1.0
	Nokia UI API
	Wireless Media API
	Mobile Media API

Browsing

WAP	WAP 1.2.1
XHTML (MP)	Yes
Browser available	Yes (third-party)

Sendo X

OS Version	Symbian OS v6.1
UI/Category	Series 60
Memory available to user	12 MB
Storage media	Yes; MMC and SD
Screen	176×220; 65 536 colors
Pointing device	No
Camera	Yes; 640×480 resolution
GSM/HSCSD/GPRS/3G	
GSM 900	Yes
GSM 1800	Yes
GSM 1900	Yes
HSCSD	No
GPRS	Yes, Class 8 (4 + 1)
3G	No
Connectivity	
Infrared	Yes
Bluetooth	Yes
USB	Yes
Serial	Yes
Java APIs	MIDP1.0
	Wireless Media API
	Bluetooth API
	Nokia UI API
	Mobile Media API
Browsing	
WAP	WAP 2.0
XHTML (MP)	Yes
Browser available	Yes (third-party)

BenQ P30

OS Version	Symbian OS v7.0
UI/Category	UIQ 2.1
Storage media	Yes; MMC and SD
Screen	208×320; 65 536 colors TFT
Pointing device	Yes
Camera	Yes; 640×480 resolution

GSM/HSCSD/GPRS/3G

GSM 900	Yes
GSM 1800	Yes
GSM 1900	Yes
HSCSD	No
GPRS	Yes (4 + 2, class 10)
3G	No

Connectivity

Infrared	Yes
Bluetooth	Yes
USB	Yes
Serial	No

Java APIs	MIDP 2.0
	PersonalJava 1.1.1
	Wireless Media API

Browsing

WAP	Yes 2.0
XHTML (MP)	Yes
Browser available	Yes

Sony Ericsson P800

OS Version	Symbian OS v7.0
UI/Category	UIQ
Memory available to user	12 MB
Storage media	Yes; Sony MS Duo
Screen	208×320 (Flip Open); 208×144 (Flip Closed); 4096 colors
Pointing device	Yes
Camera	Yes; 640×480 resolution
GSM/HSCSD/GPRS/3G	
GSM 900	Yes
GSM 1800	Yes
GSM 1900	Yes
HSCSD	Yes
GPRS	Yes (4 + 1)
3G	No
Connectivity	
Infrared	Yes
Bluetooth	Yes
USB	Yes (high speed serial connector with a USB->Serial adapter built into the desk stand)
Serial	Yes
Java APIs	CLDC 1.0
	MIDP 1.0
	PersonalJava 1.1.1
Browsing	
WAP	WAP 2.0
XHTML (MP)	Yes
Browser available	Yes (inbuilt and third-party)

Motorola A920/A925

OS Version	Symbian OS v7.0
UI/Category	UIQ
Memory available to user	8 MB
Storage media	Yes; MMC and SD
Screen	208×320; 65 536 colors TFT
Pointing device	Yes
Camera	Yes
GSM/HSCD/GPRS/3G	
GSM 900	Yes
GSM 1800	Yes
GSM 1900	Yes
HSCD	Yes
GPRS	Yes
3G	Yes
Connectivity	
Infrared	Yes
Bluetooth	A920 No/A925 Yes
USB	Yes
Serial	Yes
Java APIs	MIDP 1.03
	PersonalJava 1.1.1a
Browsing	
WAP	No
XHTML (MP)	Yes
Browser available	Yes (third-party)

Sony Ericsson P900

OS Version	Symbian OS v7.0 (+ security updates and MIDP2.0)
UI/Category	UIQ 2.1
Memory available to user	16 MB
Storage media	Yes; Sony MS Duo
Screen	208 × 320 (Flip Open); 208 × 208 (Flip Closed); 65 536 colors TFT
Pointing device	Yes
Camera	Yes; 640 × 480 resolution

GSM/HSCSD/GPRS/3G	
GSM 900	Yes
GSM 1800	Yes
GSM 1900	Yes
HSCSD	Yes
GPRS	Yes
3G	No

Connectivity	
Infrared	Yes
Bluetooth	Yes
USB	Yes (high speed serial connector with a USB->Serial adapter built into the desk stand)
Serial	No

Java APIs	MIDP 2.0
	PersonalJava 1.1.1
	Bluetooth API
	Wireless Media API

Browsing	
WAP	WAP 2.0
XHTML (MP)	Yes
Browser available	Yes

Nokia 6600

OS Version	Symbian OS v7.0s
UI/Category	Series 60 (v2)
Memory available to user	6 MB NOR flash user data storage
Storage media	Yes; MMC
Screen	176×208; 65 536 colors TFT
Pointing device	No
Camera	Yes; 640×480 resolution

GSM/HSCSD/GPRS/3G

GSM 900	Yes
GSM 1800	Yes
GSM 1900	Yes
HSCSD	Yes
GPRS	Yes (2 + 2, 3 + 1, class B and C)
3G	No

Connectivity

Infrared	Yes
Bluetooth	Yes
USB	No
Serial	No

Java APIs	MIDP 2.0
	CLDC 1.0
	Nokia UI API
	Mobile Media API
	Wireless Media API
	Bluetooth API

Browsing

WAP	WAP 2.0
XHTML (MP)	Yes
Browser available	Yes

Nokia 6620

OS Version	Symbian OS v7.0s
UI/Category	Series 60 (v2)
Memory available to user	6 MB NOR flash user data storage
Storage media	Yes; MMC
Screen	176 × 220; 65 536 colors TFT
Pointing device	No
Camera	Yes; 640 × 480 resolution

GSM/HSCSD/GPRS/3G

GSM 850	Yes
GSM 1800	Yes
GSM 1900	Yes
HSCSD	No
GPRS	No
3G	No
EDGE	Yes

Connectivity

Infrared	Yes
Bluetooth	Yes
USB	Yes
Serial	No

Java APIs	MIDP 2.0
	CLDC 1.0
	Nokia UI API
	Mobile Media API
	Wireless Media API
	Bluetooth API

Browsing

WAP	WAP 2.0
XHTML (MP)	Yes
Browser available	Yes

Nokia 7700

OS Version	Symbian OS v7.0s
UI/Category	Series 90
Memory available to user	64 MB
Storage media	Additional memory slot
Screen	640 × 320; 65 536 colors
Pointing device	Yes
Camera	Yes; 640×480 resolution

GSM/HSCSD/GPRS/3G

GSM 900	Yes
GSM 1800	Yes
GSM 1900	Yes
HSCSD	Yes
GPRS	Yes
3G	No
EDGE	Yes

Connectivity

Infrared	Yes
Bluetooth	Yes
USB	Yes
Serial	No

Java APIs	MIDP 2.0
	CLDC 1.0
	Nokia UI API
	Wireless Media API
	Mobile Media API
	Bluetooth API

Browsing

XHTML (MP)	Yes (+ HTML)
Browser available	Yes

Index